Transition to Sustainable Energy Technologies

This book explains the challenges related to global energy transition, focusing on decarbonization, renewable technologies, and sustainable mobility. It offers a comprehensive exploration of integrated socio-economic, environmental, and technological perspectives and provides actionable solutions and practical pathways for transitioning to sustainable energy systems. Designed to be used in short academic courses, this textbook is an excellent foundational text for upper-level undergraduate and graduate students in engineering and environmental sciences taking specialized courses in energy systems, renewable technologies, climate change mitigation, and environmental impacts. It is also a great resource for professionals and policy makers involved in energy projects.

Features:

- Offers an interdisciplinary approach to energy transitions combining theoretical frameworks with practical applications.
- Explores the full spectrum of energy systems, comparing conventional to renewable sources, energy vectors, and energy-storage technologies.
- Incorporates real-world data from leading institutions and integrates key historical milestones like the Paris Agreement with forward-looking trends.
- Highlights cutting-edge innovations, including solar and wind energy, hydrogen, energy-storage systems, synthetic fuels, and smart grids, and emphasizes their integration into energy systems and sustainable mobility solutions.
- Bridges the gap between socio-economic, environmental, and technological perspectives.

Transition to Sustainable Energy Technologies

Pathways, Sources, Mobility

Carlo Villante, Sonia Dell'Aversano,
and Stefano Ranieri

CRC Press
Taylor & Francis Group
Boca Raton London New York

CRC Press is an imprint of the
Taylor & Francis Group, an **informa** business

Designed cover image: iStock

First edition published 2026
by CRC Press
2385 NW Executive Center Drive, Suite 320, Boca Raton FL 33431

and by CRC Press
4 Park Square, Milton Park, Abingdon, Oxon, OX14 4RN

CRC Press is an imprint of Taylor & Francis Group, LLC

Funding Body: Università degli Studi dell'Aquila

ISBN: 978-1-041-05046-9 (hbk)
ISBN: 978-1-041-05047-6 (pbk)
ISBN: 978-1-003-63112-5 (ebk)

DOI: 10.1201/9781003631125

Typeset in Times
by Newgen Publishing UK

Dedication

To my sons, Gabriele and Laura,
and to their generation, inheriting the responsibility to make our world a place where innovation harmonizes with nature and sustainable energy fuels development.
May you lead with greater courage, wisdom, and compassion, building a brighter and more sustainable future for all.

Carlo Villante

Contents

PART III Sustainable Mobility: Sector Evolution, Propulsion Systems

Preface

Energy is the lifeblood of modern society, powering homes, industries, transportation, and communication systems. However, the ways we produce and consume energy are deeply connected to some of the most pressing challenges of our time. Dependence on fossil fuels, environmental degradation, and geopolitical tensions highlight the urgent need for a global energy transition.

The global energy landscape is undergoing a monumental transformation. Confronted with crises such as climate change, pollution, energy dependency, and resource depletion, humanity finds itself at a critical juncture where transitioning to sustainable energy systems is no longer optional but essential.

This book, "Transition to Sustainable Energy Technologies: Pathways, Sources, Mobility", serves as a comprehensive resource to understand and address the challenges and opportunities associated with this energy revolution. It is the result of extensive research and analysis, and integrates socio-economic insights, environmental considerations, and technical expertise to create a holistic guide for students, educators, policymakers, and professionals.

By bridging the gap between sustainability theory and the practical realities of energy systems, the book emphasizes both the urgency and feasibility of transitioning to cleaner, more resilient energy models.

Our motivation for writing this book stems from recognizing that many literature resources focus narrowly on specific technologies or policies, often overlooking the broader context of energy sustainability.

Therefore, by integrating historical perspectives, real-world data, and forward-looking solutions, this book aims to empower readers with the knowledge and tools to navigate the complexities of the global energy transition.

Our ambition is for this book to serve as both an introduction and a practical guide for those passionate about building a sustainable energy future. By fostering a deeper understanding of energy systems, we hope to contribute to the global dialogue on harmonizing development with environmental stewardship.

The book is organized into three interrelated yet independently readable sections:

- **Part I – Energy Transition**
  The first section of the book explores the historical evolution of energy systems and the challenges posed by dependence on fossil fuels. It highlights issues such as energy dependency, resource depletion, and climate change. The motivation behind this section is to provide a comprehensive understanding of how energy systems must evolve towards sustainability by integrating renewable energy sources and implementing policy-driven decarbonization strategies.
- **Part II – Sustainable Energy Sources**
  This section delves into various renewable energy sources, such as solar, wind, hydro, geothermal, and biomass. It evaluates their technological characteristics, potential environmental benefits, and the challenges of

integrating them into current energy systems. The main motivation is to highlight how renewable energy can replace fossil-based systems while ensuring long-term sustainability and economic feasibility.

- **Part III – Sustainable Mobility**
  The third section focuses on transforming the transportation sector, which is a major contributor to global greenhouse gas emissions. It examines both conventional and advanced solutions like vehicle electrification, fuel cell technology, and Mobility-as-a-Service (MaaS). The motivation behind this section is to address air pollution, GHG emissions, and the need for alternative propulsion systems that can support a sustainable future.

At the end of each part, a list of additional bibliographic materials is provided, categorized by subject, allowing readers to explore specific aspects of these broad topics in greater depth.

The book is also available as PowerPoint slides for use in teaching activities. These files can be requested directly from the authors.

In general, the authors are available to support professors who wish to use the book as teaching material in their courses. They encourage direct engagement with users to gather feedback and suggestions for improvements in future editions.

Acknowledgements

Drawing on his extensive Academic expertise, Prof. Villante structured the book and authored the majority of its content. The remaining editorial effort was undertaken by the other two authors, who contributed more significantly to specific chapters and thoroughly revised the entire book to enhance its fluency and effectiveness.

Here below, the overall structure of the book is briefly recalled, both detailing the authoring contributions to the single Chapters, and also providing a compact but comprehensive overview of book content.

Part I – Energy Transition

1 Sustainable Development – *C. Villante*
2 Energy Systems – *C. Villante*
3 Energy Dependency – *C. Villante*
4 Energy Sources' Exhaustibility – *C. Villante and S. Dell'Aversano*
5 Greenhouse Effect and Global Warming - *C. Villante*
6 Global Warming Effects and Related Issues – *C. Villante*
7 Decarbonization Policies Design: The Kaya Identity – *C. Villante and S. Dell'Aversano*
8 Decarbonization Pathways – *C. Villante and S. Dell'Aversano*
9 Small Space–Time Scale Environmental Impacts – *C. Villante*
10 Energy Transition to Sustainable Energy Systems – *C. Villante*
11 Key Enabling Technologies for Energy Sources Conversion – *C. Villante*
12 Sustainable Choice of Energy Carriers – *C. Villante and S. Dell'Aversano*
13 Key Enabling Technologies for Final Energy Uses – *S. Ranieri and C. Villante*

Part II – Sustainable Energy Sources

1 Sustainable Energy Sources' Characteristics – *C. Villante*
2 Solar Energy – *C. Villante and S. Ranieri*
3 Wind Energy – *C. Villante and S. Ranieri*
4 Hydro Energy – *C. Villante*
5 Geothermal Energy – *C. Villante*
6 Biomasses, Biofuels, and e-Fuels – *S. Dell'Aversano and C. Villante*

Part III – Sustainable Mobility

1 Sustainable Mobility Issues – *C. Villante*
2 Vehicular Energy and Power Requirements – *C. Villante*
3 ICEs and Their Local and Global Impacts – *C. Villante and S. Ranieri*
4 Alternative Fuels for ICEs – *C. Villante, S. Dell'Aversano and S. Ranieri*
5 Vehicle Electrification – *C. Villante*
6 Fuel Cell Vehicles (FCEVs) – *C. Villante*
7 Battery Electric Vehicles (BEVs) – *C. Villante*
8 Hybrid Electric Vehicles (HEVs) – *C. Villante*

About the Authors

Carlo Villante is the Lead and Corresponding Author of this book. He serves as an Associate Professor at the University of L'Aquila, Italy, where he has been teaching since 2000. Currently, he teaches Sustainable Energy Technologies (since 2005) and Sustainable Mobility and Propulsion Systems (since 2015). From 2009 to 2012, he worked as a researcher at ENEA (the Italian National Agency for Energy, New Technologies, and Sustainable Development). In 2012 and 2013, he was an Associate Professor at Sannio University, Italy. Since 2014, he has been officially qualified to serve as a Full Professor in Italian universities. In 2018, Prof. Villante founded CITraMS, the Italian Research Centre on Sustainable Mobility and Transportation, serving as Vice-Director and Board Member. The Centre comprises approximately 50 academic professors and researchers. He is an active member of the Society of Automotive Engineers (SAE) and the International Federation of Automatic Control (IFAC) Technical Committee on Automotive Control. Additionally, he serves on the Editorial Board of the *World Electric Vehicle Journal* (WEVJ), published by MDPI. Prof. Villante has authored approximately 80 technical papers, which have been published in leading international journals or presented at major international conferences.

Sonia Dell'Aversano is a young researcher who earned her B.Sc. and M.S. degrees in Mechanical Engineering at the University of L'Aquila in 2019 and 2022, respectively.

She is currently a Ph.D. researcher at the University of L'Aquila, focusing mainly on e-fuels synthesis and experimental testing in collaboration with ENEA, Italian National Agency for New Technologies, Energy and Sustainable Economic Development.

Stefano Ranieri earned his B.Sc. and M.S. degrees in Mechanical Engineering at the University of L'Aquila in 2016 and 2019, respectively. He completed a Ph.D. in Internal Combustion Engines in 2022 at the University of L'Aquila in collaboration with the Internal Combustion Engines Institute of the CNR (Italian National Research Centre). He has published about 20 technical papers, most of them in leading journals or international conferences. He is an Energy Expert Operator at ESC (Energy Systems Consulting) in Italy.

Detailed Contents

PART I Energy Transition: History, Issues, Policies and Pathways, KETs

Part I

Energy Transition

History, Issues, Policies and Pathways, KETs

INTRODUCTION

In Part I of this book, we examine the historical evolution of energy systems, focusing on how reliance on fossil fuels has shaped economies and contributed to global issues such as greenhouse gas emissions and climate change. Readers will gain insights into the socio-political dynamics of energy dependency, the environmental and economic consequences of resource depletion, and the pathways for decarbonization through global initiatives like the Paris Agreement.

- **Chapter I.1: Sustainable Development**
  This chapter defines sustainable development and introduces key indicators such as GDP and HDI to measure economic and social progress. It traces historical perspectives, emphasizing milestones like "The Limits to Growth" and the UN's "Our Common Future" report, which introduced sustainability as a balance between economic, social, and environmental systems. The chapter also examines the UN's 2030 Agenda for Sustainable Development Goals (SDGs), which serve as a blueprint for global progress. By integrating these frameworks, the chapter sets the foundation for understanding how development can be harmonized with environmental stewardship to ensure a sustainable future.

DOI: 10.1201/9781003631125-1

- **Chapter I.2: Energy Systems**
  This chapter explores the concept of energy systems, encompassing technologies, infrastructures, and policies that enable energy production, distribution, and consumption. It examines the economic and environmental aspects of energy demand and supply, emphasizing the need for efficiency, reliability, and sustainability. Components like energy sources, conversion technologies, storage systems, and energy carriers (e.g., electricity and chemical fuels) are analysed. The chapter also discusses the evolving role of renewables, the distinction between grey, blue, and green energy carriers, and the importance of smart energy management in addressing global challenges like climate change.
- **Chapter I.3: Energy Dependency**
  Energy dependency is examined through the lens of global reliance on fossil fuels, highlighting geopolitical and economic consequences. The chapter differentiates between fuel reserves and resources, analysing their implications for energy security. It provides a historical and contemporary analysis of oil, natural gas, and coal dependencies, supported by data on production, consumption, and reserves. The chapter concludes by discussing the inefficiencies and vulnerabilities in energy systems caused by these dependencies, emphasizing the urgency of diversifying energy sources and adopting sustainable alternatives.
- **Chapter I.4: Energy Sources Exhaustibility**
  This chapter delves into the finite nature of fossil fuels, such as oil, coal, and natural gas, and their long-term availability. It highlights the rate of consumption relative to resource regeneration, emphasizing the urgent need to transition to renewable energy. Case studies and global data are used to illustrate the regional disparities in fossil fuel availability and the challenges of resource extraction. The chapter underscores the critical role of technological innovation and policy reform in addressing the exhaustibility of energy sources and ensuring sustainable energy access.
- **Chapter I.5: Greenhouse Effect and Global Warming**
  This chapter provides an in-depth analysis of the greenhouse effect, focusing on the role of greenhouse gases (GHGs) in global warming. It explains the mechanisms of GHG emissions, their global warming potential, and their impact on climate systems. Simplified models for evaluating the greenhouse effect are presented, enabling readers to understand the interplay between anthropogenic activities and atmospheric changes. The chapter serves as a foundation for later discussions on decarbonization and climate mitigation strategies.
- **Chapter I.6: Global Warming Effects and Related Issues**
  This chapter examines the environmental, social, and economic consequences of global warming. It explores issues such as rising sea levels, extreme weather events, biodiversity loss, and their cascading impacts on food security, migration, and public health. The chapter also discusses the unequal burden of global warming on vulnerable populations and regions, emphasizing the need for global cooperation in climate adaptation and mitigation efforts.

- **Chapter I.7: Decarbonization Policies Design – The Kaya Identity**
  Using the Kaya identity, this chapter analyses the interplay between population growth, economic development, energy intensity, and emissions. It presents a structured framework for designing decarbonization policies, addressing factors like energy efficiency, emission intensity, and renewable energy adoption. The chapter integrates case studies and emission scenarios to illustrate the effectiveness of various policy interventions in achieving global and national climate goals.
- **Chapter I.8: Decarbonization Pathways**
  This chapter explores global decarbonization pathways, focusing on the Paris Agreement's goals and strategies for limiting global temperature rise. It presents emission scenarios, highlighting the role of renewable energy, energy efficiency, and carbon capture technologies. The chapter emphasizes the importance of international collaboration, equitable policy frameworks, and technological innovation in achieving a sustainable and low-carbon future.
- **Chapter I.9: Small Space–Time Scale Environmental Impacts**
  This chapter focuses on the localized environmental impacts of energy systems, particularly air pollution and urban energy technologies. It examines the diffusion mechanisms of primary and secondary pollutants, their health implications, and mitigation strategies. By analysing urban energy trends, the chapter underscores the importance of clean technologies and sustainable urban planning in minimizing localized environmental damage.
- **Chapter I.10: Energy Transition to Sustainable Systems**
  This chapter outlines pathways for transitioning from fossil-fuel-based systems to sustainable energy systems. It explores the technological, economic, and policy challenges of integrating renewable energy into existing infrastructures. The chapter highlights best practices, such as decentralized energy systems and smart grid technologies, as solutions to overcome barriers to sustainability.
- **Chapter I.11: Key Enabling Technologies for Energy Sources Conversion**
  This chapter delves into the technological advancements in renewable energy conversion, including solar, wind, hydro, geothermal, and biomass systems. It discusses the role of nuclear energy and emerging technologies in enhancing energy efficiency and reducing emissions. The chapter provides insights into the potential of these technologies to transform the global energy landscape.
- **Chapter I.12: Sustainable Choice of Energy Carriers**
  This chapter evaluates energy carriers, such as electricity, hydrogen, and biofuels, based on their environmental and economic impacts. It discusses the transition from grey to blue and green carriers, emphasizing the role of renewable energy in creating sustainable energy systems. The chapter highlights the importance of energy storage and distribution infrastructures in supporting this transition.

- **Chapter I.13: Key Enabling Technologies for Final Energy Uses**
  This chapter explores the application of key enabling technologies in final energy use, focusing on transportation, residential sectors, and smart cities. It discusses innovations that enhance energy efficiency and reduce emissions, providing a vision for integrating these technologies into sustainable urban and rural environments.

1.1 Sustainable Development

1.1.1 ON THE DEFINITION OF DEVELOPMENT AND ON DEVELOPMENT INDICATORS

There is widespread agreement that all individuals worldwide have the right to pursue their development aspirations, provided these efforts do not hinder others from doing the same. However, defining "development" and consistently measuring it within and across various national borders still remains a challenge.

First of all, can we establish a clear definition of a population's level of development? The most common interpretation of "development" refers to it as an "incremental capacity to meet the collective needs of a population".

"Needs" in this context primarily encompass basic necessities such as: nutrition; access to drinking water; basic education; personal hygiene services; access to electricity; public health; employment opportunities; minimum subsistence income; and others.

In addition, secondary needs also have to be considered: advanced education; access to recreational and sports facilities; access to broadband connectivity; availability of public and private transportation services; and many more.

The prioritization and significance of all these needs vary depending on a population's current development level and on the availability of goods and services. This approach is therefore effective only when evaluating the progress of a single population over time but may be less suitable for comparing vastly different populations.

So, is it possible to measure nations' development in a precise and effective way?

To address the complexities of defining and measuring development, various indicators have been introduced. Two of the most commonly used are the Gross National Product per Capita (GDPpc) and the Human Development Index (HDI).

1.1.1.1 Gross National Product per Capita – GDPpc

Gross Domestic Product (GDP) is one of the most widely used metrics to estimate a population's development level. GDP represents the total dollar value of all goods and services produced within a nation over a year. Policymakers, economists, and researchers frequently use it to assess economic health and compare economies.

DOI: 10.1201/9781003631125-2

GDP can be calculated using three primary approaches:

- Production Approach: This method sums up the value added at each stage of production for all goods and services. It focuses on output levels and the efficiency of production.
- Income Approach: This approach considers all income earned by individuals and businesses in the economy, including wages, profits, rents, and taxes (minus subsidies). It effectively measures the earnings generated by economic activity.
- Expenditure Approach: This is the most common method and sums up all expenditures made in an economy:
  C, "Consumption": Spending by households on goods and services.
  I, "Investment": Spending on equipment, structures, residencies.
  G, "Government Spending": spending on goods and services.
  NX, "Net Exports": Exports (X) minus imports (M).

To summarize, GDP increases if more goods and services are produced and sold, and roughly represents the cumulated values in the current currency of the goods and services produced and consumed within a nation, both on private and/or public sides.

It is essential to distinguish between real GDP and nominal GDP. Nominal GDP reflects the economy's output without adjusting for inflation or deflation, while real GDP adjusts for these factors, providing a more accurate view of economic growth over time. This distinction is crucial for policymakers analysing economic performance across different periods.

For global comparisons, real GDP is often expressed in "International Purchasing Power Parity Dollars" (PPP$). PPP$ adjust for cost-of-living differences between countries, allowing for a more meaningful comparison, because 1$ may buy different goods in different countries. Virtual exchange rates between the local currencies and PPP$ are commonly defined and updated by international organizations (like the OECD or World Bank).

GDP serves as a key indicator of an economy's overall health. A rising GDP suggests economic growth, which may lead to more jobs and higher income levels, while a declining GDP can indicate recessionary periods. As already said, GDP is mainly used to compare economic performance between countries.

When GDP is divided by the number of residents in a nation, GDP per capita (GDPpc) offers insights into the living standards of individuals within different nations. It is certainly a valuable development and economic indicator, but still has several limitations about:

- Income Distribution: It does not account for internal income inequality.
- Informal Economy: Informal economic activities are not included.
- Environmental Impact: It does not account for the cost of environmental impacts.
- Social Welfare and Quality of Life: It does not measure quality of life, well-being, or poverty levels.

As depicted in Figure I.1.1, global income distribution has changed dramatically over the last century. In 1950, most of the world's population had extremely low incomes (consider that the x-axis is logarithmic), with a stark divide between regions.

By 1980, rapid industrialization had raised incomes in Europe and the Americas, yet Asia and Africa largely remained near the extreme poverty line. The income poverty line (IPL) is defined depending on the mean income level of a nation and is currently set, respectively, to about 70 $/month in low-income; 115 $/month in middle-income; 215 $/month in high-income countries.

By 2020, income distribution had shifted much more to the right, as many countries in Asia experienced rapid economic growth, reducing extreme poverty worldwide. The world may not be still divided in two distinct parts ("developed" vs. "developing" countries), even though African nations still have an exceptionally low daily income per capita. Substantial disparities persist, particularly in Africa, where daily income per capita remains much lower on average.

Economic progress has lifted many regions out of extreme poverty. However, significant inequalities persist, with millions still living below the local extreme poverty lines. As development continues to evolve, in fact, it is essential to consider both economic growth and the equitable distribution of wealth and well-being.

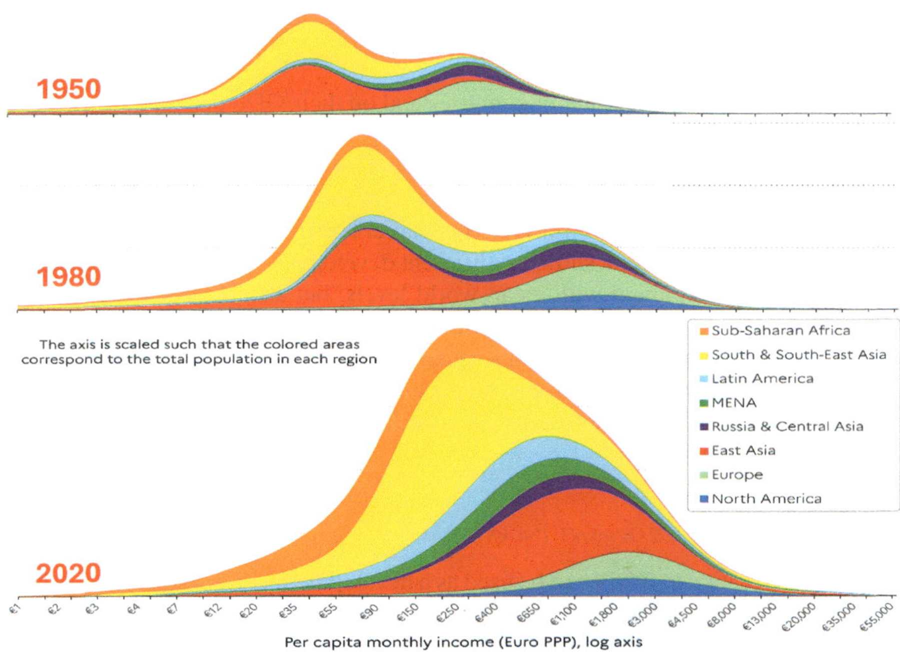

FIGURE I.1.1 Income inequalities: People vs. Monthly GDPpc in 1950, 1980, and 2020. Source: Chancel, L., Piketty, T., Saez, E., Zucman, G. et al. – World Inequality Report 2022, World Inequality Lab. wir2022.wid.world. Licensed by C.C. 4.0: https://creativecommons.org/licenses/by-sa/4.0/deed.en

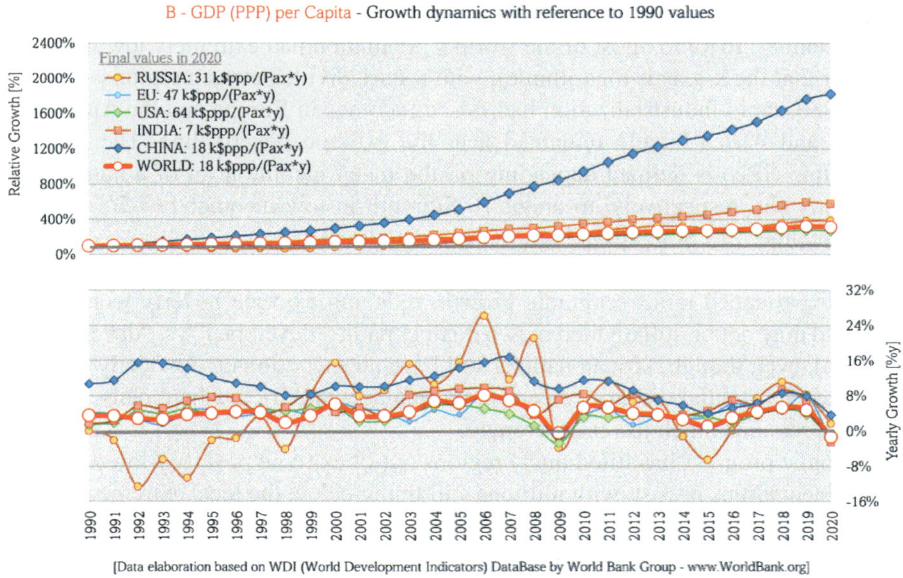

FIGURE I.1.2 GDP per capita time dynamics in world's most populous countries and aggregates.

Figure I.1.2 focuses on world major countries and on their GDP per capita dynamics over the last three decades.

While all countries show growth, China's rapid increase stands out, particularly over the past 15 years, narrowing its income gap with developed nations. India, though growing, remains at a lower GDP per capita level.

Data about GDP per capita show the different development levels reached in these countries and compare them with mean world levels, but also show their dynamics vs. time. According to their GDPpc trends, all the countries increased their development in this 30-year period, but China is certainly the country which has been better performing in the last 15 years, recovering only partially the gap it had with most developed countries. India also started its growth but still is characterized by much lower GDP per capita levels.

I.1.1.2 Human Development Index – HDI

A valuable alternative to GDPpc to be used as a development indicator is the so-called Human Development Index (HDI). HDI is a composite statistic used to rank countries based on human development levels. Introduced by the United Nations Development Programme (UNDP) in 1990. HDI incorporates three key dimensions (health, education, and standard of living) as reported below:

- The health dimension is measured by life expectancy at birth, reflecting the overall health and longevity of a population.

Population	[Millions]	8025	1403	3060	2811	723	1429	1411	335	278	240	224	216	173	146
	[World %]	100%	17%	38%	35%	9%	18%	18%	4%	3%	3%	3%	3%	2%	2%
GDPpc	[k$ PPP]	23	63	23	9	2	10	25	82	16	6	6	21	9	44
	[Global Ratio]	1,00	2,75	1,01	0,40	0,10	0,44	1,07	3,55	0,68	0,27	0,27	0,89	0,39	1,92
Electricity	[Pop %]	86	100	99	78	32	85	100	100	97	90	54	100	62	100
Lfe Exepctancy	[Years]	72	80	75	67	61	69	77	79	70	65	52	74	70	71
Literacy	[Pop>15 %]	85	..	95	73	59	..	..	..	95	57	..	92	61	..
Drinking Water	[Pop %]	88	99	93	85	52	89	92	99	88	89	67	98	97	96
Hand Washing	[Pop %]	..	..	..	56	21	60	..	..	77	61	30	..	38	..
HDI (*100)	1990	60	79	56	44	35	43	48	88	53	39	..	62	40	74
	2000	65	83	63	50	40	49	59	89	60	43	..	67	49	73
	2010	70	87	70	57	47	57	70	92	67	50	49	72	56	80
	2020	74	90	76	63	52	64	78	92	71	54	54	76	66	83
	2022	74	90	76	64	52	64	79	93	71	54	55	76	67	82
ΔHDI (*100)	1990-2010	10	9	14	13	12	14	22	4	14	10	..	10	16	6
	2010-2022	4	3	6	7	4	7	9	1	5	4	6	4	11	2
ΔHDI (*100)	1990-2022	14	12	20	20	16	21	31	5	19	15	..	14	27	8

FIGURE I.1.3 GDP per capita, HDI components, and HDI values vs. time.

- The education dimension considers both the mean years of schooling for adults and expected years of schooling for children.
- Finally, the standard of living is normally assessed using GDPpc.

Properly weighting the various indicators and dimensions, countries are assigned an HDI value ranging from 0 to 1, with higher values indicating greater human development. This index serves as a valuable tool for policymakers and researchers alike, as it encourages a broader understanding of well-being beyond mere economic indicators (like GDPpc). By focusing on health, education, and income, the HDI promotes a holistic view of progress and helps identify areas for improvement to enhance human welfare globally.

HDI can be used to address national policy choices, trying to understand if and how two countries with the same level of GDP per capita can have different human development outcomes, and therefore correctly influencing government policy priorities.

However, HDI is also a too simplified description of the actual development of a nation, capturing only some of the aspects that human development entails, and, like GDP, it still does not reflect inequalities, poverty, human security, empowerment, and other aspects.

Figures I.1.3 and I.1.4 report values for both GDP per capita and HDI among the biggest countries of the world, respectively, in a tabular and a graphical way. A description of the variation over time of the two indicators is also shown:

- The upper part of Figure I.1.4 shows global HDI maps in 2005 and 2022, highlighting that HDI increased at a mean of about 0.04, but inequalities among countries largely persist.
- The middle part reports a histogram representation of the data in Figure I.1.3, for the most populous countries and aggregates.
- The lower part depicts the correlation between HDI and GDP per capita for the world's most populous countries. Interestingly, the relationship appears more quadratic than linear, suggesting diminishing returns to HDI improvement as GDP per capita increases.

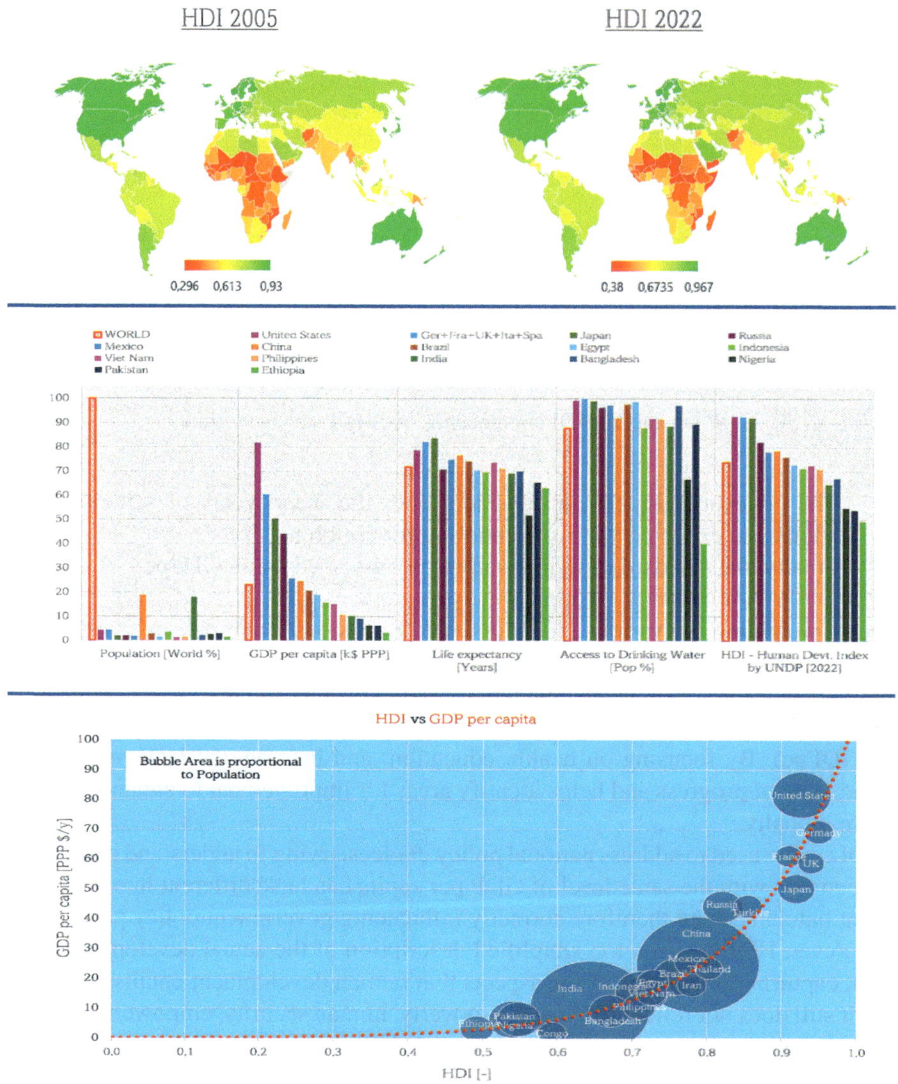

FIGURE I.1.4 GDP per capita, HDI components, and HDI values. Upper part: Global HDI maps in 2005 and 2022. Middle part: Histogram comparing values for the most populous countries and aggregates. Lower part: Correlation between HDI and GDP.

HDI and GDP per capita each offer unique insights into development, with GDP capturing economic output and HDI highlighting well-being through health, education, and living standards. Together, these indicators provide a more comprehensive picture of development, guiding policies that address both economic growth and human welfare.

1.1.2 BRIEF HISTORY OF INDUSTRIAL DEVELOPMENT

Following our discussion on the definitions and indicators of development, a natural question arises: how fast have global populations and economies evolved over the past centuries?

Understanding the history of development may provide valuable lessons for the future as we consider humanity's potential for continued growth.

Global income per capita has experienced significant growth since 1800, reflecting substantial worldwide development. But how did this transformation occur? What key events and innovations drove this progress?

The remarkable development observed over the past two centuries can be attributed to a series of at least four rapid growth periods, commonly referred to as "Industrial Revolutions".

1.1.2.1 Industry 1.0 – The First Industrial Revolution (1760)

The first Industrial Revolution began in the late 18th century, driven by the invention of steam engines and thermodynamic cycles powered by coal. These technologies industrialized manufacturing processes, particularly in the textile sector across Europe. Often referred to simply as "The Industrial Revolution", this period marked the advent of large-scale industrial production.

1.1.2.2 Industry 2.0 – The Electricity Revolution (1870)

The late 19th century witnessed the emergence of electricity as a transformative energy carrier, ushering in the second Industrial Revolution. Electricity enabled the creation of innovative technologies and services, particularly in urban areas, fundamentally reshaping industrial and societal structures. This era is often termed the "Electricity Revolution".

1.1.2.3 Industry 3.0 – The ICT Revolution (1970)

Following the global recovery from two World Wars, the 1970s brought the third Industrial Revolution. Key enablers were Information and Communication Technologies (ICTs), particularly the invention of microchips, which facilitated the development of industrial computers and IT systems. Automation of production processes, including the use of Programmable Logic Controllers (PLCs), further boosted productivity by partially replacing human labour in assembly lines. This period is known as the "ICT Revolution".

1.1.2.4 Industry 4.0 – The Internet and Robotics Revolution (2000)

At the close of the 20th century, advancements in robotics and high-speed connectivity fuelled the fourth Industrial Revolution. These technologies dramatically increased industrial productivity, giving rise to what is known as the "Internet & Robotics Revolution".

1.1.2.5 Industry 5.0 – Cybernetics, IoT, and AI Revolution (2020)

Today, we stand on the brink of a fifth Industrial Revolution – often called "Industry 5.0". This era builds upon the digital foundation of Industry 4.0 and is characterized by advanced technologies such as cyber-physical systems, the Internet of Things (IoT), artificial intelligence (AI), machine learning, big data, and cloud computing. Industry 5.0 emphasizes decentralized, intelligent, and adaptive production systems, enabling greater customization and efficiency.

However, the full potential of Industry 5.0 faces critical challenges, including energy availability, climate change, and resource limitations. Addressing these challenges will be essential to ensuring that technological advancements can be sustainably integrated into the global economy.

1.1.3 BRIEF HISTORY OF EARLIER DEVELOPMENT ECONOMICS

Before delving into the intrinsic link between development potential and energy availability, it is useful to explore the evolution of development economics over the past centuries.

Development economics has evolved in tandem with the Industrial Revolutions, as economists responded to emerging challenges and opportunities. Although a comprehensive history of economic thought is beyond the scope of this book, highlighting key milestones since the 18th century will illuminate how economic theories have shaped modern perspectives on development.

1.1.3.1 Classical Economics

Classical economic thought, embodied by Adam Smith (1723–1790), laid the foundation for capitalist economies. Smith's seminal work, *An Inquiry into the Nature and Causes of the Wealth of Nations* (1776), often referred to as *The Wealth of Nations*, remains influential. In this text, Smith introduced concepts such as productivity and free markets, which continue to underpin modern economic policies.

According to Classical Economics, a nation's incremental productivity (and thus its development) is determined by the availability of four factors of production: land, labour, entrepreneurship, and capital. These inputs were seen as essential for creating goods and services, with those controlling them often enjoying the greatest wealth. However, this perspective primarily considered resources as static inputs, overlooking the dynamic implications of resource depletion and environmental constraints.

1.1.3.2 Neoclassical Economics

In the late 19th century, neoclassical economists such as Alfred Marshall (1842–1924), John Maynard Keynes (1883–1946), and Arthur Pigou (1877–1959) reframed development as synonymous with "incremental productivity" rather than mere wealth accumulation. This era embraced a technocratic and anthropocentric vision, placing full faith in markets to adopt the best available technologies (BATs). While interactions between economic, social, and environmental systems were acknowledged, they were often underestimated.

One significant contribution from Pigou was the concept of "negative externalities" and the introduction of mechanisms, such as Pigouvian taxation, to compensate for environmental and social damages caused by economic activities. Despite advancements in economic thought during the 20th and 21st centuries, many governmental economic interventions still draw heavily from the foundations laid by Keynes and Pigou.

1.1.3.3 Evolutionary Economics and Sustainability

A major shift in economic thought came with Kenneth Boulding (1910–1993), a pioneer of evolutionary economics and a prominent figure in the discourse on sustainable development.

Kenneth Boulding was among the first economists to seriously explore the economic behaviour of finite systems with limited resources, a concept he called "Spaceman Economics". He contrasted this with traditional, unconstrained economic systems, which he referred to as "Cowboy Economics". Remarkably, Boulding introduced these ideas in the 1960s, a time when the world seemed almost boundless in its resources and technological possibilities, coinciding with the era of US and Soviet space exploration. Despite the optimism of that period, Boulding foresaw critical issues that have since become increasingly relevant.

In contrast to the expansive mindset of "Cowboy Economics", Boulding theorized a constrained economic system, the "Spaceman Economy", characterized by the efficient management of limited resources. He highlighted the risks of overexploitation and pollution, anticipating what we now term a "circular economy". In this model, resources are finite, and the surrounding environment imposes narrow limits on economic activities.

This raises a fundamental question for economists to address: Can our planet sustain continuous, exponential economic growth?

> Boulding's influential paper, *The Economics of the Coming Spaceship Earth* (1966), reshaped economic perspectives on development, emphasizing the finite nature of Earth's resources and the necessity of sustainability. This work remains a cornerstone of modern sustainable development theory, marking a pivotal transition in economic thinking. Its reading is certainly interesting, since it sets an important milestone for what happened later, and it is still happening nowadays. An extract from the paper is therefore reported here below.
>
> Extract from "The Economics of the Coming Spaceship Earth by K. E. Boulding, 1966":
>
> > Man has been accustoming himself to [...] a closed sphere of human activity. [...] In a closed system, the outputs of all parts [...] are linked to the inputs of other parts. [...] Three important classes are matter, energy, and information. The present world economy is open in regard to all three. [...]

The econosphere involves inputs of available energy in the form, say, of water power, fossil fuels, or sunlight, which are necessary in order to create the material throughput and to move matter from the noneconomic set into the economic set [...] In advanced societies this is supplemented very extensively by the use of fossil fuels, which represent as it were a capital stock of stored-up sunshine.

Because of this capital stock of energy, we have been able to maintain an energy input into the system, particularly over the last two centuries. [...] This supplementary input, however, is by its very nature exhaustible. [...] The large energy inputs which we have obtained from fossil fuels are strictly temporary. [...]

If the rest of the world were to rise to American standards of power consumption, and still more if world population continues to increase, the exhaustion of fossil fuels would be even more rapid. [...]

The time is not very far distant, historically speaking, when man will once more have to retreat to his current energy input from the sun. [...] The closed earth of the future requires economic principles which are somewhat different from those of the open earth of the past. [...]

I am tempted to call the open economy the "cowboy economy," the cowboy being symbolic of the illimitable plains and also associated with reckless, exploitative, romantic, and violent behavior, which is characteristic of open societies. The closed economy of the future might similarly be called the "spaceman" economy, in which the earth has become a single spaceship, without unlimited reservoirs of anything, either for extraction or for pollution, and in which, therefore, man must find his place in a cyclical ecological system which is capable of continuous reproduction of material form even though it cannot escape having inputs of energy. [...]

If there are infinite reservoirs from which material can be obtained and into which effluvia can be deposited, then the throughput is at least a plausible measure of the success of the economy. [...]

By contrast, in the spaceman economy, [...] the essential measure of the success of the economy is [...] the nature, extent, quality, and complexity of the total capital stock. [...]

What we are primarily concerned with is stock maintenance, and any technological change which results in the maintenance of a given total stock with a lessened throughput [...] is clearly a gain.

This idea that both production and consumption are bad things rather than good things is very strange to economists. [.... .]

The problems which I have been raising in this paper are of larger scale and perhaps much harder to solve than the more practical and immediate problems. [...]

One can hope, therefore, that as a succession of mounting crises, especially in pollution, arouse public opinion and mobilize support for the solution of the immediate problems, a learning process will be set in motion which will eventually lead to an appreciation of and perhaps solutions for the larger ones. My neglect of the immediate problems, therefore, is in no way intended to deny their importance, for unless we at least make a beginning on a process for solving the immediate problems we will not have much chance of solving the larger ones. On the other hand, it may also be true that a long-run vision, as it were, of the deep crisis which faces mankind may predispose people to taking more interest in the immediate problems and to devote more effort for their solution. This may sound like a rather modest optimism, but perhaps a modest optimism is better than no optimism at all.

I.1.4 ON "THE LIMITS TO GROWTH"

To answer this question, the Club of Rome (a nonprofit organization founded in 1968 at the Accademia dei Lincei in Rome) commissioned a groundbreaking study in the early 1970s. This group, comprising intellectuals, business leaders, former heads of state, UN officials, scientists, and economists, tasked a team of 17 researchers, led by Donella H. Meadows, Dennis L. Meadows, Jørgen Randers, and William W. Behrens III, with modelling the socio-economic dynamics of the world.

Using the computational power of centralized computers at MIT, the team developed a mathematical model of the global system, incorporating key variables such as population, industrialization, resource use, and pollution. Their findings were published in 1972 in a seminal report titled *The Limits to Growth*.

This was one of the earliest attempts to model the planet's interconnected systems, made possible by solving a set of interrelated differential equations. The goal was to determine whether growth trends could continue or if they would lead to overshooting the Earth's carrying capacity within the next century.

The right part of Figure I.1.5 shows the results of the report about the trends of the main variables defining world growth. Analysing them, the report concluded that:

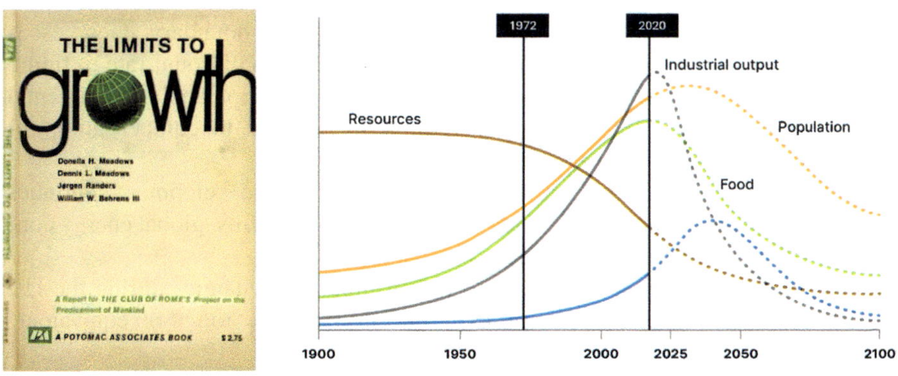

FIGURE I.1.5 Original cover and key graphical projections from *The Limits to Growth* report (1972), illustrating potential scenarios for global development based on different policy paths. Source: Meadows et al. (1972). Licensed under Creative Commons BY-NC license.

- If current trends in population growth, industrialization, pollution, food production, and resource depletion remain unchanged, the limits to growth will be reached within 100 years, likely resulting in a sudden and uncontrollable decline in population and industrial capacity.
- It is possible to alter these trends and establish ecological and economic stability that is sustainable over the long term. This would ensure that basic material needs are met, and individuals can realize their potential.
- Achieving this sustainable outcome requires prompt and decisive action. The sooner societies work toward this goal, the greater their chances of success.

Their conclusions were substantially right, even though the time previsions for model overshooting were wrong, due to errors in the estimation of the initial conditions of some variables, including energy fossil resources, as will be shown in detail in the next chapters of this book.

Nevertheless, the value of the message of the book still remains an unchanged acquisition in development economics: the world cannot develop without limitations and constraints, trying to "auto-solve the problems it generates both in terms of incremental exhaustion rapidity of many of its natural resources, and in terms of the incremental amplitude of externalities that the economic systems generates and reverts on the mutually interacting social and environmental eco-systems."

Development must be moderated and controlled to overcome these challenges, maintaining and sustaining a durable continuous development.

I.1.5 ON THE ROLE OF ENERGY IN DEVELOPMENT ECONOMICS

Traditional neoclassical economics simplifies reality by treating the economy as an almost self-regulating system, relying on a limited set of productive factors. In contrast, Romanian economist Nicholas Georgescu-Roegen (1906–1994), a founder of "Ecological Economics", argued that the economy is a subsystem of the broader, interconnected biosphere.

Ecological economics considers long-term, intergenerational issues, emphasizing the preservation of humanity without a temporal boundary. Georgescu-Roegen criticized neoclassical economics for applying a high discount rate to future generations' well-being, ignoring the costs of resource depletion and environmental degradation. He argued that future generations are entitled to compensation for inheriting a polluted and impoverished planet.

Energy is a dominant factor in this dynamic, especially as development continues to depend on finite fossil fuels. Despite recognition of these limits, global energy consumption has grown exponentially.

Figure I.1.6, for example, depicts energy consumption dynamics during the last 60 years. Yearly energy consumption is expressed in *Gtoe/y*. A "ton of oil equivalent" (*toe*) is equal to about 41.87 *GJ* or 11.63 *MWh*.

Global energy consumption in 2023 reached approximately 15 *Gtoe/y*, which means about 1840 *ktoe/y* per capita. This equates to an average daily per capita consumption of around 5 *kg* of oil equivalent, underscoring the scale of resources required to support current development levels.

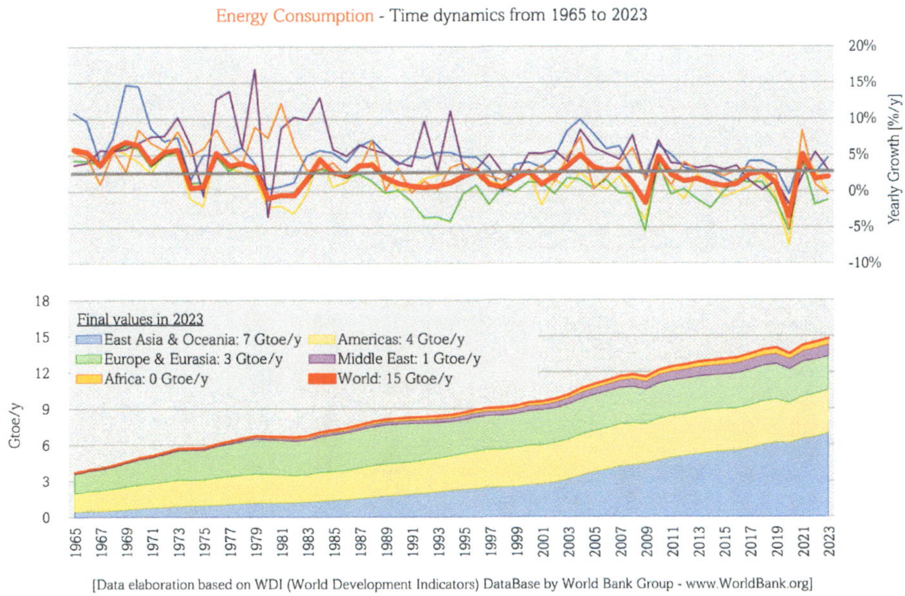

FIGURE I.1.6 Energy consumption time dynamics during the last 60 years (1965–2023), showing global and regional trends in total energy use.

As shown, energy consumption continuously increased, substantially ignoring any exhaustibility in energy sources used, nor considering the cost of the increasing environment externalities which are reverted and accumulated on the environment.

However, even though development economists, for decades, all agree in confirming this assumption, the evolution of the world economic system seems instead to absolutely underestimate the amplitude of any existing limit to growth, leading to almost exponential energy consumption growth.

These considerations are confirmed by further data reported data in Figure I.1.7 which disaggregate energy consumption, respectively, by world regions (left part of figure) and by energy sources.

As shown, especially in the Asia Pacific region, which has been the most rapidly developing area in the last two decades, the utilization of high-impact fossil energy sources (especially coal) is very relevant.

Lastly, Figure I.1.8 reports the per capita energy consumption evolution, confirming that it is still rapidly increasing in most of the world regions (especially in Asia), substantially evolving freely without limits or constraints.

I.1.6 ON "OUR COMMON FUTURE"

Another milestone in sustainable development was the 1987 United Nations report Our Common Future by the Brundtland Commission. This report introduced the concept of "Sustainable Development," defined as meeting present needs without compromising the ability of future generations to meet theirs.

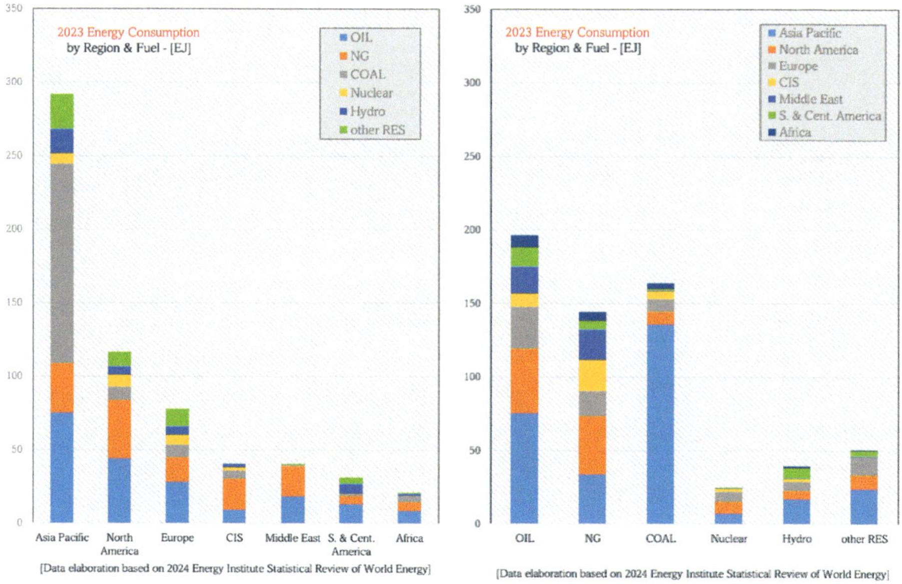

FIGURE I.1.7 Energy consumption disaggregated by region and fuel source in 2023, highlighting the reliance on fossil fuels in rapidly industrializing regions.

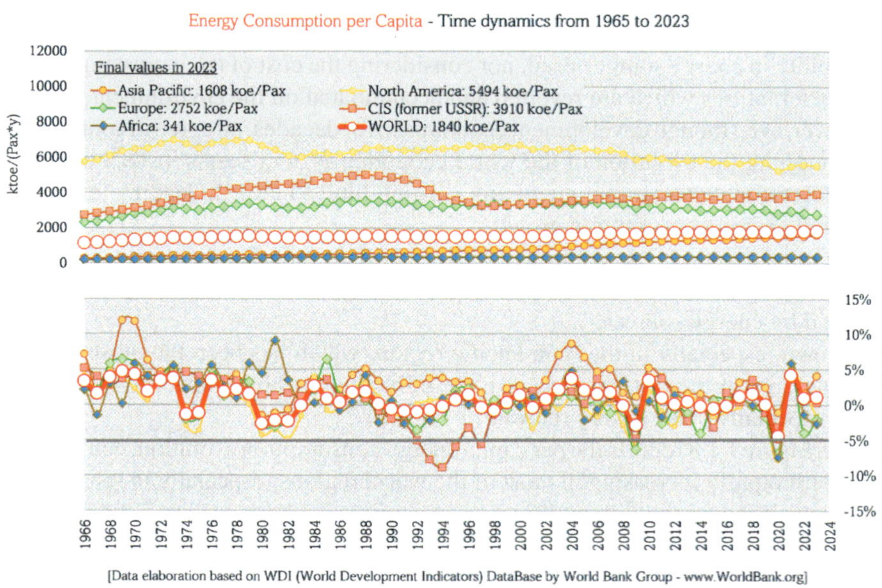

FIGURE I.1.8 Per capita energy consumption dynamics over the last 60 years, illustrating rising energy demand across different world regions.

FIGURE I.1.9 Left part: "Our Common Future" report cover, 1987. Source: World Commission on Environment and Development (1987). Right part: A sketch describing sustainable development characteristics.

The report reinforced ideas from Boulding and Georgescu-Roegen, emphasizing that sustainable development requires balancing economic, social, and environmental priorities.

The right part of Figure I.1.9 reports a currently widely accepted graphical representation of sustainable development, showing the inter-dependencies between the three essential subsystems deifying our eco-sphere: Economics, Society, and Environment. According to the report, development can only be sustainable if equal importance and dignity are given to all three systems, and if their interactions are considered and balanced. This approach ensures that economic growth, social progress, and environmental protection are integrated in a way that supports long-term, intergenerational well-being.

In fact, each of these systems plays a crucial role in shaping the dynamics of the biosphere. Only by recognizing the interconnectedness of these domains, can nations create policies that allow for long-lasting, resilient development.

Since its publication, Our Common Future has served as a guiding document for sustainable policymaking, influencing environmental, economic, and social strategies worldwide.

I.1.7 THE UN 2030 AGENDA FOR SUSTAINABLE DEVELOPMENT

In 2015, the UN General Assembly adopted Agenda 2030, a global framework endorsed by all 193 member states. The agenda sets several Sustainable Development Goals (SDGs). These goals address diverse issues, from eradicating poverty to promoting clean energy, while emphasizing the interconnected nature of social, economic, and environmental priorities.

FIGURE I.1.10 UN Agenda 2030 Sustainable Development Goals (SDGs), illustrating the 17 goals that guide global efforts towards sustainable development.

The SDGs provide a comprehensive roadmap for addressing today's challenges while safeguarding the well-being of future generations. By promoting integrated strategies, Agenda 2030 aims to build a sustainable, equitable, and prosperous future for all.

SDGs are categorized in 17 Sustainable Development macro-Goals (SDGs), shown in Figure I.1.10, which serve as thematic areas within Agenda 2030, each encompassing specific targets, not only addressing energetic and/or environmental domains.

These goals address only global issues (with a global, shared, and undividable responsibility) and range from ending poverty, energy, promoting equality, and fostering peace and partnerships, to more energy-related issues, like protecting ecosystems (Goal #13: Climate Action), and ending hunger to ensuring sustainability (Goal 7: Clean and Affordable Energy).

The SDGs promote a holistic approach to sustainable development, recognizing that progress in one area often supports advancement in others. This integrated approach encourages countries to pursue strategies that balance economic, social, and environmental priorities. Agenda 2030 provides a comprehensive blueprint for global progress, uniting nations in a shared commitment to build a more sustainable, equitable, and prosperous future. The SDGs offer a roadmap for addressing today's challenges while ensuring that future generations can thrive.

In summary, the key concepts introduced in this chapter highlight that sustainable development must strike a balance between economic growth, social well-being, and environmental protection. Therefore, achieving long-term sustainability should require integrated policies that address these interdependent factors, ensuring that progress today does not compromise the well-being of future generations.

1.2 Energy Systems

In the previous chapter, we discussed the critical role of energy availability in driving and sustaining societal development. The broader concept of "energy system" encompasses the technologies, policies, and practices that determine how energy is produced, distributed, and consumed. So, exactly what is an "energy system"? How can we define it?

An energy system may be defined as: "the results of all the technological, political, economic, and social choices which are made to determine a territory/population energy demand and find the most viable ways to fulfil it".

Therefore, an energy system includes all those interconnected components and processes that produce, convert, store, distribute, and consume energy. It involves all the technologies, infrastructures, and practices used to generate and manage energy, whether from renewable, non-renewable, or other sources. Energy systems are designed to meet the energy demands of households, industries, transportation, and other sectors, while maintaining efficiency, reliability, and sustainability.

About the first part of the given definition, we must consider that the concept of energy demand is an economic one: not all the "desired" energy quantities, in fact, would be produced and delivered to final users, but only those that could be produced and delivered at a reasonably low price. Therefore, if a consumer is not willing to pay a sufficient price, his energy requirements will not be considered within national demand. Obviously, the production and delivery price of energy also depends on how the national energy system is realized, so that better national energy systems are also those which may increase energy demand, maintaining its price sufficiently low by making prices low.

The local price of energy, however, may strongly be affected by Pigouvian taxations, if a nation only should consider all the local and global damages created by the various phases of energy use. Therefore, the many political and social choices connected to all those aspects (as will be clearer in the next chapters of this book) also have to be considered a crucial part of an energy system.

An energy system can be local (e.g., a building's heating system or a district's power supply) or national/global (e.g., a country's electricity grid or the global oil market). Energy systems are critical to modern society and are a focus of innovation

DOI: 10.1201/9781003631125-3

and transformation, especially as the world transitions towards more sustainable and low-carbon energy solutions to address climate change.

Now, narrowing our attention to a more technological dominion, the key components of an energy system typically include:

- **Energy Sources**
  An energy system definition includes choices about the energy sources mix. These, in turn, could be:
  - Available at a territorial level or imported from elsewhere.
  - Non-renewable (e.g., fossil fuels like coal, oil, and natural gas [NG, mostly made by methane], or of limited availability like uranium and other fissile materials) or renewable energy sources (RES: solar, wind, biomasses, hydropower, geothermal, etc.).
  - Potential energies (chemical, both fossil and renewable, gravitational, nuclear), or flowing energies (solar, wind, river waters, water currents, tidal waves, geo-thermal fluxes). This characteristic is strictly connected with the programmability of energy production, which is needed to fulfil a demand which is not fully predictable in time and space.
- **Energy Conversion Technologies**
  The process of converting raw energy sources into usable forms, such as converting coal into electricity through combustion in a power plant, or harnessing wind or sunlight to generate electricity.
- **Energy Storage Systems (ESSs)**
  Technologies that store energy for later use, such as batteries, pumped hydro storage, or compressed air energy storage. As we will see, the requirement for ESSs becomes much higher if the energy sources mix moves towards non-programmable RESs.
- **Energy Carriers (also referred to as Vectors)**
  These are the energy forms in which we chose to transfer energy from the place it is produced to where it will have subsequent transformation processes and/or final uses. Some important aspects about energy carriers must be underlined:
  - Some carriers are chemical: energy is transferred in a potential form, as chemical bounds present in specific fuel substances: gasoline, diesel, naphtha, kerosene, compressed or liquified methane (CNG and LNG), liquified petroleum gases (LPGs), compressed and liquified H_2 (CH_2 and LH_2) are among these. They have a limited need to be stored, and normally only require flexible but high energy-consuming infrastructures.
  - The most used energy carrier is electricity, which has been since the 19th century. This is mainly because its transmission and distribution grids are very low energy-consuming, because of the high transmission (and distribution) voltages.
  - The prevalence of electricity over chemical carriers is expected to increase in the future, for at least two reasons: using chemical carriers in final-use technologies produces much higher air pollution issues (especially in

urban contexts); conversion technologies for renewable sources are more oriented towards electricity than chemical carriers' production.

- A further distinction among energy carriers, which is increasingly used, is in grey, blue and green carriers: green carriers are those produced by RES; blue carriers are produced from non-RES capturing and sequestrating CO_2 (by means of so-called CCS – Carbon Capture and Sequestration technologies); grey carriers are produced from non-RES releasing CO_2 into the atmosphere.
- Chemical carriers may also be fully "synthetic": high-energy chemical bonds may be produced using raw materials like H_2O and CO_2, and electricity (which are both known as e-fuels, or synfuels). They are, therefore, a way to store and transfer electricity to some final uses which have difficulty in using electricity (usually referred as "hard-to-abate sectors"): among these, long-distant mobility applications are the most relevant example.

- **Energy Final Uses**
  These are the final uses of energy by consumers, whether in homes, factories, transportation, or other sectors. Figure I.2.1 illustrates the distribution of energy consumption across key sectors in IEA countries (which are the main global energy consumers [excluding China]). As of 2023, the breakdown has remained stable over the past two decades, with industry and services consuming 43%, residential sectors accounting for 22%, and transportation at 34%. Notably, transport is further divided into passenger transport (23%) and freight (11%). These patterns highlight very stable trends, and the persistent dominance of industrial and transportation energy demands.
- **Energy Management and Regulation**
  Policies, practices, and technologies that ensure the efficient, safe, and equitable delivery and use of energy. These include smart electric grid technology, pricing structures, energy efficiency standards, and environmental regulations.

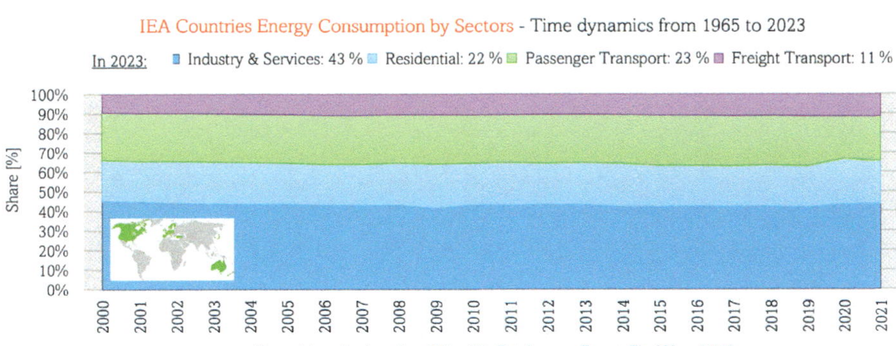

FIGURE I.2.1 Subdivision of world energy consumption trends by final uses.

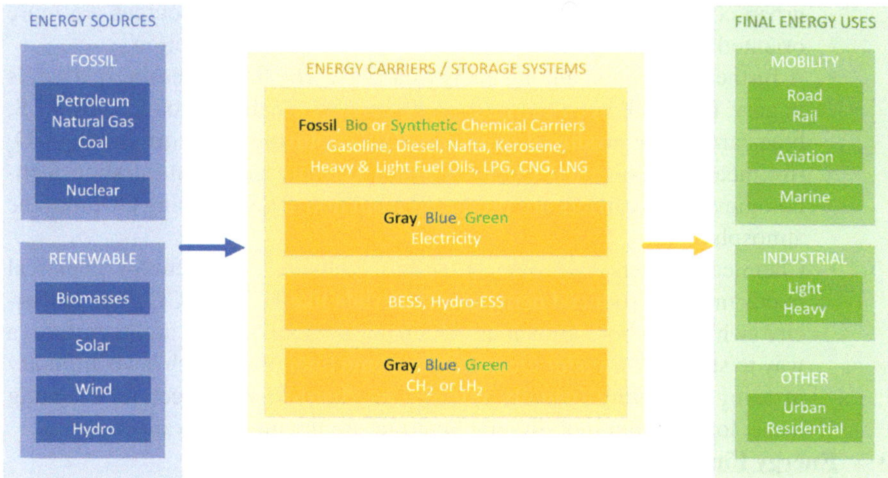

FIGURE I.2.2 Graphical representation of the technological parts of an energy system.

Figure I.2.2 provides a schematic view of an energy system, illustrating the flow from energy sources to carriers and final uses. It highlights the diversity of sources (both fossil and renewable) and the growing importance of energy storage and clean carriers like green hydrogen.

Having established the foundational components and challenges of energy systems, we are now equipped to explore the pressing sustainability and short-term issues they face. The following chapters will delve into key topics:

- Energy dependency (Chapter 3).
- Energy sources exhaustibility (Chapter 4).
- Global warming and climate change (Chapters 5–8).
- Small space–time scale environmental impacts and clean air issues (Chapter 9).

These discussions will provide the framework for understanding the energy transition towards more sustainable energy systems.

1.3 Energy Dependency

An energy system can ensure sustainable development only if it is based on energy sources whose availability in space and time can be assured, at stable and convenient prices for a long period, therefore not prejudging future generations' ambitions to maintain the same approach.

As we will see in detail in this chapter, sources unavailability was the first sustainability issue that most of the biggest and more economically competitive energy systems had to face during their development. As a consequence, our attention here will be prevalently dedicated to actual energy systems and their inter-dependency issues in recent decades, but it is noteworthy to underline that this problem is new.

Relevant conflicts were fought, for example, within Central Europe, during World War II to get control of the Ruhr area and related coal reserves. Still now, the Ruhr basin (in Germany) is one of the most important industrial areas in Europe, although its production is not predominantly based on coal extraction and utilization.

As is well known, fossil fuel sources are those that may cause greater dependency issues, being present and therefore produced only in restricted areas of the globe. However, how big are these dependencies? Are energy systems evolving to solve their inter-dependencies, or are the problems becoming increasingly critical as time passes by? Is the dependency only regarding oil? Or natural gas and coal also?

All these questions will be the subject of our detailed attention within the following sections. We examine and discuss data about fuel consumption, production, and reserves: data come from a public historical database made available for the scientific community by the Energy Institute (in particular from its 2024 Statistical Review on World Energy).

All the data elaboration shown in this and the following chapters were performed by the authors, who developed most of the graphical representations reported in this book.

Before going in depth into data analysis and discussion, the not-obvious distinction between fuel reserves and resources is introduced and discussed.

Fuel Reserves and Resources

Understanding the distinction between reserves and resources is critical in evaluating energy dependency. While resources represent the total potential availability of a fuel, reserves represent that part of the resources which are "recoverable". This

DOI: 10.1201/9781003631125-4

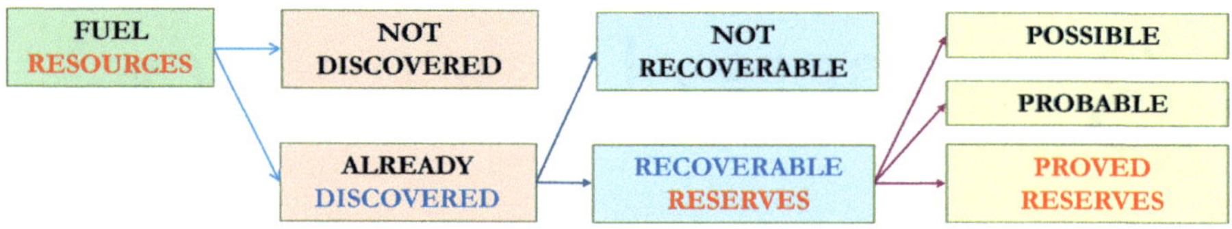

FIGURE I.3.1 Definition of the concepts of reserves and resources of non-renewable energy sources.

"recoverability" not only depends on technical advancements but also on geopolitical stability and economic feasibility, which are often unpredictable.

Figure I.3.1 defines these same concepts schematically according to commonly applied sectorial regulations. As illustrated, the term reserves therefore only addresses the portion of total fuel resources that have been discovered and are classified as recoverable, based on technical, economic, and political considerations. Figure I.3.1 provides a detailed breakdown of how resources are classified into reserves, highlighting the layers of technical and commercial uncertainty that influence their "recoverability".

Unfortunately, most world fuel reservoirs are almost inevitably in unstable conditions, which are generated, also, by the high economical value of these prime materials and on the consequent relevance for the consumer countries to somehow gain efficient control and/or good economic relations with the governments installed in the areas where the reservoirs are located. Due to all these considerations, the resources of each fuel are expected to be much larger than reserves.

So, how are fuel reserves and resources expected to evolve with time? Reserves of fossil sources are obviously expected to diminish over time because their consumption rate is much faster than the rate with which biomass fossilization creates new fuels. On the other hand, they may also increase due to new discoveries, diffusion of new and/or more economical technologies for their extraction, conversion, and final use, or political stabilization of geographic areas which may increase their recoverability.

These factors will be explored in greater detail in the next chapter, which focuses on the exhaustibility of energy sources. For the purpose of discussing energy dependency, instead, this chapter will rely solely on the actual data about proven fuel reserves.

I.3.1 OIL: PRODUCERS, CONSUMERS, AND DEPENDENCY ISSUES

Oil dependency is a critical challenge for modern energy systems, deeply tied to global energy security, economic stability, and environmental sustainability. As a finite resource with highly uneven geographical distribution, oil has long been a focal point of geopolitical strategies and international trade. This section examines global oil consumption, production, and reserves, highlighting the inherent dependency issues and their implications for energy systems worldwide.

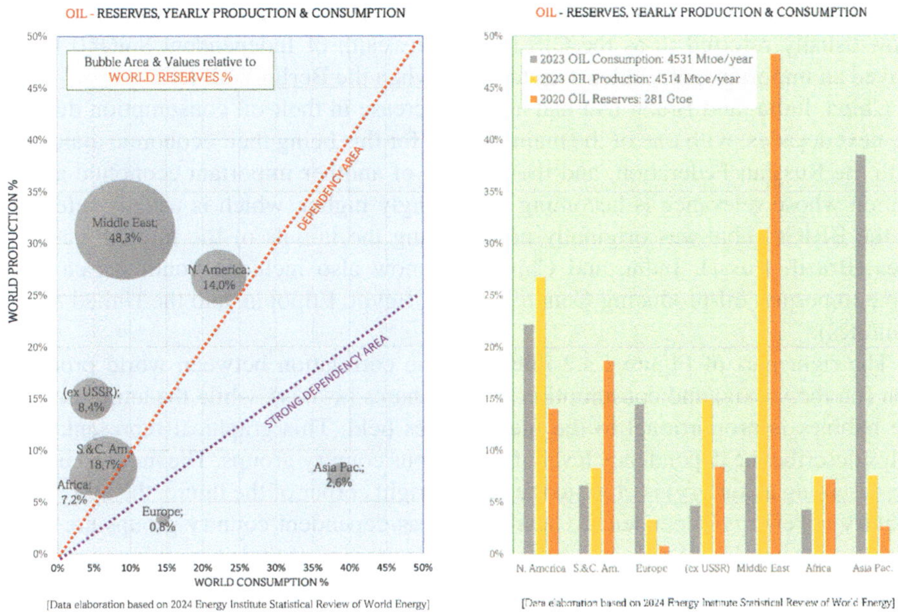

FIGURE I.3.2 2023 Data about oil consumption, production, and reserves in all world areas.

Figure I.3.2 reports data about oil consumption, production, and reserves in all world areas. The left part shows a histogram for three different sets of data, respectively, about oil consumption (in grey), production (in yellow), and proved reserves (in orange). Data are shown as percentages of total world values which are reported in the legend.

Data show that Asia Pacific countries (including China, India, Japan, South Korea, and Australia) accounted for approximately 38% of global oil consumption in 2023, driven by rapid industrialization and population growth. North America, led primarily by the United States, consumed 22%, while Europe contributed nearly 15%. Almost half (48%) of world reserves, however, are present in the very restricted Middle East area, which includes Iran, Iraq, Israel, Kuwait, Oman, Qatar, Saudi Arabia, the United Arab Emirates (UAE), Yemen, and other small countries. Other important oil reserves (19%) are present in South America (mainly in the Orinoco Belt in Venezuela, where some of the biggest world oil reservoirs, more than 15%, are located).

Coherently, countries owning most of the natural oil reserves are also major producers, even though the rhythm of oil production depends also on the commercial agreements stipulated with the major consuming partners. The situation, as is clear, is very bifurcated, with a few countries having the resources on one side (mainly grouped in the economic association named OPEC [Organization of the Petroleum Exporting Countries]), and the most consuming economies on the other side. For a long time, most of oil-buying countries following a capitalistic economy are grouped in an association called OECD (which evocatively stands for Organization for Economic Co-operation and Development).

Among the others, former USSR (Union of Soviet Socialist Republics) countries (now usually referred to as the CIS [Commonwealth of Independent States]) have played an important role, at least up to 1989, when the Berlin Wall fell.

China, India, and Brazil had had a huge increase in their oil consumption during the next decades, with one of the main reasons for this being their economic dialogue with the Russian Federation, and the creation of another important economic association whose relevance is becoming increasingly higher, which is usually referred as the BRICs. This was originally named using the initials of the founding countries (Brazil, Russia, India, and China), but now also includes South Africa and some important oil-producing Countries (Iran, Egypt, Ethiopia, and the United Arab Emirates).

The right part of Figure I.3.2 illustrates the correlation between world production (on the y-axis) and consumption market shares (x-axis), while the amplitude of the bubbles is proportional to the fuel reserves held. This graphical representation helps describe the dependency level of the various country groups. The more a bubble (representing a country) is close to the bottom right corner of the figure, the more that country is "energy-dependent". Clearly, the less-dependent country groups are also those represented by the biggest bubbles, due to the size of their remaining reserves. Figure I.3.3 further graphically crystallizes this situation, being relative to the import (blue bars, lower part of figure) and export (orange bars, upper part) percentage market shares.

Saudi Arabia and other Middle Eastern countries account for over one-third of global oil exports. Canada has also emerged as a significant exporter in recent years. (The specific role of Canada in the oil sector will be clarified in the next chapter.)

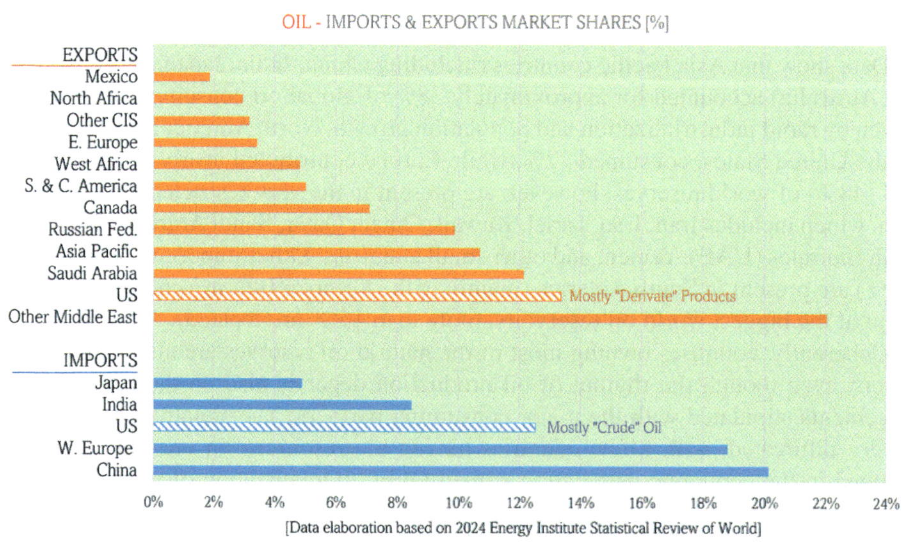

FIGURE I.3.3 Oil import and export main market shares for various countries.

Currently, the major oil-importing countries are China, Western Europe, the United States, India, and Japan, in that order. The United States occupies a unique position as both one of the world's largest oil exporters and importers. This dual role arises from the development of some of the world's largest oil-refining companies in the US. Over the years, the country has imported substantial amounts of crude oil while exporting refined oil products, particularly to Western Europe and other OECD member nations.

Regarding Western Europe, an additional important aspect is noteworthy to be discussed, which is not evidenced by Figure I.3.3: Norway's rise as a major exporter of oil and natural gas over the last two decades. In fact, Norway currently extracts oil from its sub-Arctic offshore reservoirs and supplies more than one-third of the oil consumed in EU countries, remaining outside of the European Union, so not being subjected to its environmental and climate regulations.

A significant portion of oil revenues in countries like Saudi Arabia, the UAE, and Norway is channelled into sovereign wealth funds. These funds are invested in acquiring controlling stakes in major multinational corporations worldwide, extending far beyond the fuel sector. As a result, oil revenues permeate global economics. Since the globalization era of the 1990s, these interconnected economic phenomena have expanded beyond the traditional scope of fuel trade, creating complex interdependencies that go far beyond the dynamics of simple fuel buying and selling (in which Figure I.3.3 is limited).

For example, the Norwegian Sovereign Wealth Fund is a prime case in point. It currently holds a capitalization of approximately $3 trillion, which is more than three times Norway's GDP. Although Norway has one of the world's most sustainable domestic energy systems (largely powered by renewable energy, even in challenging sectors such as passenger mobility) its energy transition was financed entirely through revenues from oil exports to the EU.

This highlights one of the many inherent contradictions in the energy transition strategies pursued by different nations.

The oil market remains highly inefficient due to its rigidity on both the supply and demand sides, dominated by a small number of key players. This structural inefficiency has persisted since the end of World War II, contributing to recurring crises.

Figure I.3.4 illustrates the volatility of global oil prices from 1970 to the present. These price fluctuations are often disconnected from changes in the actual cost of industrial production, underscoring the complex dynamics of the oil market.

Oil price was, in fact, influenced by a number of external socio-political and even sanitary phenomena (as highlighted in the figure), such as the Yon Kippur War and Iranian Revolution in the 1970s, or the diffusion of revolutionary movements within Middle Eastern countries known as the Arab Spring at the start of this century, coming to the visible effects of the COVID-19 pandemic more recently (mostly due to consumption contraction).

Regardless of the numerous socio-political and ethical considerations surrounding this unreasonable situation, it is clear that oil dependency remains a significant unresolved issue and a primary driver of the unsustainability of modern energy systems worldwide.

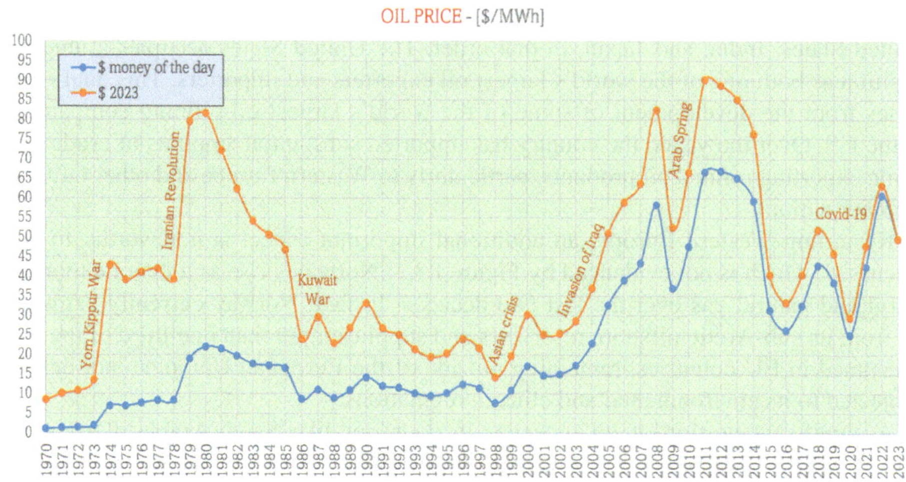

FIGURE I.3.4 Mean world oil buying price oscillations during the last 55 years.

I.3.2 NATURAL GAS: PRODUCERS, CONSUMERS, AND DEPENDENCY ISSUES

Let's now use the same instruments to analyse the natural gas sector.
Figure I.3.5 reports data about NG consumption, production, and reserves in all world areas. Cumulative Data here are in conventional cubic metres of standard natural gas, each of which is equivalent to 8.816 kWh.

Data show that North America (mostly the US alone) is the biggest NG consumer (about 28%), followed by Asia Pacific countries (24%) and former USSR countries (about 15% vs. less than 5% in the oil sector). This situation is the result of a distribution of NG resources which privileges ex-USSR countries (holding 30%). The Middle East also has the biggest NG world reserves (more than 40%) but only started to extract it from its soil during the last few decades (after the Berlin Wall fell).

Before the 1990s, in fact, one could say, in short, that capitalistic economies were mainly basing their energy systems on oil, while communist countries were using natural gas and coal more widely. Only after the Berlin Wall fell did the huge availability of natural gas coming from the Russian Federation become a substantial occasion for energy independence for EU countries which started diverting their energy systems more and more towards methane.

The transition to natural gas as a key energy source brought with it logistical challenges. Unlike oil, natural gas transportation demands specialized infrastructure. Compressed natural gas (CNG) requires pressurized pipelines and compressor stations, necessitating significant investment and long lead times. Liquefied natural gas (LNG) offers flexibility in transportation, and can be shipped globally.

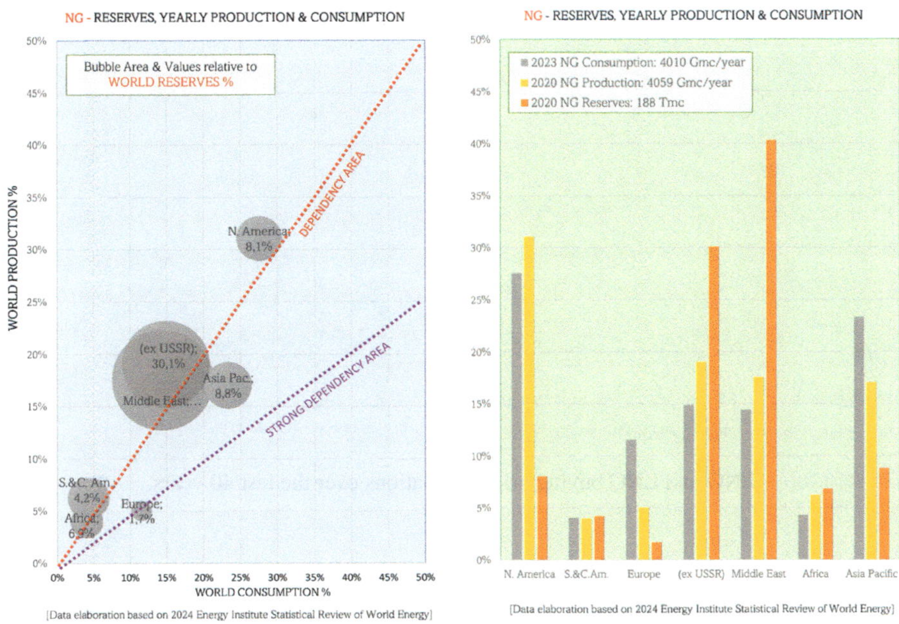

FIGURE I.3.5 2023 Data about natural gas consumption, production, and reserves in all world areas.

However, the cryogenic processes required for liquefaction are energy-intensive, requiring temperatures below –163°C, alongside costly insulated storage and regasification systems. Consequently, LNG tends to be more expensive than CNG, particularly in regions with established pipeline networks.

The geopolitical implications of infrastructure reliance became evident during the Ukraine crises in 2008 and 2022. As a critical transit country for Russian gas to Europe, disruptions in Ukraine highlighted the risks of dependency on singular supply routes.

Figure I.3.6 illustrates the volatility of natural gas prices over the past 40 years, with sharp increases during periods of geopolitical tension, such as the annexation of Crimea in 2008 and the escalation of the Russia–Ukraine war in 2022.

Europe's dependency on Russian natural gas has driven significant policy changes. The development of alternative pipeline routes, such as those through the Balkans or Turkey, aims to connect Europe with Middle Eastern gas suppliers, bypassing Ukraine and Russia. Simultaneously, Europe has expanded its LNG import capacity, sourcing gas from the United States (enabled by advancements in fracking) and Middle Eastern producers such as Qatar. Despite these efforts, the transition remains incomplete, as pipeline infrastructure inherently lacks the flexibility of LNG shipping.

Figure I.3.7 reports the situation in import and export market shares in the natural gas sector in compressed (CNG) and liquified (LNG) forms.

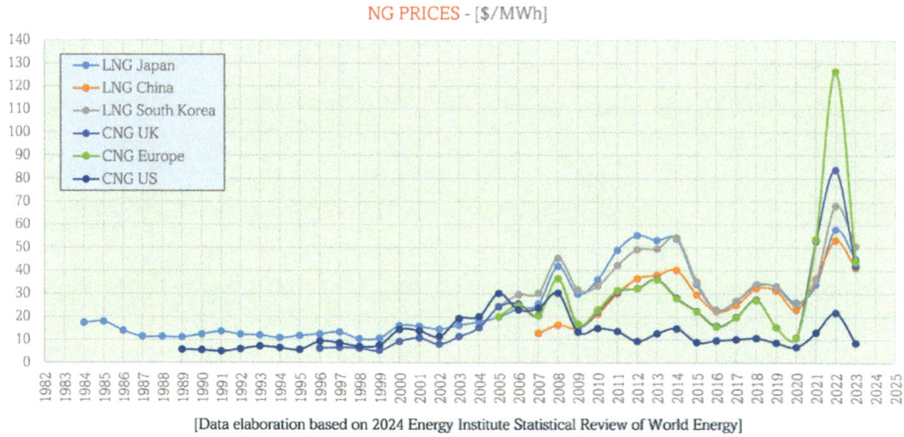

FIGURE I.3.6 CNG and LNG buying price oscillations over the last 40 years.

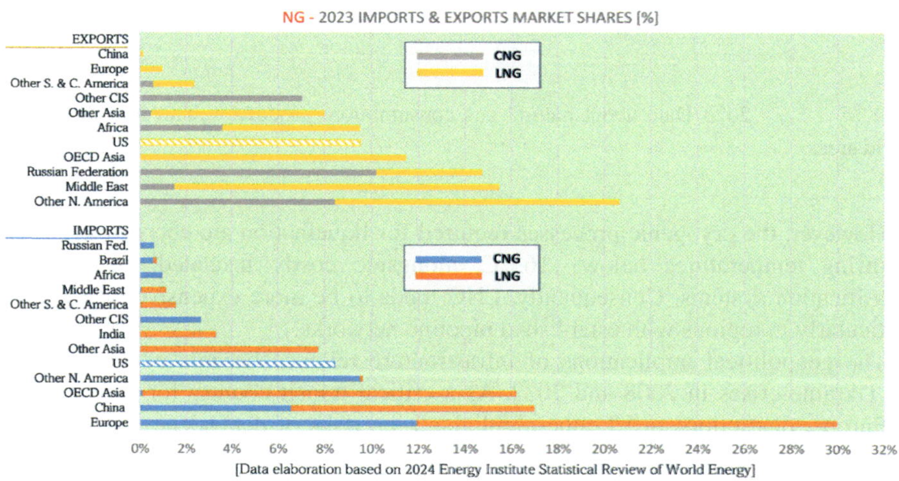

FIGURE I.3.7 CNG and LNG import and export main market shares for various countries.

I.3.3 COAL: PRODUCERS, CONSUMERS, AND DEPENDENCY ISSUES

Let's now consider the coal sector. Coal remains a critical component of the global energy system, particularly in rapidly industrializing nations. However, what do the data reveal about global coal consumption, production, and reserves? How dependent are energy systems worldwide on this resource, and what does that mean for energy security and sustainability?

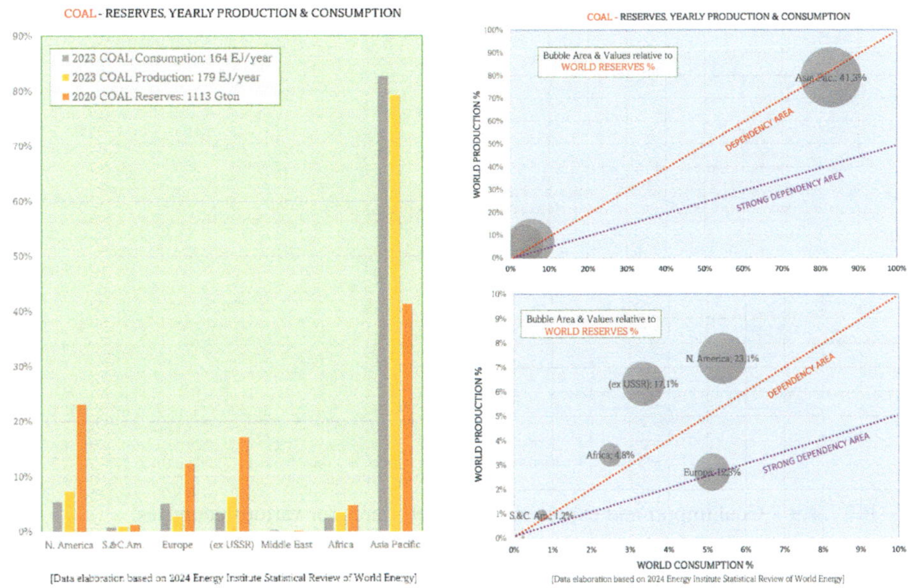

FIGURE I.3.8 2023 data about coal consumption, production, and reserves in all world areas.

To discuss this issue, again, we will use the same types of data and approach analysis used for oil and NG.

Figure I.3.8 reports data about coal consumption, production, and reserves in all world areas. Cumulative data here are in conventional tons of coal, each of which is equivalent to 8.141 MWh.

This figure clearly shows how large the most important actors in the coal sector are, with Asia Pacific countries accounting, in 2023, for 84% of world consumption, 79% of production, and still holding more than 40% of coal world energy reserves.

When India and China started to develop their economics very rapidly (from the 1990s onwards, as already described in Chapter 1), they did it mainly using their own fuel reserves, so they used mainly coal (also producing very severe local pollution, as we will briefly discuss later). This has not produced strong dependency issues for their energy systems but is a huge problem from other viewpoints, as we will show clearly later, especially if global warming issues are included in the discussion.

Figure I.3.9 depicts the global trade in coal, revealing a distinct pattern of exporters and importers. Indonesia and Australia stand out as the largest exporters, together commanding a significant share of the international coal market. These two nations are followed by other prominent coal-exporting countries such as Russia, South Africa, and the United States, which also contribute considerably to global coal supplies.

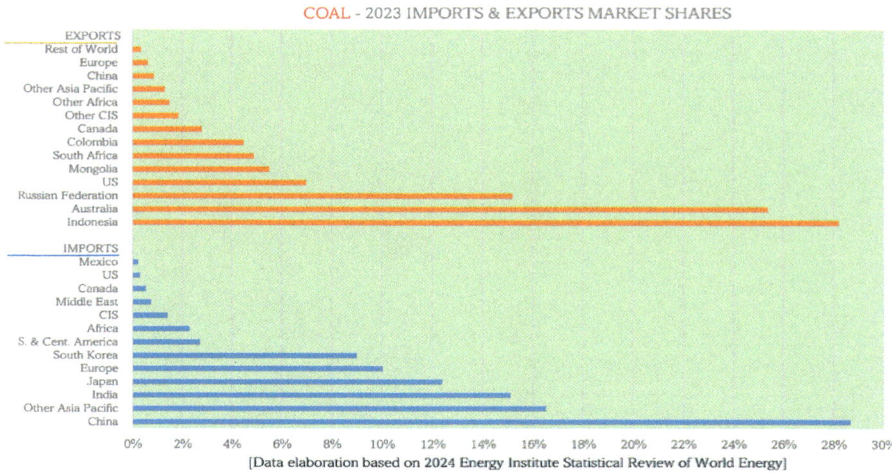

FIGURE I.3.9 Coal import and export main market shares for various countries.

On the importing side, China emerges as the dominant player, reflecting its immense demand for coal to sustain its industrial activities. Other major importers include India, Japan, South Korea, and various European countries, all of which rely on coal imports to meet their energy and industrial requirements. These trends highlight the complexity of the trade relationships underpinning the global coal market.

1.4 Energy Sources' Exhaustibility

As underlined in 1972 by the famous Report on "The Limits to the growth", cited in Chapter 1, one of the main sustainability issues of an energy system could come from the exhaustibility of the energy sources they rely upon. Was the report right in its prevision? Is the time horizon to fossil energy resources exhaustion reducing as they foresaw?

In this chapter we will discuss these questions: anticipating some of the conclusions that will arise from this discussion, we can say that the report's qualitative predictions were and remain valid, while its quantitative forecasts were partially wrong and have not fully materialized. In fact, fortunately, during the last 50 years our time horizon to exhaustion on a world scale still has not substantially reduced: how was this possible? And again, is this situation expected to be maintained, or will fossil fuels effectively finish in the coming decades?

To explore these issues, it is vital to revisit a key concept introduced in the previous chapter. There, we focused primarily on the short-term availability of fossil fuels, emphasizing proven reserves – quantities that are immediately available for exploitation. Instead, on a wider time-term, it would be more correct to refer to whole fuel resources, these probably being much bigger than reserves. Unfortunately, accurately forecasting overall fuel resources is not possible, making it necessary to rely on reserve values for both mid- and long-term analyses.

Anyhow, we should keep in mind that reserves of fossil sources not only are expected to diminish over time because of their consumption, but also, on the other hand, they may increase due to new discoveries, diffusion of new and/or more economical technologies for their extraction, conversion and final use, or due to the political stabilization of geographic areas which may increase their recoverability.

1.4.1 OIL: RESERVES, RESOURCES, AND EXHAUSTIBILITY ISSUES

Let us now look in some detail at what has happened during the last few decades to proved oil reserves and try to draw some conclusions about their possible exhaustibility time-horizon. To do that we will make critical use of a very simple time-to-exhaustion

DOI: 10.1201/9781003631125-5

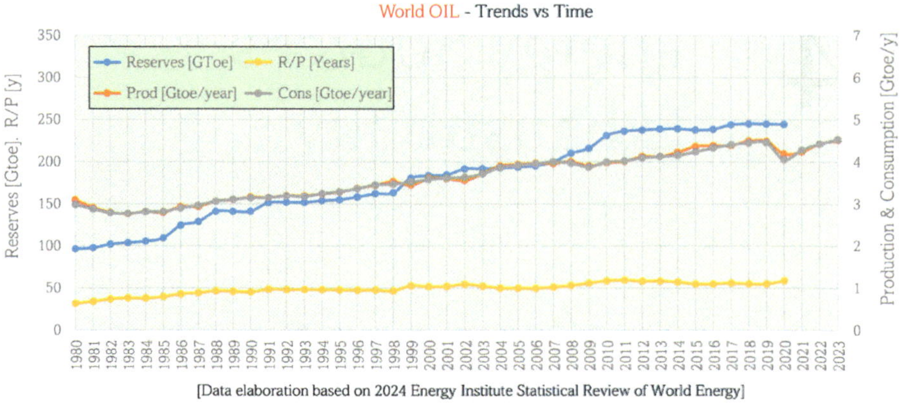

FIGURE I.4.1 World oil reserves, production, and consumption & R/P ratio vs. time.

indicator, which is called the "R-on-P ratio" (or simply R/P). This is calculated by the simple division (expressed in years) between time-varying reserves of a fuel and yearly production of the same fuel.

Both the numerator and denominator of this ratio vary as time goes by, with their own trends. However, if we make the hypothesis to maintain fuel production constant, and that new fuel quantities are not discovered nor made "recoverable" for a certain period, the R/P would be expected to decrease linearly with a –1 slope (decreasing by one year for each year that passes).

Instead, as is clearly shown by the data plotted in Figure I.4.1 referring to the oil sector, what happened in the real world, in a somewhat surprising way, was that both R and P increased very rapidly. R increased more rapidly than P, so that the R/P ratio not only did not decrease but increased as time passed.
Referring to Figure I.4.1, in particular, in the time span 1980–2020, production increased by about 35%, but reserves increased by almost 154%, so that R/P passed from about 31 years as estimated in 1980, to 58 years in 2020.

What was the reasons for this huge and somewhat unpredictable increment in available and proved reserves? As already underlined, the reasons why reserves can increase are new discoveries, diffusion of new and/or more economical technologies for their extraction, and/or the political stabilization of geographic areas.

All these things have happened in the last four decades. As shown also quantitively by data in Figure I.4.2, new relevant reservoirs were discovered in Saudi Arabia, Libya, the UAE, and Nigeria, and new production technologies were applied both in Canada and in Venezuela to extract petroleum oil from so-called oil sands, particularly available in Canada at high-latitude terrains, making those already known resources now recoverable, although at a higher industrial cost.

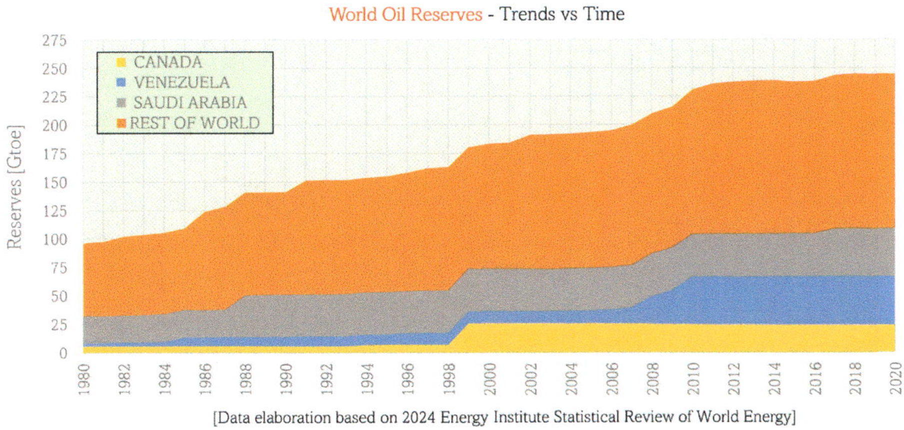

FIGURE I.4.2 World oil reserves in the main world sites vs. time.

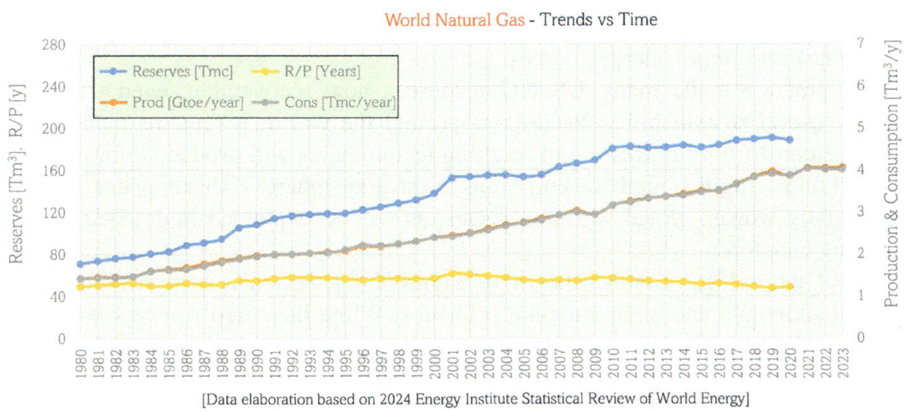

FIGURE I.4.3 World natural gas reserves, production, and consumption & R/P ratio vs. time.

I.4.2 NATURAL GAS: RESERVES, RESOURCES, AND EXHAUSTIBILITY ISSUES

Let us now take into consideration world natural gas (NG) trends for the same variables.

The trends in global natural gas production, reserves, and R/P ratios, as shown in Figure I.4.3, parallel those observed for oil, albeit with distinct differences in scale.

Production increased by about 171%, confirming a relevant movement of world energy systems towards NG, but reserves correspondently grew by almost the same amount (165%), so that R/P remained stable, increasing from 49 years in 1980, to 50 years in 2020.

This stability reflects advancements in exploration technologies, more efficient extraction methods, and the utilization of unconventional resources, such as shale

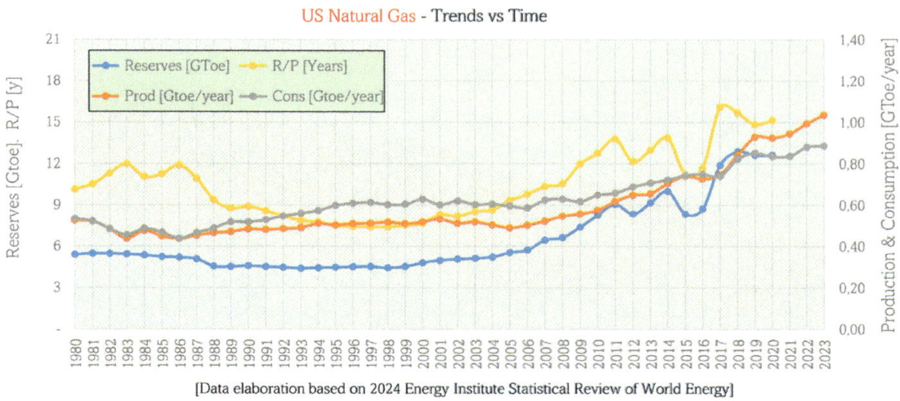

FIGURE I.4.4 US natural gas reserves, production, and consumption & R/P ratio vs. time.

gas. Figure I.4.4 highlights the transformative impact of fracking technology in the United States, which has not only bolstered domestic production but also reduced US dependency on foreign energy sources.

This practice made many US NG resources now recoverable, even if it has created several sustainability discussions around the world. It remains unclear, in fact, whether diffused utilization of fracking technologies will produce relevant present and future fugitive methane emissions in the atmosphere. If this happens, it may substantially worsen global warming issues, as will be explained in detailed in the following chapters.

Global natural gas trends, shown in Figure I.4.4, reveal a complex interplay between economic and environmental priorities. While increased reserves and production underscore the resource's pivotal role in meeting energy demands, they also raise questions about their compatibility with long-term climate goals.

As the world seeks to transition towards sustainable energy systems, the role of natural gas, particularly in tandem with carbon capture technologies and renewable substitutes like biogas, will probably be critical in shaping the energy landscape.

I.4.3 COAL: RESERVES, RESOURCES, AND EXHAUSTIBILITY ISSUES

Lastly, let us look at coal trends for the same variables. Coal, long the backbone of industrial energy systems, exhibits unique trends in reserves, production, and exhaustibility when compared to oil and natural gas.

Global coal trends from 1981 to 2020 are depicted in Figure I.4.5. A distinct scenario emerges, where coal production nearly doubled (+98%) globally during this period, even though most OECD countries notably reduced their coal use. Reserves increased at a much slower rate (+16%).

The result of these trends was a very notable decrease in the R/P ratio, which dropped from 237 years in 1981, to 139 years in 2020. The R/P ratio therefore

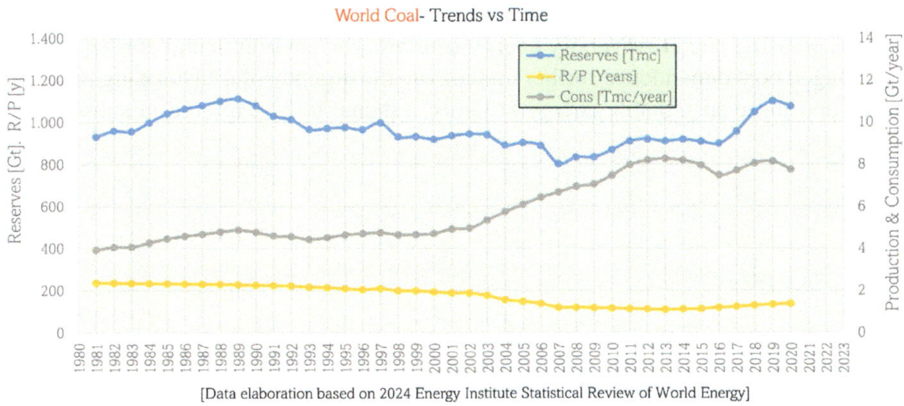

FIGURE I.4.5 World coal reserves, production, and consumption & R/P ratio vs. time.

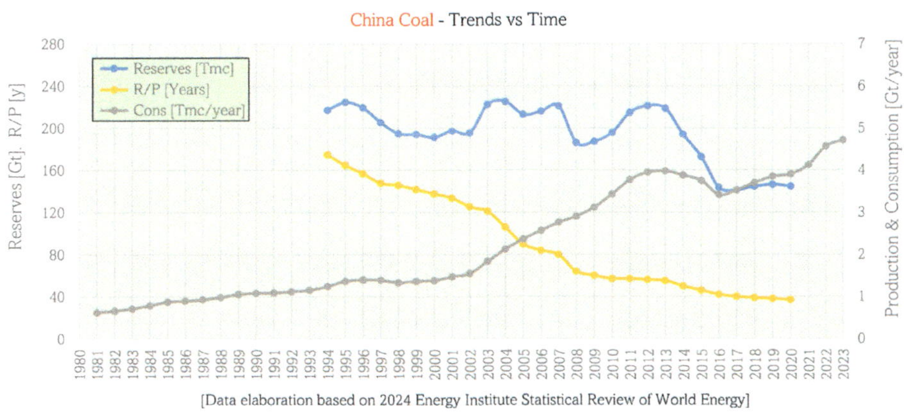

FIGURE I.4.6 China coal reserves, production, and consumption & R/P ratio vs. time.

decreased much more than linearly, losing a mean of more than 2.5 years each year. This probably means that residual coal will last much less than the forecast 139 years.

The primary driver of these trends has been China's remarkable economic expansion, heavily reliant on its domestic coal resources (Figure I.4.6). Over the last two decades, China's reliance on coal has not only alleviated pressure on world oil and natural gas reserves but also created unprecedented environmental challenges. Locally, this reliance has led to widespread air pollution and degraded ecosystems, while globally, it has significantly amplified CO_2 emissions, a critical factor in accelerating climate change.

Figure I.4.5 highlights the accelerated consumption of coal reserves worldwide, while Figure I.4.6 underscores China's pivotal role in these trends, as its reserves deplete to sustain soaring production levels. China's trajectory reflects the broader trade-offs between rapid economic development and long-term environmental sustainability.

While coal remains abundant compared to oil and natural gas, its declining R/P ratio signals a pressing need to diversify energy sources. The environmental consequences of prolonged coal dependence further underscore the urgency of transitioning towards cleaner and more sustainable energy systems, a theme to be explored in subsequent chapters.

1.5 Greenhouse Effect and Global Warming

1.5.1 SPACE–TIME SCALES OF ANTHROPIC INTERACTIONS WITH THE BIOSPHERE

As already introduced, the main sustainability issues of energy systems may depend on the way they interact with the external environment producing damage to the soil, water, and air.

Each energy process, in fact, generates environmental interactions, but these can be more or less relevant for the unit energy source chosen, energy carrier distributed, and energy utilization for its final uses.

Some of these interactions produce damage at a limited space distance and for a limited time with respect to the time and point where and when the interaction was effectively put in place. Even though these could be very dangerous and harmful, those small time-scale interactions are implicitly not creating sustainability issues, but local problems only impact on current generations, that are still very relevant in most cases but have to be solved locally by administrative authorities.

These problems will therefore not be the object of our attention now, but will be treated later in this book.

Instead, let us now look at those interactions that produce damage at a wider scale both in space and time. Obviously, the wider the scale, the more those interactions could create very dangerous and harmful sustainability issues. Interactions, in fact, can be cumulated in space and time and produce cumulative effects which may easily become uncontrollable on a global scale, affecting future generations.

Significantly, the more harmless a single interaction is, the more it tends to be neglected by the community, so that it cumulates in space and time and become more harmful from a sustainability point of view.

This is certainly the case with some atmospheric interactions produced by the emission of gaseous substances in air by industrial processes. Especially those emissions which are not toxic, nor chemically reactive or soluble by atmospheric water vapour, can remain there for decades. During this time, they accumulate and slowly diffuse into the whole atmosphere so that after a long period of time they can produce significant modifications to the overall thermo-physical properties of the atmosphere, and, consequently, of the whole biosphere.

DOI: 10.1201/9781003631125-6

These simple considerations generate a number of questions to be solved to understand the possibly generated issues. Some of these are listed here. We will try to answer only the first of them in this chapter, while the others will be the subject of our attention in the following chapters.

- What are the main gases emitted by human activities which are expected to create substantial changes in the thermo-physical behaviour of the atmosphere? How does this happen?
- What are the main modifications we are producing in the atmosphere and what are the effects of these modification on the whole biosphere? (Chapter I.6)
- Is it possible to design the interventions on energy systems we must implement to control and limit the future incremental impact we will produce on the biosphere? (Chapter I.7)
- Are these interventions being implemented effectively? What pathways can solve these problems, and are we currently on the right track? (Chapter I.8)

I.5.2 ON THE THERMO-PHYSICAL BEHAVIOUR OF THE ATMOSPHERE

The so-called greenhouse effect (GH in the following) is the process by which the Earth's temperature rises due to the presence of certain gases in the atmosphere (such as water vapour, carbon dioxide, methane, nitrous oxide, and ozone) that can absorb energy at specific wavelengths. These gases are known as greenhouse gases (GHGs) because they cumulatively make the atmosphere produce the same effects on the planet that the glass panels of a greenhouse produce on its content. Summarizing, our atmosphere allows sunlight to enter our biosphere but prevents a part of the generated heat escaping from it.

In fact, most of the sunlight passes through the atmosphere almost unaltered and reaches the Earth's surface, it is absorbed by the land and oceans, and is then re-emitted back into the atmosphere. In this process of absorption and re-irradiation, this energy amount changes its electro-magnetic characteristics.

Incoming sunlight, in fact, is mostly constituted by visible electromagnetic waves (coming from a high-temperature emitting source, the sun), while the energy re-emitted by low-temperature emitting bodies (on the Earth's surface) is mostly made by waves in the invisible infra-red (IR) spectrum. While some of these re-emitted waves escape into space, many of them are absorbed by atmospheric GHGs.

This natural process keeps the Earth's average temperature warm enough to sustain life. Therefore, GH is definitely a positive effect, since without the presence of GHGs in the atmosphere, the planet would be more than 30°C colder, making water mostly ice and the world uninhabitable for most organisms.

However, if GH intensifies due to an increase in GHGs, it can lead to excessive warming, with potential negative impacts on ecosystems and human societies. Even small shifts in global temperature, in fact, can cause disruptions to climate patterns, agriculture, and biodiversity.

I.5.3 ON GREENHOUSE GASES AND THEIR GLOBAL WARMING POTENTIAL

As explained, GHGs are natural or human-made gases in the Earth's atmosphere that may absorb IR (emitted by the Earth) and/or visible (emitted by the Sun) electromagnetic energy waves, contributing to and eventually intensifying atmospheric GH.

The main species among GHGs include carbon dioxide (CO_2), methane (CH_4), nitrous oxide (N_2O), and water vapour (H_2O), with ozone (O_3) also playing a role.

A simplified, though physically consistent, explanation for this phenomenon is sketched in Figure I.5.1, together with a visual representation of the absorption characteristics of these gases and their influence on solar and terrestrial radiation.

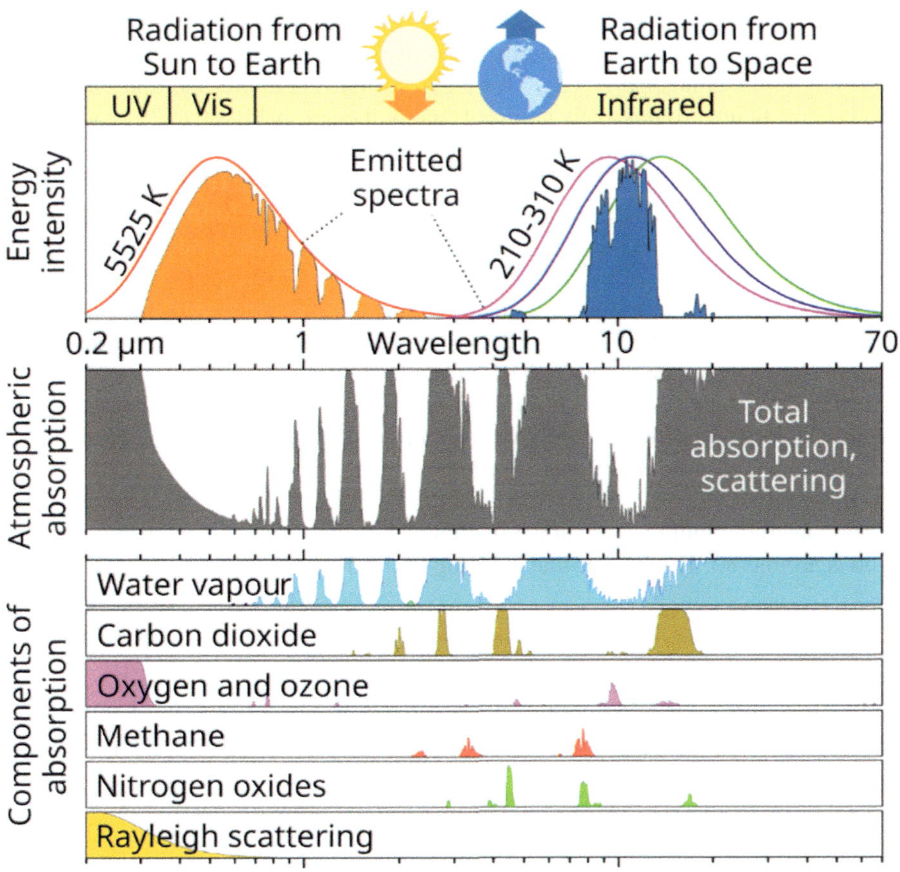

FIGURE I.5.1 GHGs and their absorption bands (middle panel) and the effect that this has on both solar radiation and upgoing thermal radiation (top panel). Individual absorption spectrum for major greenhouse gases plus Rayleigh scattering are shown in the lower panel. Source: This figure was prepared by Robert A. Rohde for the Global Warming Art project.

As is well known, human activities, particularly the burning of fossil fuels and deforestation, have significantly increased the concentration of some of GHGs, and so amplified the greenhouse effect, leading to significant global warming (GW) of the planet (we will evaluate in detail the amplitude of this effect in the following chapters).

Each GHG has a different Global Warming Potential (GWP), which is a measure of how much heat a mass unit of that gas can contribute to GW, "trapping" energy in the atmosphere over a specific time frame.

GWP, in fact, is both dependent on the chemical and thermo-physical properties of the considered GHG (which determine how much radiation the gas absorbs), and on its atmospheric lifetime (which governs how long the gas typically remains in the atmosphere). Therefore, from this point of view, non-toxic, non-reactive, non-soluble gases are much more harmful than others, having a much higher GWP (remaining in the atmosphere 100 years or more).

To evaluate the GWP of each GHG, a time reference period of 100 years was commonly used, but this led to an underestimation of the harmfulness of some species which stay in the atmosphere "only" for a generation. Therefore, the time-reference period is now being progressively adjusted, and will be probably set to 20 years in the next official reports on the subject.

The official evaluation of GHGs' GWP is made by UN IPCC (we will talk diffusively about this UN-based scientific organization in the following chapter) in its Assessment Reports (ARs). The current revision of the AR is number 6, usually referred as AR6.

Table I.5.1 reports the values evaluated for the GWP of the main GHGs by AR4, AR5, and AR6, both on 100-year and 20-year time reference periods. As shown, both CH_4 and N_2O have much higher GWP values than CO_2.

These values must be used as multipliers of GHG quantities to make their effect on GW equivalent to a correspondently equivalent quantity of "virtually" emitted CO_2. This is the method used to compare the GW responsibility of single species among the GHGs.

Please keep in mind that, as shown in Table I.5.1, the GWP of CH_4 over a 20-year period is around 80. This means that emitting in the atmosphere a single kg of CH_4 is

TABLE I.5.1
Main GHGs and their GWP values according to the last assessment reports by UNIPCC, evaluated both on a 20- and a 100-year time reference period

	100-year time period			20-year time period		
GHG	AR4 2007	AR5 2014	AR6 2021	AR4 2007	AR5 2014	AR6 2021
CO_2	1	1	1	1	1	1
CH_4 (fossil)	25	28	29.8	72	84	82.5
CH_4 (non-fossil)			27.2			80.8
N_2O	298	265	273	289	264	273

equivalent to 80 kg of CO_2 emissions, while burning the same quantity of CH_4 would produce a CO_2 emission of 2.75 kg, according to the mass balance of its combustion reaction:

$$CH_4 + 2\ O_2 = CO_2 + 2\ H_2O$$

This means that we should be particularly careful when extracting natural gas from reservoirs, limiting as far as possible CH_4 fugitive emissions in the atmosphere, and eventually burning those CH_4 quantities which could not be captured during the extraction (and would instead be emitted into the atmosphere).

This problem has been traditionally neglected in normal extraction processes of methane from gaseous reservoirs (leading to so-called gas-flaring) but is becoming increasingly important when new fracking technologies are considered (especially in the US) to extract CH_4 from shale rocks.

I.5.4 A SIMPLIFIED MODEL FOR GREENHOUSE EFFECT EVALUATION

As already explained, the Earth's climate is governed by a delicate balance between incoming solar radiation and outgoing heat. As shown in Figure I.5.2, some of the atmospheric components, like CO_2 and water vapour, allow short-wavelength solar radiation to pass through but are much less transparent to longer-wavelength IR radiation. This means that while sunlight can easily enter the atmosphere, the heat trying to leave is partially trapped by these gases.

According to Kirchhoff's law of thermal radiation, these gases release the energy they absorb, emitting further IR radiation in all directions (both upwards into space and downwards back towards the Earth's surface). Over time, the Earth's temperature stabilizes when the amount of incoming energy equals the amount leaving the planet, achieving a new thermal equilibrium. In this equilibrium, greenhouse gases cause the surface to be warmer than it would be otherwise.

In 1896, the Swedish scientist Svante Arrhenius described a model explaining the fundamental principles of atmospheric thermal behaviour. Since then, his model has become a cornerstone of climate science. The model simplifies the Earth's energy flow into distinct "layers", with a surface at a constant temperature (denoted as T_S) and an atmospheric layer at a slightly different temperature (denoted as T_a).

In this simplified view, the surface and atmosphere exchange energy, with strong oceanic and atmospheric currents ensuring that temperature values remain relatively stable across the globe. The temperatures used in the model are typically long-term averages, smoothing out daily with seasonal fluctuations (the model, in fact, does not consider the effect of Earth's revolution and rotation).

The interaction of various gases with solar and infrared radiation is essential in understanding how energy flows through the Earth's system. This model, although simplified, helps explain how the greenhouse effect regulates the planet's temperature and its potential for change in response to increased greenhouse gas concentrations.

The model aims to find the temperatures T_S (surface temperature) and T_a (atmospheric temperature) that balance the incoming and outgoing energy. Specifically, it adjusts these temperatures so that the outgoing radiative power (the heat leaving Earth's atmosphere) equals the absorbed solar energy.

When applied to Earth, the radiation involved is divided into two types: shortwave (mostly visible waves) radiation from the Sun (incoming) and longwave (invisible, mostly distant-IR waves) radiation from the Earth (outgoing). These two forms of radiation have distinct characteristics in terms of emission and absorption.

In the idealized model, we assume that the atmosphere is completely transparent to sunlight. The planetary albedo α_p is the fraction of incoming solar energy that is

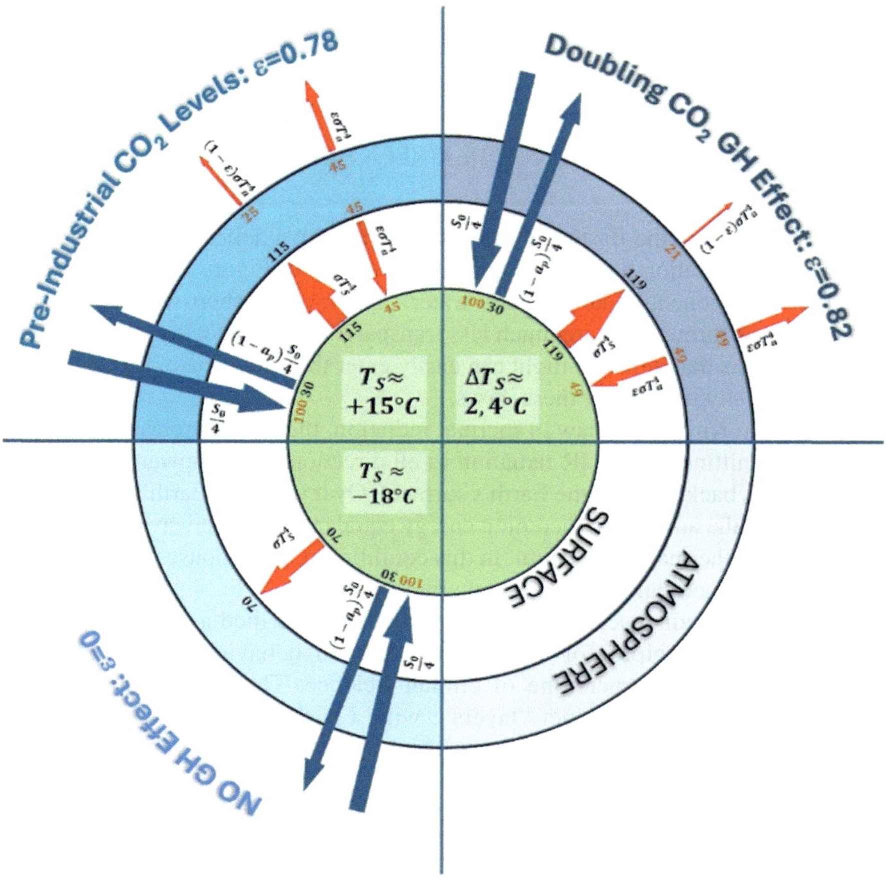

FIGURE I.5.2 Simplified sketch of the thermal fluxes defining thermal balance of Earth's surface and its atmosphere according to the Arrhenius model: Bottom-left quadrant represents an atmosphere without a greenhouse effect (ε=0); upper-left quadrant shows the natural pre-industrial greenhouse effect (ε=0.78); upper-right quadrant shows the CO_2 doubling scenario greenhouse effect (ε=0.82).

reflected back into space. Since sunlight passes freely through the atmosphere, it does not matter whether this reflection occurs on the surface or the top of the atmosphere. For Earth, we normally use these values:

- Solar constant: S_0=1366 W/m^2
- Planetary albedo: a_p=0.30

To account for the fact that a planet's surface area is four times larger than the area of its "shadow" (as seen by the sun, the so-called "Earth discus"), we average the incoming solar radiation, giving a value of $S_0/4$ hitting Earth's surface.

For outgoing longwave radiation, the Earth's surface is treated as a perfect emitter, meaning it has an emissivity of 1 (a "black body" in the infrared spectrum). The surface emits radiation according to the Stefan–Boltzmann law, where the radiative flux density F is related to the surface temperature by the formula $F = \sigma T_S^4$, where σ is the Stefan–Boltzmann constant.

A crucial assumption of the model is Kirchhoff's law of thermal radiation, which states that the absorptivity of a material at a given wavelength is equal to its emissivity. In this model, in particular, we assume that atmospheric emissivity (the ability to emit radiation) and absorptivity (the ability to absorb radiation) are identical for both the incoming and outgoing infrared radiation streams.

Thus, we use the symbol ε to represent both the emissivity and absorptivity of the atmosphere, simplifying the interaction of longwave radiation as it passes through the atmosphere.

Now, we will write two energy balance equations, respectively referring to the Earth's surface and to the atmospheric layer energy, referring to the graphical identification of the various thermal fluxes involved, reported in Figure I.5.2.

These balances consider a generic atmosphere, having its characteristic ε value.

Under these assumptions, the following equation can be written for the infrared flux density out of the top of the atmosphere:

$$F^{\uparrow} = \varepsilon\sigma T_a^4 + (1-\varepsilon)\sigma T_S^4$$

With ε being the fraction of upward longwave radiation from the surface that is absorbed (the absorptivity of the atmosphere), the remaining fraction $(1 - \varepsilon)$ is transmitted to space through an atmospheric window. In the first term on the right, ε is the emissivity of the atmosphere, according to the Stefan–Boltzmann law (to account for the fact that the atmosphere is not opaque to sunlight). Thus, ε plays the role of neatly blending, or averaging, the two streams of radiation in the calculation of the outward flux density.

Zero net radiation leaving the top of the atmosphere requires:

$$\frac{S_0}{4}\left(1-a_p\right)+\varepsilon\sigma T_a^4+(1-\varepsilon)\sigma T_S^4$$

Zero net radiation entering the surface requires:

$$\frac{S_0}{4}\left(1-a_p\right)+\varepsilon\sigma T_a^4-\sigma T_S^4$$

Energy equilibrium of the atmosphere can be either derived from the two above equilibrium conditions, or independently deduced:

$$2\varepsilon\sigma T_a^4-\varepsilon\sigma T_S^4=0$$

Note the importance that the atmosphere radiates both upward and downward. Thus, the ratio of T_a to T_S is independent of ε:

$$\frac{T_S}{T_a}=2^{\frac{1}{4}}\approx 1.189$$

Thus, T_a can be expressed in terms of T_S, and a solution is obtained for T_S in terms of the model input parameters:

$$T_S=\left[S_0\frac{1-a_p}{4\sigma}\frac{1}{1-\frac{\varepsilon}{2}}\right]^{\frac{1}{4}}$$

Let us now solve these balances in three different conditions (also graphically represented in the various parts of Figure I.5.2), to calculate:

- The temperature the Earth would have without any GHG content (therefore putting ε =0, scenario represented in the bottom-left part of Figure I.5.2).

The solution to the previous equation can also be expressed in terms of the effective emission temperature, T_e, which is the temperature that characterizes the outgoing infrared flux density F, as if the radiator were a perfect radiator. T_e is also the solution for T_S, for the case of ε=0, or no atmosphere:

$$T_e=\left[S_0\frac{1-a_p}{4\sigma}\right]^{\frac{1}{4}}$$

With the definition of T_e

$$T_S=T_e\left[\frac{1}{1-\frac{\varepsilon}{2}}\right]^{\frac{1}{4}}$$

For a perfect greenhouse, with no radiation escaping from the surface, or ε = 1:

$$T_S = 2^{\frac{1}{4}}\,T_e \approx 1{,}189\,T_e,\quad T_a = T_e$$

Using the parameters defined above to be appropriate for Earth,

$$T_e \approx 255\,K = -18°C$$

For ε = 1, it would be:

$$T_S \approx 300\,K = +27°C$$

Which estimates the temperature the Earth would have with a complete GH.

- The temperature the Earth had before industrialization (ε = 0.78, scenario represented in the upper-left part of Figure I.5.2).

$$T_S \approx 288.3\ K = 15.2°C,\quad T_a \approx 242.5K,\quad GH = T_S - T_e = 33.2°C$$

This value of T_S happens to be close to the published 287.2 K of the average global "surface temperature" based on measurements. ε = 0.78 implies 22% of the surface radiation escapes directly to space, consistent with the statement of 15% to 30% escaping in the greenhouse effect.

- The temperature the Earth would have for a CO_2-doubling concentration scenario (ε = 0.82, scenario represented in the upper-right part of Figure I.5.2).

If a CO_2-doubling scenario is applied, the value of ε should be increasing to approximately 0.8 due to the simple presence of the additional CO_2, but it must be considered that Earth's temperature increase would produce an amplifying effect on ε values due to the greater presence of water vapour suspended in the atmosphere.

Experimental measurements show that this effect almost doubles the effect on the ε increase, so that a value of 0.82 would be more appropriate and coherent with atmospheric behaviour. In these conditions:

$$T_S \approx 290.7\ K = 17.6°C,\quad GH = 35.6°C$$

Therefore, doubling atmospheric CO_2 with respect to its pre-industrial value concentration would produce an additional GW of about 2.4 °C.

I.6 Global Warming Effects and Related Issues

Global warming (GW) could be an extremely important limitation to development. This directly depends on the amplitude and severity of the issues that are expected to happen because of the level of GW and consequent climate change. Therefore, since the 1980s, the United Nations has taken charge of this problem, as it is a globalized issue which can be controlled and solved only by global and coordinated actions.

That is why, in 1988, the United Nations Environment Programme (UNEP) and the World Meteorological Organization (WMO) together decided to establish the so-called Inter-Governmental Panel on Climate Change (IPCC), an institution bringing together the world's leading scientists and policymakers to assess the latest scientific findings related to climate change and its effects.

The IPCC plays a vital role in shaping international climate policy, guiding the direction of global climate negotiations, and informing national and local decision-making on how to mitigate and adapt to climate change.

The primary mission of the IPCC, in fact, is to provide comprehensive, objective, and policy-relevant information on climate change, including its physical science, impacts, vulnerabilities, and the potential for adaptation and mitigation. The IPCC, whose structure is depicted in Figure I.6.1, operates through three Working Groups (WG):

- WG I focuses on the physical science basis of climate change, exploring how the climate system functions and how human activity, particularly the burning of fossil fuels, is influencing global temperatures and weather patterns.
- WG II examines the impacts of climate change on ecosystems, human societies, and economies, analysing vulnerabilities and identifying options for adaptation.
- WG III addresses mitigation strategies, exploring pathways to reduce greenhouse gas emissions and transition to low-carbon economies.

All the work by the WGs is synthesized in periodic assessment reports (Ars), which are considered the most authoritative sources of climate knowledge globally. In 2023, three IPCC WGs together released their last report, named "AR6 Synthesis Report: Climate Change".

DOI: 10.1201/9781003631125-7

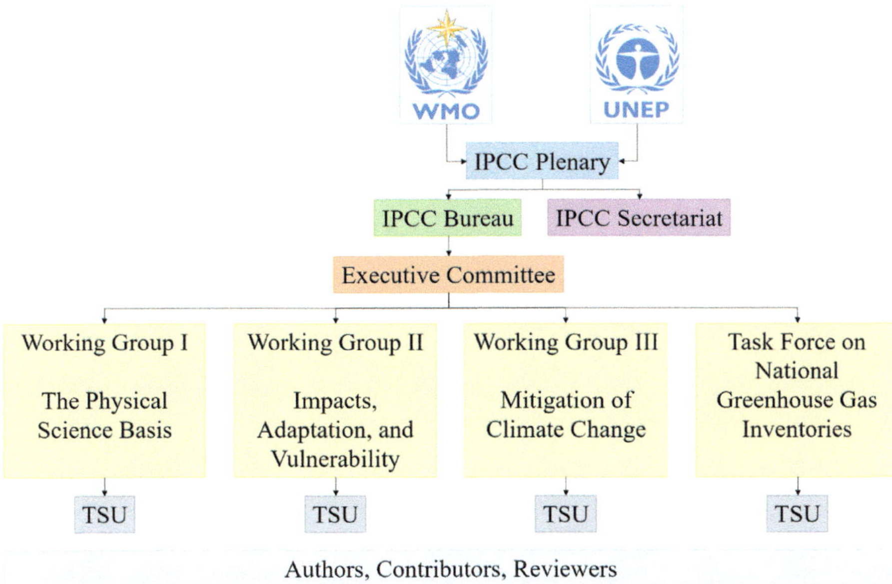

FIGURE I.6.1 IPCC structure.

Each AR draws on the work of thousands of scientists, researchers, and experts from around the world. The IPCC's findings are reviewed by governments and scientific communities before being published, ensuring a consensus-driven approach to climate knowledge.

In this chapter, some of the output of AR6 is reported, which has therefore been fully confirmed by world scientific community.

In particular, according to AR6, emissions of GHGs have increased rapidly over recent decades and led to confirmed values in 2019 of:

- 410 ppm of CO_2, against a pre-industrial (in 1850) value of about 280 ppm
- 1866 ppb of CH_4, against a pre-industrial value of about 800 ppb
- 332 ppb of N_2O, against a pre-industrial value of about 270 ppb.

Regarding cumulative CO_2 emissions in the period 1850–2019, and therefore the most measured GW, AR6 states that the highest responsibility is from North America (26%), followed by Europe (16%), and Eastern Asia (12), as depicted in Figure I.6.2.

In regard to GW, it states that "Global surface temperature was around 1.1°C above 1850–1900 in 2011–2020 (1.09°C [0.95°C–1.20°C]), with larger increases over land (1.59 [1.34 to 1.83]°C) than over the ocean (0.88°C [0.68°C–1.01°C]). Observed warming is undoubtedly human-caused, with warming from greenhouse gases (GHG), dominated by CO_2 and methane (CH_4)".

GW determines a much higher amount of energy resident in the atmosphere, together with a greater amount of water vapour. It is difficult to foresee exactly where

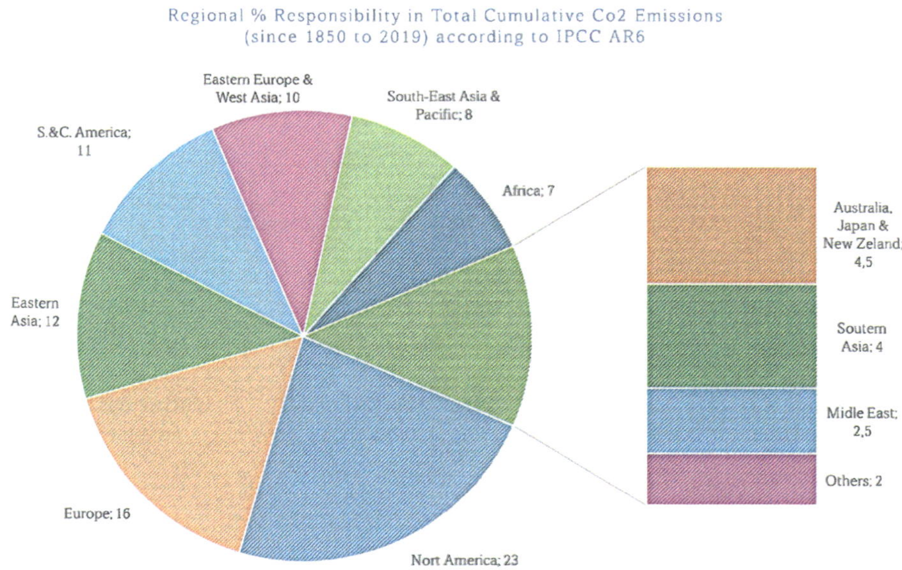

FIGURE I.6.2 Cumulative CO_2 emissions in the period 1850–2019. Data source: IPCC AR6.

this energy will be concentrated, and where and when the water vapour will aggregate to produce more frequent clouds and precipitation, but all these phenomena are likely to happen and to increase in severity, as long as GW increases.

Also, ice and snow covering the highest mountains are progressively diminishing. Ice contents at the polar zones are also diminishing so that sea level is progressively increasing and is already creating huge problems in low-lying zones of the globe.

Permafrost is also thawing in some areas (such as Mongolia), generating relevant structural problems to buildings, because the terrain is losing its normal structural properties.

The cascading effects of GW on both ecosystems and human societies are visualized in Figure I.6.3, summarizing the interconnectivity between climate change drivers, feedback loops, and resulting impacts on the environment and human life.

The IPCC AR6 report provides a detailed and clear assessment, emphasizing that human-caused climate change is already driving many weather and climate extremes across all regions of the globe.

These changes have resulted in adverse impacts on food and water security, human health, economies, and societies, leading to substantial losses and damage to both nature and people. Notably, the most vulnerable communities (those that have historically contributed the least to climate change) are disproportionately affected.

There is strong evidence of observed changes in extreme weather events, such as heatwaves, heavy precipitation, droughts, and tropical cyclones. The attribution of these phenomena to human influence is increasingly clear.

It is virtually certain that since the 1950s, hot extremes (including heatwaves) have become more frequent and intense across most land regions, while cold extremes

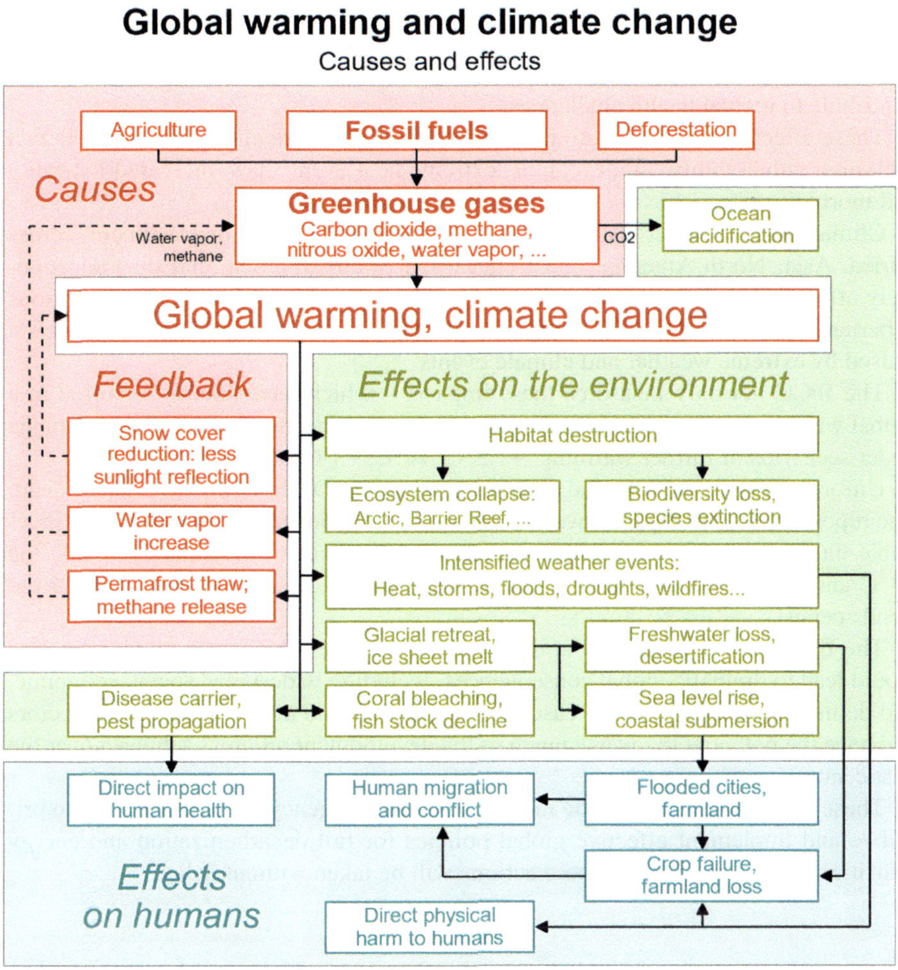

FIGURE I.6.3 Cascading effects of GW on both ecosystems and human societies. Source NASA – License: https://creativecommons.org/licenses/by-sa/4.0/deed.en. (No changes were made).

(including cold waves) have become less frequent and severe. Marine heatwaves have also roughly doubled in frequency since the 1980s.

Global warming has intensified agricultural and ecological droughts in some regions due to increased land evapotranspiration. Additionally, the proportion of major tropical cyclones (Categories 3–5) has risen over the past four decades.

The IPCC has examined the effects of climate change on agriculture and water availability, finding that global warming has reduced food and water security. These challenges stem from altered precipitation patterns and the greater frequency and intensity of climate extremes, making it increasingly difficult to achieve the UN Agenda 2030 Sustainable Development Goals (SDGs).

In urban areas, global warming has exacerbated heatwaves, worsened air pollution events, and adversely affected human health, livelihoods, and critical infrastructure. Cities have experienced intensified heatwaves that directly impact physical health and contribute to mental health challenges.

These effects have also led to humanitarian crises where climate hazards interact with high vulnerability. In all regions, extreme heat events have increased mortality and morbidity rates.

Climate and weather extremes are increasingly driving displacement across Africa, Asia, North America, and Central and South America, and disproportionately affecting small island states in the Caribbean and South Pacific. These regions experience heightened vulnerability through displacement and involuntary migration caused by extreme weather and climate events.

The IPCC not only measures these impacts – which have already occurred at a global warming level of +1.1°C – but also provides projections of escalating damage under scenarios of further warming: +1.5°C, +2°C, +3°C, and +4°C.

Unfortunately, current global policies to reduce CO_2 emissions are insufficient. The report warns that the first two scenarios (+1.5°C and +2°C) are now nearly inevitable unless extremely stringent policies are implemented immediately. While the +3°C and +4°C scenarios are avoidable, they remain disturbingly likely if current trends persist.

The IPCC AR6 highlights that the realization of these higher-warming scenarios would lead to dramatic global consequences, including widespread social, economic, and demographic crises. These cascading impacts would affect all economic sectors and have the potential to reverse much of the development progress achieved over the past century.

These findings should provide more than enough urgency for world leaders to prioritize and implement effective global policies for full decarbonization and energy transition. We sincerely hope these actions will be taken without delay.

1.7 Decarbonization Policies Design – The Kaya Identity

In this chapter we will try to understand if and how it is possible to design the interventions on energy systems which are needed to control and limit the future incremental impact we will produce on the biosphere. To this aim, it is necessary to preliminarily identify the main socio-economic and technological variables which mostly contribute to the definition of each nation's yearly CO_2 emissions.

One of the simplest but still the most effective way to reach this aim was firstly introduced by the Japanese energy economist Yoichi Kaya in 1990 and is therefore usually referred as the Kaya Identity.

"Kaya Identity" is basically a framework used to decompose and analyse the factors influencing CO_2 emissions. It was developed to help understand the relationship between economic growth and CO_2 emissions. The identity, in fact, expresses global carbon emissions as a quite simple four-term factorization, according to the algebraic and textual formulations reported below:

$$CO_2\,Emissions = Population * GDP\ per\ capita * Energy\ intensity * Carbon\ intensity$$

$$\mathrm{Em(t) = A(t)\,B(t)\,C(t)\,D(t)}$$

$$\mathrm{En(t) = A(t)\,B(t)\,C(t)}$$

$$\Delta \mathrm{Em(t)} = \int_{1965}^{t} Em(\tilde{t})\,d\tilde{t}$$

$$\Delta[\mathrm{CO_2}](\mathrm{t}) \approx \Delta \mathrm{Em}(\mathrm{t})/7.82 \text{, in ppm, with } \Delta \mathrm{Em(t)} \text{ in } \mathrm{Gt_{CO2}}$$

$$\Delta \mathrm{T_{GW}}(\mathrm{t}) \approx \Delta\left[\mathrm{CO_2}(\mathrm{t})\right]/120$$

where:

- **A(t)** describes **Population** dynamics
  It is a demographic index, referring to the total number of inhabitants vs. time, and is an indicator of the effectiveness of many national social policies.

DOI: 10.1201/9781003631125-8

- **B(t)** describes **GDP per capita** dynamics
  It expresses the economic output per person, and, as already explained, is one of the main indicators used to evaluate national development.
- **C(t)** is **Energy intensity**
  It is the amount of energy used per unit of GDP and is an effective indicator of energy-efficient policies.
- **D(t)** is **Carbon intensity**
  It is the amount of CO_2 emissions produced per unit of energy, and directly reflects energy sustainability policies, especially those oriented towards RES penetration.

The Kaya Identity helps policymakers and researchers identify the specific areas where emissions reductions can be targeted. For instance, emissions can be lowered by reducing carbon intensity (e.g., using cleaner energy) or improving energy efficiency (reducing energy intensity), while controlling population growth, and granting a designed increase in GDP per capita to sustain development.

In essence, the Kaya Identity provides a very immediate and powerful tool to understand the drivers of carbon emissions and helps guide efforts towards reducing them.

It is noteworthy to consider that in an unconstrained economy the first two terms, A and B, of the Kaya Identity are expected to grow exponentially at given yearly rates τ_A and τ_B expressing the % growth of the variable per year. In a "sane" world, increasing wealth and development at a reasonable rhythm (not considering revolutionary phases) τ_A is expected to vary between 1 and 2%, while τ_B should be maintained around 3 and 4%.

Since Kaya Identity is a simple factorization, the growing rates of A and B add to one another. It is therefore very simple to conclude that energy efficiency and decarbonization policies, acting respectively on C and D, will have to produce negative yearly growing rates in C and D (τ_C and τ_D) to balance (τ_A and τ_B). Substantially, yearly emissions will decrease only if:

$$\tau_A + \tau_{B+}\tau_C + \tau_D < 0$$

It is obvious to conclude that if we want at least to maintain yearly emissions constant in such a developing world, we should at least intervene to decrease C and D cumulatively at around 5% each year. Acting like this, would be already a significant action, but unfortunately it would be largely insufficient to solve or maintain our sustainability issues.

Those, in fact, as clearly shown by the simplified correlations reported in the bottom part of Table I.7.1, would worsen if CO_2 emissions are not decreased but simply stabilized. In fact, since CO_2 emissions generally remain in the atmosphere for a significant number of decades (which is very difficult to estimate), its effect on GW is not dependent simply on their yearly emissions, calculated through Kaya Identity, but on the "moving" integral of those emissions evaluated over a given number of decades.

TABLE I.7.1
Kaya Identity Variables Definitions and Their World Values in 2020 (Data elaborated from the Energy Institute 2020 Statistical Review of World Energy)

KAYA IDENTITY VARIABLES	WORLD (2020 Values)	
A(t) = Population	7821	[MPax]
B(t) = Yearly GDP per capita	18	[k$/Pax]
C(t) = Energy intensity	1265	[Wh/$]
D(t) = Emission intensity	91	[gCO_2/kWh]
En(t) = Yearly energy consumption	15	[Gtoe/y]
Em(t) = Yearly CO_2 emissions	34	[$GtCO_2$/y]
ΔEm(t) = Cumulative emissions since 1965	1272	[$GtCO_2$]
Δ[CO_2](t) = Cumulative CO_2 emissions since 1965	163	[ppm]
ΔT_{GW}**(t)** = Cumulative global warming since 1965	1.4	[°C]

Therefore, the only way not to worsen GW is to start immediately emitting at most the same amount of CO_2 as we introduced in the atmosphere decades ago. This would mean implementing polities to reduce instantaneously CO_2 emissions by about 90% with respect to our actual emissions.

Any other decarbonization policy and scenario, unfortunately, will only further worsen GW and the sustainability of our development. That is the huge level of criticality at which we have now reached. That is the direct effect of our actions and of the delay in the implementation of policies that the whole scientific community has been designing and proposing for the attention of political decision makers for decades, as shown in previous chapters of this book.

I.7.1 KAYA FACTORS: A – POPULATION AND DEMOGRAPHIC CONSIDERATIONS

Let us now have a look at the individual Kaya Identity factors and their dynamic behaviour in recent decades, to draw some conclusions on their possible short-term evolution. We first focus on population and demographic considerations.

Figure I.7.1 depicts population dynamics in the world's major nations: curves in the upper part are normalized by their initial values in 1990 and therefore depicts variable relative growth; the bottom part, instead, in the same colours, reports yearly growth rates of the single curves above. The same type of graphs will be used in the following paragraphs for the other Kaya factors and for all the variables reported in Table I.7.1.

The data analysed cover the period 1990–2020, with the final 2020 values shown in the legend. This figure clearly shows the huge increase in the global population over the last 30 years, now reaching more than 8 billion people (also indicated as GPax). India's population grew by more than 60%, reaching almost 1.4 GPax. Also, the populations of China and the US increased a lot, while the populations of the European Union and Russian Federation were much more stable. The global

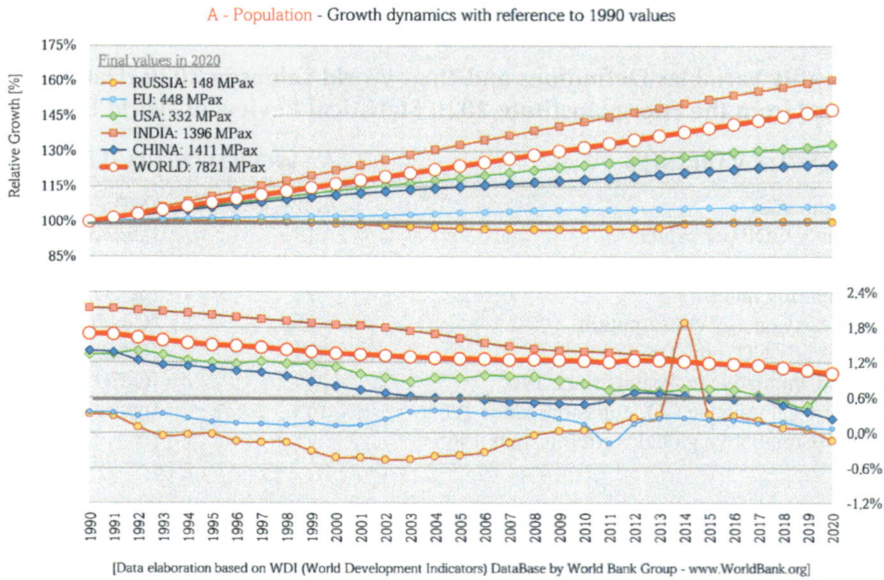

FIGURE I.7.1 Population dynamics for the world and the major nations: curves in the upper part are normalized by their initial values in 1990 and therefore depict variable relative growth; the bottom part, in the same colours, reports yearly growth rates of the single curves above.

population has overall grown by more than 45% in the last three decades, having a mean annual growth rate in the period of about 1.5%/year.

This increase in population has been very intense and started long before the 1990s, as shown in Figures I.7.2 and I.7.3. Figure I.7.2 is similar to Figure I.7.1, but represents data starting from 1950, disaggregating them depending on national income (in four groups: low; lower-middle; upper-middle; high). As shown, the world's population is increasing at about 1 GPax every 22 years.

This increment was largely due to low-income countries (mostly in Africa and Asia), whose populations have been growing steadily at a rate greater than 2.5%/year for 50 years.

Figure I.7.3 is particularly impressive and depicts so-called world "demographic pyramids", showing population sub-divisions by sex and age, starting from 1950 (upper left corner), 1980 (upper right), and 2020 (lower left corner).

In the lower-right corner, the expected distribution for population in 2055 is shown. Values reported are absolute, so the areas under the curves are indicative for the total world population.

Starting from a broad-based pyramid in 1950, indicative of high birth rates and short life expectancy, the shape evolves into a more rectangular form by 2055, reflecting longer life expectancy and declining fertility rates.

The future pyramid suggests a significant increase in the working-age population, particularly in low- and middle-income regions, potentially driving increased energy demand and economic activity.

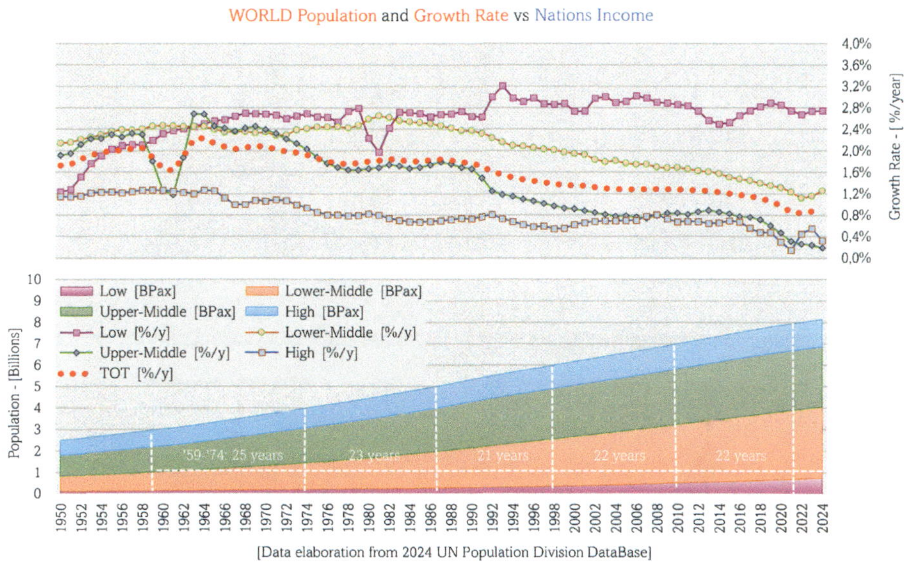

FIGURE I.7.2 World population dynamics starting from 1950: data are disaggregated by level of national income (low; lower-middle; upper-middle; high): curves in the lower part show absolute values; the upper part reports yearly growth rates for the single national income groups.

In the context of the Kaya Identity, population growth directly contributes to increase energy demand and CO_2 emissions, especially in rapidly developing regions.

However, the demographic transition observed in high-income countries highlights the potential for stabilization in population growth as economies develop. This underscores the importance of integrating demographic insights into energy and climate policies to address the challenges posed by varying growth rates across regions.

I.7.2 KAYA FACTORS: B – GDP PER CAPITA AND ECONOMIC DEVELOPMENT

Figure I.7.4 depicts GDP per capita dynamics, and the subsequent Figure I.7.5 illustrates total GDP dynamics. The latter combines the effects of the previously discussed population dynamics with those of GDP per capita, which, as explained, is mostly used as an indicator of economic development.

Figure I.7.4 clearly shows that the world experienced substantial development in the last three decades, although some nations grew much faster than others. China's GDP per capita increased approximately 18-fold, reaching about 18 k$/y per capita, which is also the current world average. By contrast, most developed countries grew much less but still have GDP per capita levels that are more than twice the world mean value (47 k$/y in the European Union and 64 k$/y in the United States).

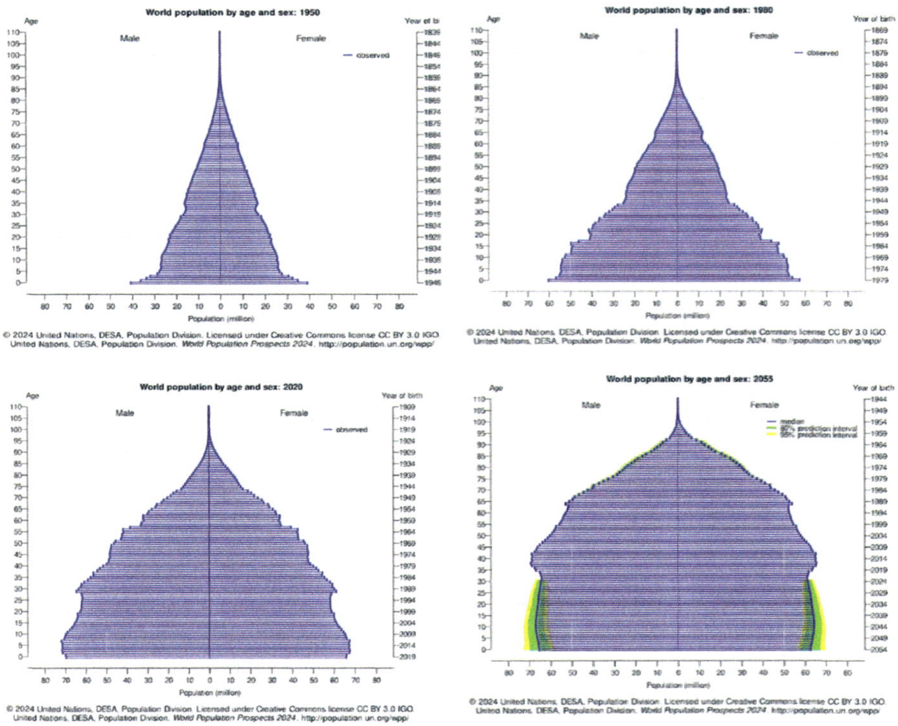

FIGURE I.7.3 World demographic pyramids showing population sub-divisions by sex and age, staring from 1950 (upper left), 1980 (upper right); 2020 (lower left). In the lower-right corner is the expected distribution for population in 2055. Values reported are absolute, so the areas under the curves are indicative of the total world population (2024 United Nations, DESA, Population Division. Licensed under Creative Commons license CC BY 3.0 IGO).

The lower panel in Figure I.7.4 highlights yearly growth rates. Notably, China displayed consistently high growth rates until recently, driving its rapid economic expansion.

Other countries, such as the EU and US, maintained relatively stable and moderate growth. These patterns are essential for decarbonization strategies, as higher GDP per capita often corresponds with increased emissions but also greater potential for investments in clean technologies.

According to Figure I.7.5, China, with a total GDP of approximately 25 trillion $/y, has overtaken both the United States and the entire European Union, which each stand at around 21 trillion $/y. Together, these three economies (China, the US, and the EU) account for more than half of the world's GDP. Particularly striking is that the US and EU, which collectively house just one-tenth of the world's population, still produce about one-third of global GDP.

The lower panel of Figure I.7.5 shows yearly growth rates for GDP. China's extraordinary growth trajectory is evident, though its pace has begun to stabilize in recent years. This shift aligns with the nation's transition to a more consumption- and

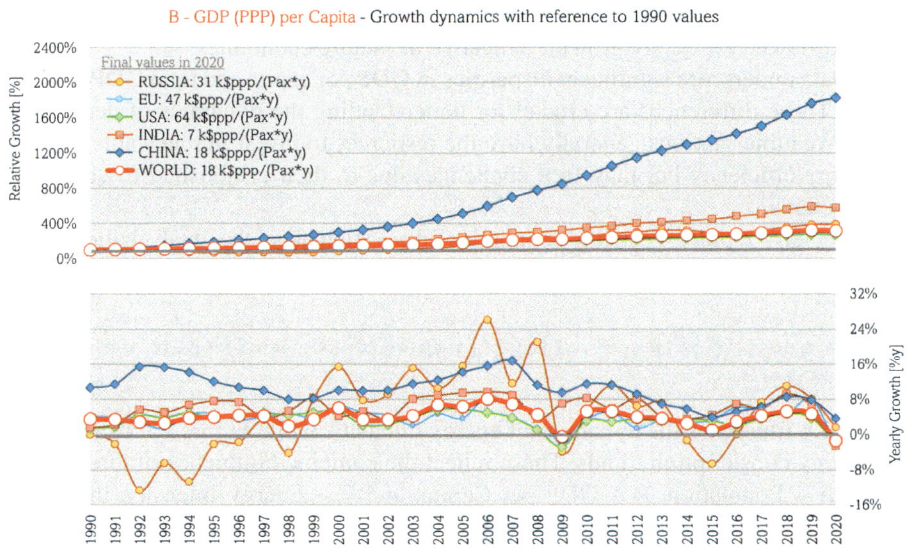

[Data elaboration based on WDI (World Development Indicators) DataBase by World Bank Group - www.WorldBank.org]

FIGURE I.7.4 GDP per capita dynamics in the world and in major nations: curves in the upper part are normalized by their initial values in 1990 and therefore depict variable relative growth; the bottom part, in the same colours, reports yearly growth rates of the single curves above.

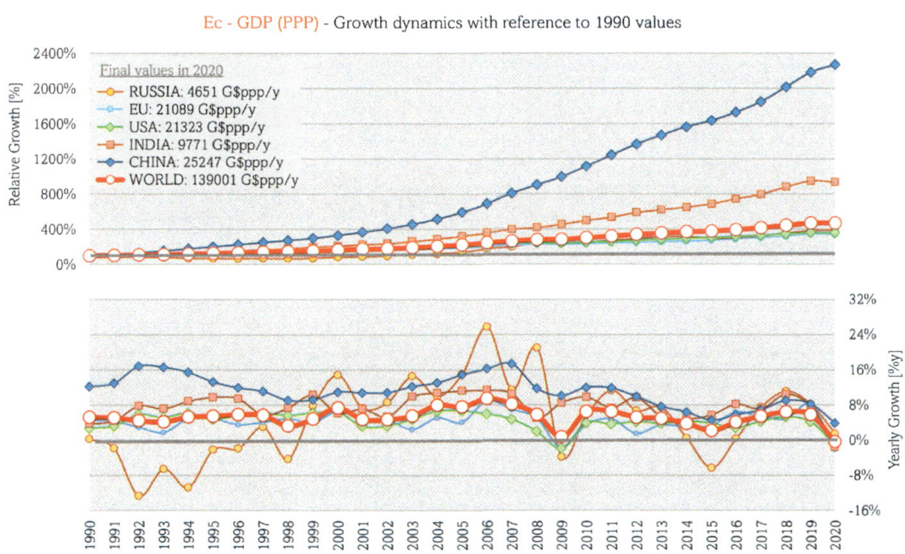

[Data elaboration based on WDI (World Development Indicators) DataBase by World Bank Group - www.WorldBank.org]

FIGURE I.7.5 GDP dynamics in the world and in major nations: curves in the upper part are normalized by their initial values in 1990 and therefore depict variable relative growth; the bottom part, in the same colours, reports yearly growth rates of the single curves above.

innovation-driven economy. Meanwhile, developed nations like the US and EU exhibited slower yet steady growth, indicative of mature economies.

The data underscore significant disparities in GDP per capita and total GDP among nations. These differences are crucial for understanding the dynamics of decarbonization. Wealthier nations generally have the resources to invest in renewable energy and energy efficiency but may face challenges due to their established fossil-based infrastructures.

Conversely, rapidly growing economies like China face the dual challenge of maintaining economic expansion while reducing their carbon intensity.

I.7.3 KAYA FACTORS: C – ENERGY INTENSITY AND EFFICIENCY POLICIES

Figure I.7.6 illustrates the dynamics of energy intensity, while Figure I.7.7 shows the total energy consumption trends. These reflect the combined influence of three Kaya factors: A – Population, B – GDP per Capita, and C – Energy Intensity, the latter serving as an indicator of energy-efficiency policies.

Figure I.7.6 shows that energy-saving policies around the world are probably much more intense than what is normally perceived by people. Whole world energy intensity, in fact, has decreased by about 60% in the last 30 years, meaning that

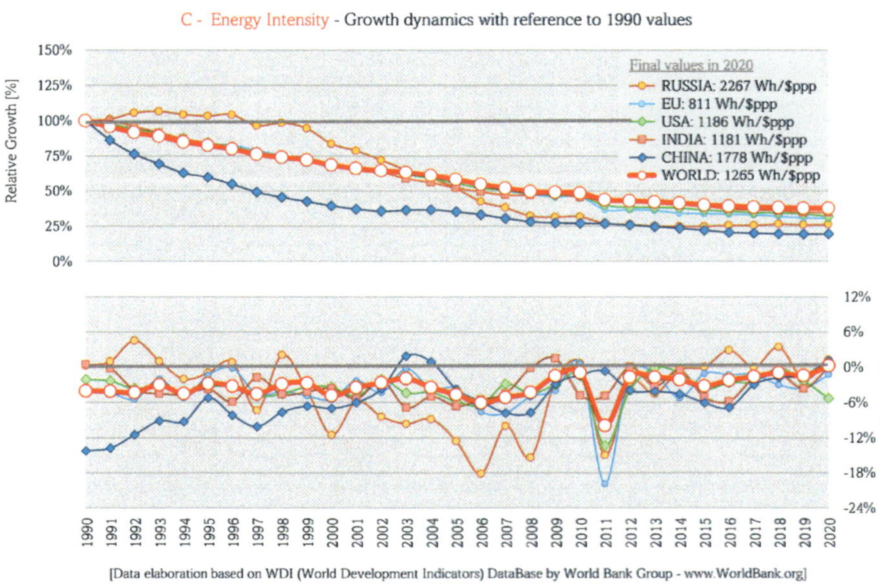

FIGURE I.7.6 Energy intensity dynamics in the world and in major nations: curves in the upper part are normalized by their initial values in 1990 and therefore depict variable relative growth; the bottom part, in the same colours, reports yearly growth rates of the single curves above.

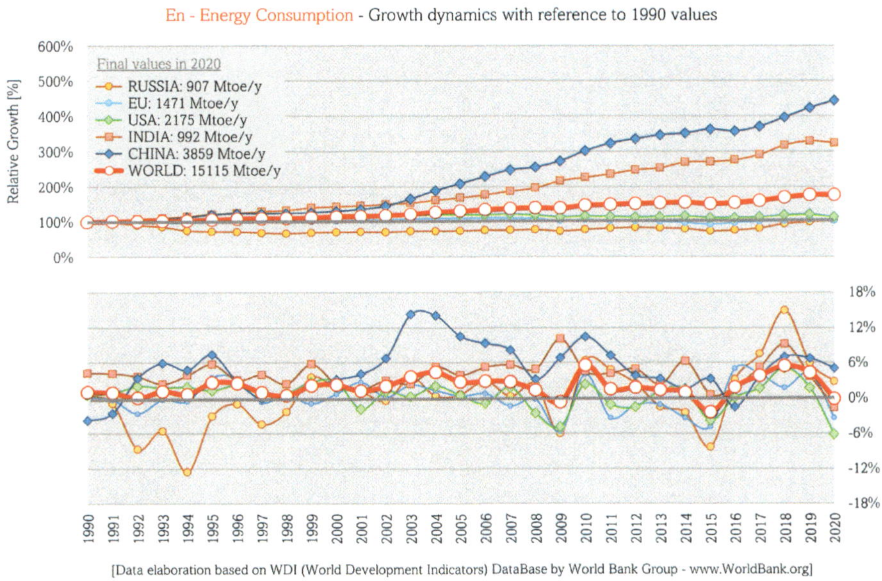

FIGURE I.7.7 Energy consumption dynamics in the world and in major nations: curves in the upper part are normalized by their initial values in 1990 and therefore depict variable relative growth; the bottom part, in the same colours, reports yearly growth rates of the single curves above.

we now use less than half the energy we used in the 1990s to produce the same GDP unit.

Energy intensity reductions were observed across all countries. This decrease, in fact, was not only due to the increased attention to sustainability issues, but mainly to market competitiveness issues, with the cost of energy being an important part of the cost of production in many industrial sectors. That is the reason why most industrialized countries followed the same paths.

Despite these positive developments, data reveal that reductions in the C factor are insufficient to offset the rapid growth of A (population) and B (GDP per capita). This imbalance is underscored by Figure I.7.7.

Total energy consumption at the global level almost doubled between 1990 and 2020. This is despite the remarkable progress in energy intensity reduction. Notably, China's energy consumption has increased by a factor of 4.5, while India's energy consumption has more than tripled (by a factor of 3.2). Conversely, energy consumption in OECD countries, including the United States and European Union, has remained relatively stable.

While improvements in energy efficiency (C factor) are encouraging, the simultaneous surge in energy consumption due to growing populations and economies (A and B factors) poses significant challenges to sustainability. These dynamics emphasize the pressing need for stronger, multifaceted policy interventions to effectively manage global energy demand while fostering sustainable growth.

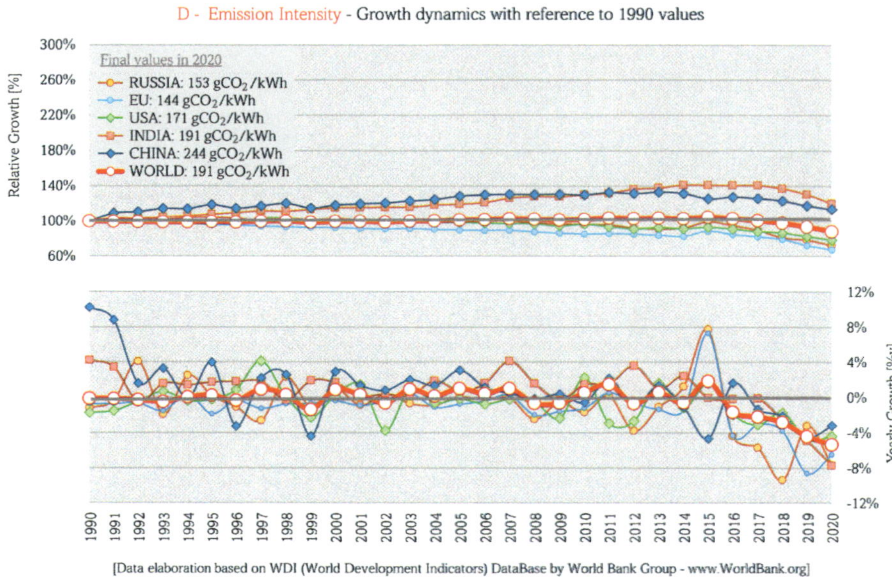

FIGURE I.7.8 Emission intensity dynamics in the world and in major nations: curves in the upper part are normalized by their initial values in 1990 and therefore depict variable relative growth; the bottom part, in the same colours, reports yearly growth rates of the single curves above.

I.7.4 KAYA FACTORS: D – EMISSION INTENSITY AND DECARBONIZATION

Figure I.7.8 depicts emission intensity dynamics and subsequently Figure I.7.9 presents data on yearly CO_2 emissions.

According with Kaya Identity, CO_2 emissions "sum up" the dynamical effects of all the four Kaya factors: A – Population; B – GDP per Capita; C – Energy Intensity; D – Emission Intensity. The latter, as introduced, is one of main indicators of decarbonization policies in the energy sector.

Trends clearly show the enormous underestimation the world is giving to GW issues. At a global level, emission intensity only began decreasing significantly in the last three years, influenced by the global economic slowdown caused by the COVID-19 pandemic.

The European Union, however, stands out for its consistent progress in reducing emission intensity, reflecting its robust and forward-thinking decarbonization policies. This contrasts sharply with China and India, where emission intensity increased due to their heavy reliance on coal to fuel economic growth.

Emission intensity is strongly tied to a nation's energy mix (the sources of energy used to meet its demand). We will talk about this in more depth in Chapters I.10 and I.11, but it is important here to recall that no energy source is totally CO_2 free on a global emissions level.

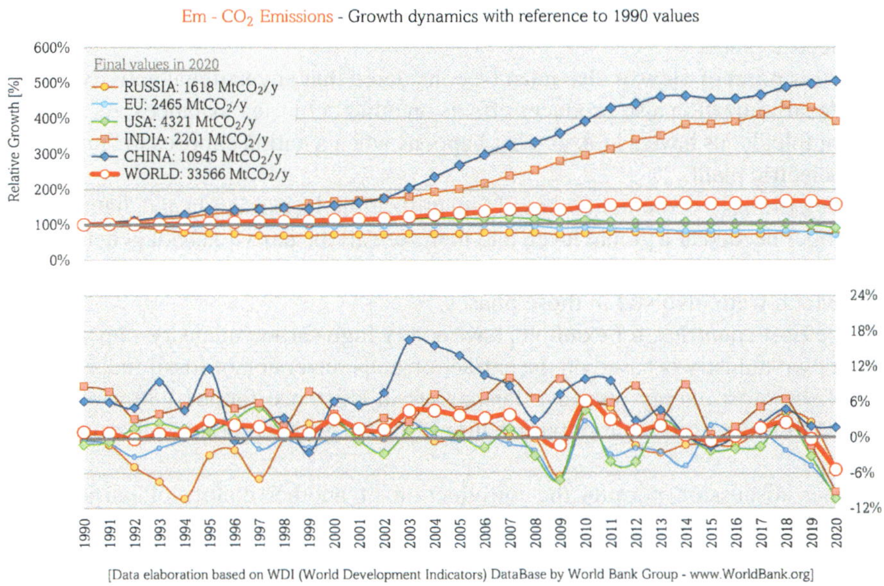

FIGURE I.7.9 CO_2 dynamics in the world and in major nations: curves in the upper part are normalized by their initial values in 1990 and therefore depict variable relative growth; the bottom part, in the same colours, reports yearly growth rates of the single curves above.

In fact, if one nation considers introducing renewable energy sources (RESs) to produce energy for its demand, it must design, build, maintain, exercise, and eventually dismantle new production plants (at the end of their life cycle). On a "life-cycle analysis" (LCA) basis, the unit of energy produced is not totally "green" as may seem, because it must consider the emissions due to all this additional energy consumption which would not have been part of energy demand otherwise.

How is it possible to evaluate this effect?

Those additional emissions could be accounted for using the data presented in Figure I.7.9, i.e. the emission intensity of the single nations and its foreseeable value for the coming years (when new RES plants would be installed and operated). The same intervention (introduction of a new RES plant) may therefore have a substantially different effect on national decarbonization depending on the nation that proposes it, having a greater effect where decarbonization is more advanced (and therefore national emission intensity is already lower).

All these effects must be considered both locally and globally and could lead to unexpected drawbacks of CO_2 balance at a national level. As an example, we cite the case of the Italian energy sector: Italy currently imports about 7% of its electricity from bordering countries. This energy is therefore almost completely CO_2 free and contributes to lowering Italian emission intensity. Therefore, if Italy decides to move towards RES (as it has been doing in recent years), paradoxically its emission intensity will increase if this electricity produced is used to limit its electricity imports, while it will produce an effective reduction only if it replaces that produced by fossil thermoelectric plants.

From this point of view, it also must be considered that any national activity about its own decarbonization also produces effects on other countries' balances. As a clarifying example, let us have a look at what happens when a nation chooses to "turn-off" a thermoelectric plant.

This determines a decrement in fossil fuel utilization. Its extraction and distribution are no longer needed at a global level, and therefore all the emissions not generated in those phases will produce automatic decarbonizations in the CO_2 balance of the other nations which were involved in those phases.

Middle East countries, for example, have a very high carbon intensity, considering that their own industry is based on the extraction and conversion of fossil fuels, which is one of the most energy-intensive businesses in existence (about 1 litre of oil is consumed to make it possible to burn 1 litre of its derivative products all around the world).

Previous discussion permits the introduction of another couple of particularly important considerations:

- The first is a direct conclusion of what has been already explained. The CO_2 emissions produced using fossil sources are much higher (even more than double) than those evaluable only considering its final usage ("only" implying a certain amount of fuel combustion).
- Also, considering only their final utilization (for example, in local thermoelectric plants), distinct types of fossil fuels are not at all equivalent in terms of their CO_2 emission intensity.

Let us concentrate on this latter aspect. To evaluate a hydrocarbon's (with a basic formula C_nH_m) specific emissions due only to its combustion process we must refer to its simplified stoichiometric combustion reaction:

$$C_nH_m + \left(n + \frac{m}{4}\right) O_2 \rightarrow n\ CO_2 + \frac{m}{2}\ H_2O$$

This formula highlights the differences in CO_2 emissions per gram of fuel burned, depending on the hydrogen-to-carbon (H/C) ratio of the fuel. Therefore, the combustion of each mole of C_nH_m, weighing $12n + m$ grams, would produce and emit in the atmosphere n moles of CO_2, weighing $n(12+2*16) = 44n$ grams. Summarizing, each gram of burnt fuel produces emissions which are given by the ratio:

$$\frac{44\,n}{12\,n+m} = \frac{44}{12+\frac{m}{n}}\ \frac{g_{CO_2}}{g_{C_nH_m}}$$

This ratio only depends on the m/n ratio of the fuel and has a minimal value of 2.75 for NG (CH_4, $m/n = 4$), but may be much higher for other fuels. Oil derivatives, having *m/n* values around 2, produce about 3.2 grams of CO_2 for each gram of burnt C_nH_m. This ratio reaches values around 3.4 for coal (having a *m/n* value around 1 or less).

What we have calculated up to now is the mass ratio, but we need to evaluate the emission intensity of CO_2 for a unit of electricity. To do that we first have to consider both the different thermal energies delivered by a unit mass of different fuels, and the gross electric efficiency of the considered thermoelectric plants.

Considering the first aspect, we will use different reference values for the H_i (lower calorific value) of the various fuels: 50 MJ/kg for methane; around 40 for liquid fuels; and 25 for solid fuels.

In regard to the second aspect, we must consider a standard thermodynamic cycle with an efficiency of about 0.4, but using both liquid fuels and especially solid fuels, a part of the energy produced is lost in exhaust cleaning technologies (needed to make local emissions compatible with the environment).

If a value of 0.40 for the electric efficiency of NG plants is used, correspondent values of around 0.36 can be used for oil plants. As regards coal plants, a maximum corresponding value of 0.32 is acceptable, but advance exhaust cleaning technologies could lead to the net efficiency going down to 0.24.

Under all these assumptions, and considering assumed parameters, values can vary somewhat around the value assumed above. The following values can be calculated for mean emission intensity in electricity production (D_{El}) using various fuels (NG, oil, coal), only considering their final use inside a thermoelectric plant:

$$D_{El-NG} \approx 450 \div 550 \frac{g_{CO_2}}{kWh_{El}}$$

$$D_{El-Oil} \approx 750 \div 850 \; g_{CO_2} / kWh_{El}$$

$$D_{El-Coal} \approx 1450 \div 1550 \; g_{CO_2} / kWh_{El}$$

$$D_{El-Clean\ Coal} \approx 1900 \div 2100 \; g_{CO_2} / kWh_{El}$$

These values show that choosing the correct energy mix is extremely important, including only considering fossil fuels among possible fuel sources: producing electricity by natural gas, for example, could have up to four times less impact on GW issues than using other types of fossil fuels.

In contrast, the figures so easily calculated explain the great global apprehension of GW scientists around the world created by the decision of such large and energy-consuming economies as China and India to fuel their development mainly through coal.

All this considered, the data in Figure I.7.9 about overall CO_2 emissions dynamics appear clear.

As shown in this figure, overall global CO_2 emissions in the last three decades were almost constantly increasing, coming to a final value in 2020 of about 33.5 $GtCO_2/y$ (58% more than the 1990 value).

This increment was primarily due to emission increments in China and India. Emissions in China, in particular, were multiplied by a factor 5, so that China alone now is responsible for about 30% of total world CO_2 emissions.

Figures I.7.8 and I.7.9 reveal the urgent need for effective decarbonization policies. Although global emission intensity has begun to decline, the pace is far too slow to counteract the effects of rising energy consumption and economic growth.

The reliance on coal in nations like China and India underscores the need for targeted international collaboration to reduce the global reliance on high-emission energy sources.

While nations like those in the EU have demonstrated the potential of decarbonization through renewable energy adoption and efficiency improvements, the global challenge remains daunting.

Without accelerated efforts to reduce coal dependence and improve emission intensities, the world is unlikely to meet its climate goals, with severe consequences for global warming and sustainability.

I.7.5 GLOBAL AND NATIONAL KAYA IDENTITIES DYNAMICS

Let us try now to gather all the considerations we have made about the individual Kaya factors, plotting them together for nations/economic systems and for the whole world.

Staring from a global perspective, Figure I.7.10 depicts world dynamics for all the Kaya factors.

As in previous figures, the curves in the upper part are normalized to their initial values in 1990, depicting variable relative growth. The bottom part, using the same colours, reports the yearly growth rates of the individual curves.

These data clearly show how both population (yellow curve) and GDP per capita (cyan curve) dynamics have generated growth (and a yearly growth rate) which was not balanced by a corresponding decrease in energy intensity (green curve), leading to a growth in energy consumption (violet curve), that has almost doubled during the last three decades.

Decarbonization policies (represented by emission intensity dynamics) have only produced some positive effects in the last three years (since the COVID-19 pandemic), leading to a final increment of CO_2 emissions of about 58% with respect to the 1990 value.

Figure I.7.11 delves deeper into world annual CO_2 emission dynamics over a longer observation period starting from 1965. The lower part of the figure disaggregates emissions by world geographical areas, highlighting their contributions to total emissions.

The growth in the responsibility of Asian countries among emitting regions is particularly striking. In 1965, their share of yearly emissions was marginal (13%), whereas it has now reached a significant 54%.

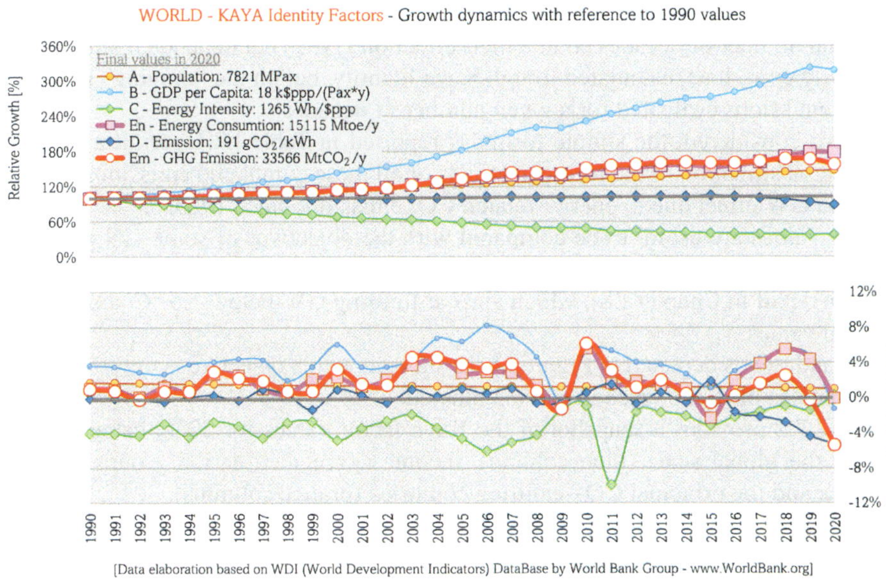

FIGURE I.7.10 World dynamics for all the Kaya factors: curves in the upper part are normalized by their initial values in 1990 and therefore depict variable relative growth; the bottom part, in the same colours, reports yearly growth rates of the single curves above.

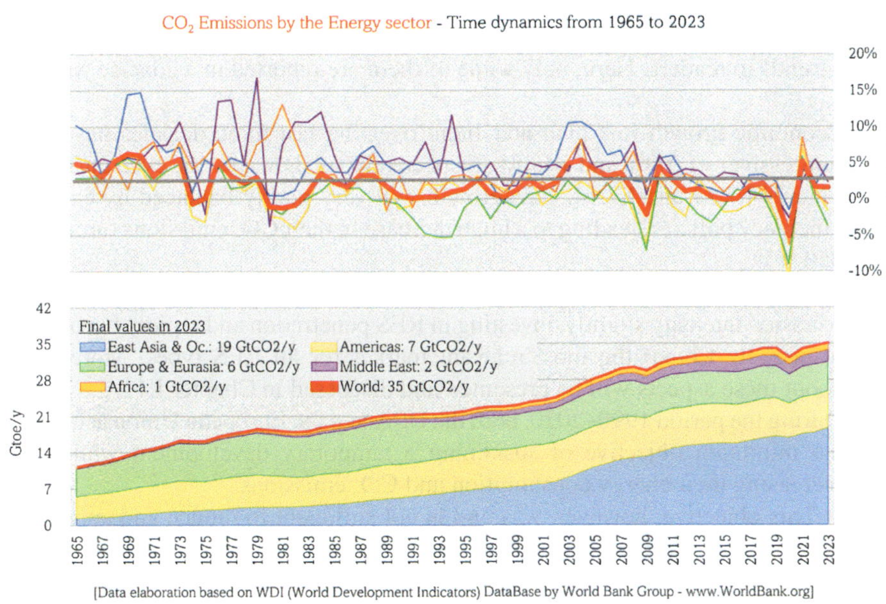

FIGURE I.7.11 World yearly CO_2 emission dynamics: curves in the lower part disaggregated for world geographical areas, identifying their responsibility for total emissions; the upper part, in the same colours, reports yearly growth rates of the single curves below.

As already described, these CO_2 emissions will remain in the atmosphere for a substantial number of decades so that their effect on GW is not dependent simply on their yearly emissions, calculated trough Kaya Identity, but on the "moving" integral of those emissions evaluated for a given number of decades.

All this considered, the simple formulas reported in Table I.7.1 permit a simple preliminary estimation of the foreseen effects of all these cumulated emissions on the additional GW effect due to energy-related CO_2 emissions since 1965.

These values are useful to be compared with the objectives of some UN policies which are under application, such as the so-called Paris Agreement (which will be treated in detail in Chapter I.8), which aims at limiting GW below 1.5 °C above pre-industrial values.

Data and trends depicted in Figure I.7.12 show how this objective is unfortunately almost impossible to be reached and maintained.

Table I.7.2 provides a snapshot of the Kaya factor values for 2020, categorizing them by the global scale (blue column), income-aggregated nation groups (violet columns), and the principal CO_2-emitting countries (orange columns).
The following discussions will explore Kaya factor evolution dynamics across key world emitters and economic aggregations. These include China, India, the USA, the European Union, and the Russian Federation.

Figures I.7.13–I.7.17 detail the behaviour of these factors at the national level, using the same variables presented for the global scale in Figure I.7.10.

Our attention is aimed at China, India, USA, the European Union, and the Russian Federation. Figures I.7.13–I.7.17 detail the behaviour at a national level of the same variables presented in Figure I-7.9 at a global level.

We do not describe all the trends in detail, leaving most considerations about the depicted trends to readers. Here, only some of them are reported in a concise way:

- Economic growth in China and India over the last three decades has been impressive: +1700% and +480%, respectively, as shown by the cyan curves in Figures I.7.13 and I.7.14. This growth was very poorly balanced by energy-efficiency policies, leading to a huge increase in their CO_2 emissions (around + 350 %).

  Over the very last years, both China and India have started lowering their emission intensity slightly, investing in RES penetration and rapidly becoming (especially China) the most relevant market for all RES types. More data about these aspects will be presented and discussed in Chapter I.11.
- During the period 1990–2020, both the USA and the European Union achieved the minimum objective of sustaining a temporary development while not increasing their energy consumption and CO_2 emissions.

  This objective, however, was and is not sufficient to reduce and solve the many sustainability issues of their energy systems. In more detail, the US is mainly focused on solving dependency issues in the mid-term using ways (such as the intensive use of fracking) which are certainly not sustainable over the long term, and, as explained, are generating increasing worries around the globe.

TABLE I.7.2
Values of the Kaya Factors in 2020 in the World and in the Main CO_2-Emitting Countries

2020 DATA Elaborated from the WDI Database by the World Bank		World	Most Relevant Nations or Aggregates China	India	USA	EU	Russia
A – Population	[MPax]	7821	1411	1396	332	448	148
	[World %]	100%	18%	18%	4%	6%	2%
B – GDP per Capita	[k$/y]	18	18	7	64	47	31
Economic Development Index	[Ratio]	100%	101%	39%	362%	265%	177%
Ec – GDP	[G$/y]	139,001	25,247	9771	21,323	21,089	4651
National Economic Indicator	[World %]	100%	18%	7%	15%	15%	3%
C – Energy Intensity	[Wh/$]	1265	1778	1181	1186	811	2267
En. Efficiency Indicator	[Ratio]	100%	141%	93%	94%	64%	179%
En – Energy Consumption	[Gtoe/y]	15115	3859	992	2175	1471	907
National Energy Use Index	[World %]	100%	26%	7%	14%	10%	6%
D – CO_2 Intensity	[g_{CO2}/kWh]	191	244	191	171	144	153
Use of RES Sources Index	[Ratio]	100%	128%	100%	89%	75%	80%
CO_{2-} Yearly CO_2 emissions	[Gt_{CO2}/y]	33,566	10,945	2201	4321	2465	1618
National GW Responsibility Index	[World %]	100%	33%	7%	13%	7%	5%
ΔCO2 – Cumulative CO_2 emissions	[Gt_{CO2}]	873	52	100	163	42	192
(cumulative emissions since 1990)	[ppm_{CO2}]	112	7	13	21	5	25
GW – Estimated GW	[°C]	0.90	0.05	0.10	0.17	0.04	0.20
Historical GW responsibility Index	[World %]	100%	6%	12%	19%	5%	22%

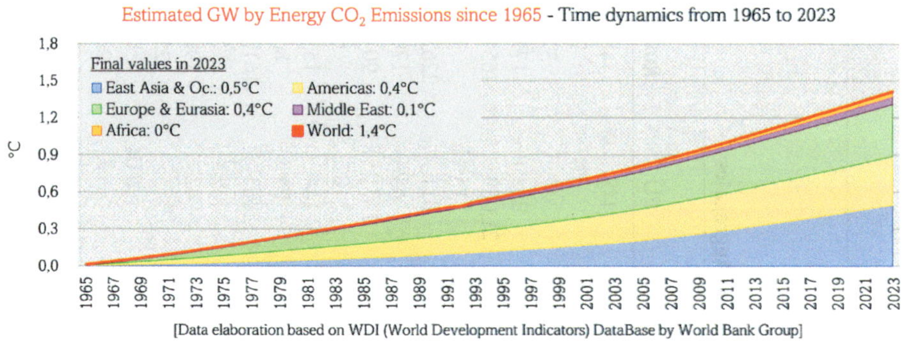

FIGURE I.7.12 World GW dynamics: curves are disaggregated for world geographical areas, identifying their responsibility for total GW.

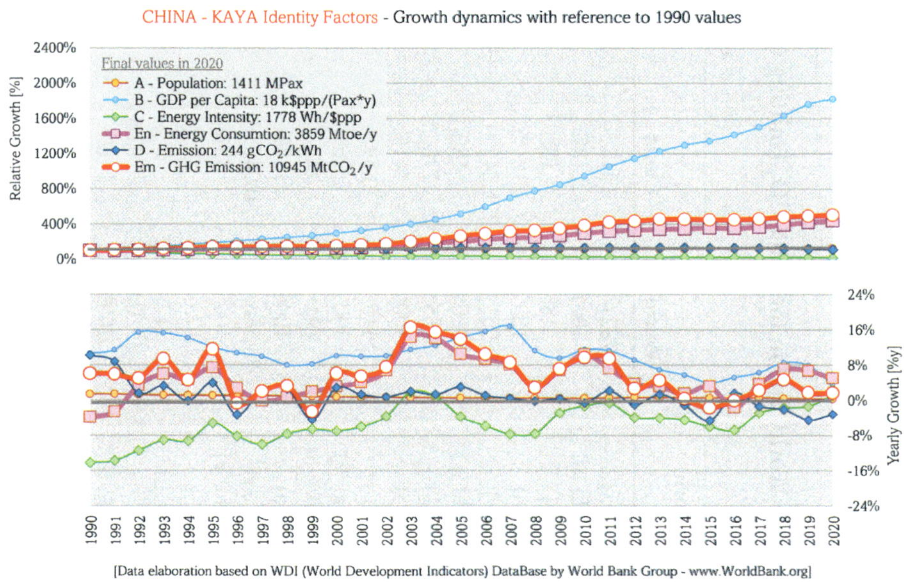

FIGURE I.7.13 China dynamics for all the Kaya factors: curves in the upper part are normalized by their initial values in 1990 and therefore depict variable relative growth; the bottom part, in the same colours, reports yearly growth rates of the single curves above.

On the other end, the European Union shows a more coherent and front-running behaviour towards decarbonization, but its results also are far from being sufficient.

Moreover, somewhat paradoxically, their efforts are becoming increasingly negligible on a global scale, as Asian economies' relevance on world overall emissions increases (already 58%, against the 7% of the European Union overall).

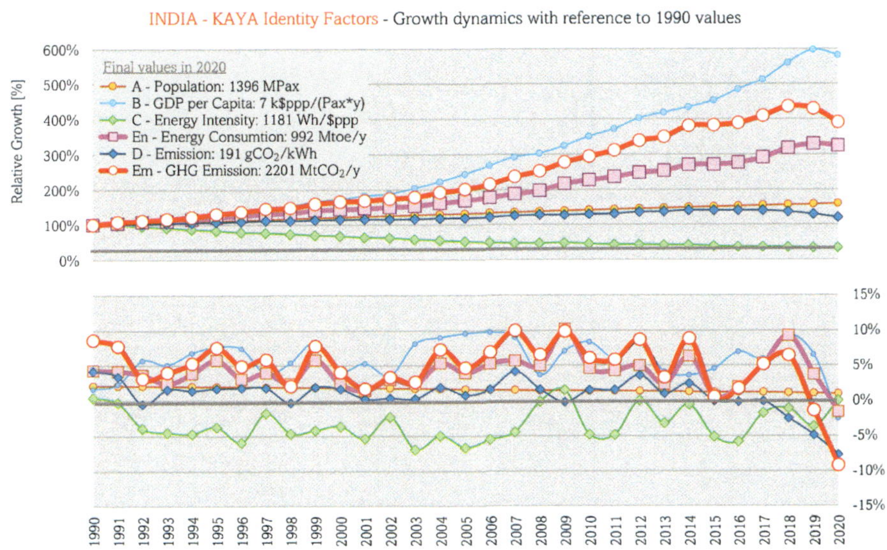

[Data elaboration based on WDI (World Development Indicators) DataBase by World Bank Group - www.WorldBank.org]

FIGURE I.7.14 India dynamics for all the Kaya factors: curves in the upper part are normalized by their initial values in 1990 and therefore depict variable relative growth; the bottom part, in the same colours, reports yearly growth rates of the single curves above.

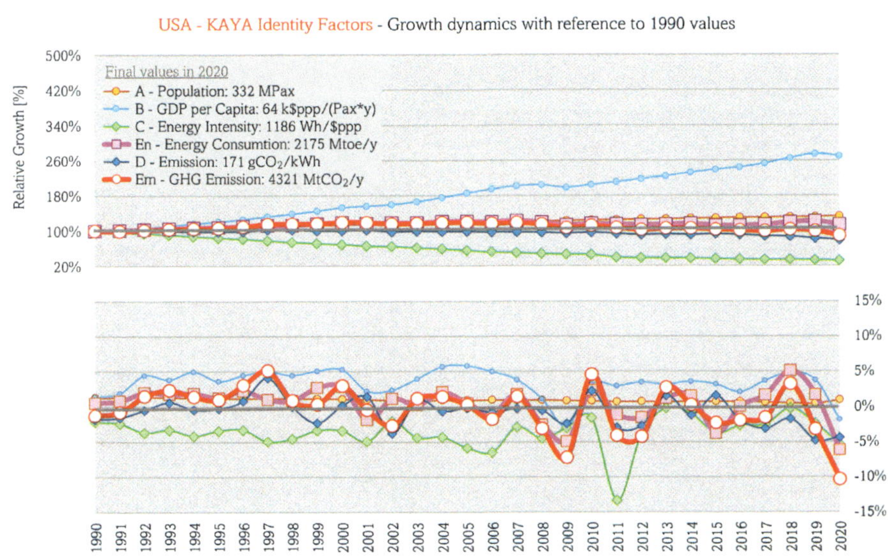

[Data elaboration based on WDI (World Development Indicators) DataBase by World Bank Group - www.WorldBank.org]

FIGURE I.7.15 USA dynamics for all the Kaya factors: curves in the upper part are normalized by their initial values in 1990 and therefore depict variable relative growth; the bottom part, in the same colours, reports yearly growth rates of the single curves above.

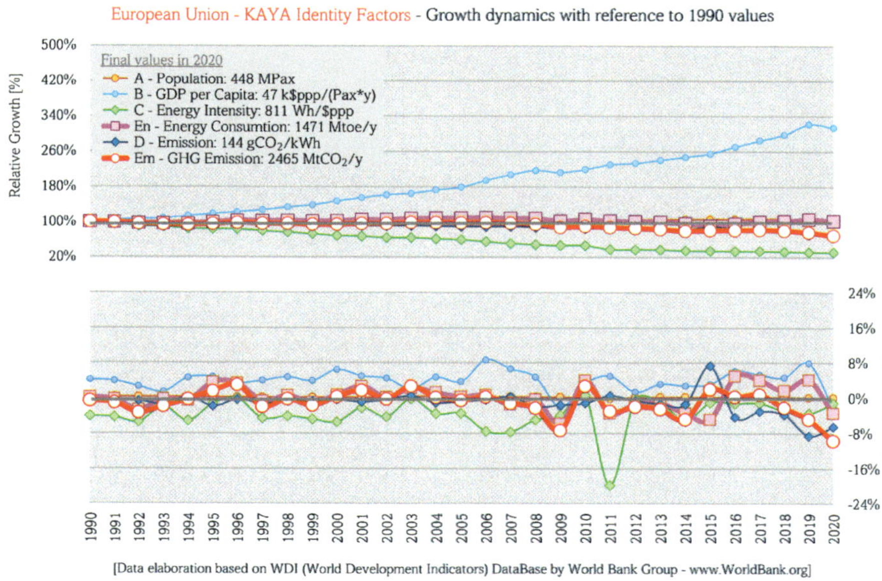

FIGURE I.7.16 European Union dynamics for all the Kaya factors: curves in the upper part are normalized by their initial values in 1990 and therefore depict variable relative growth; the bottom part, in the same colours, reports yearly growth rates of the single curves above.

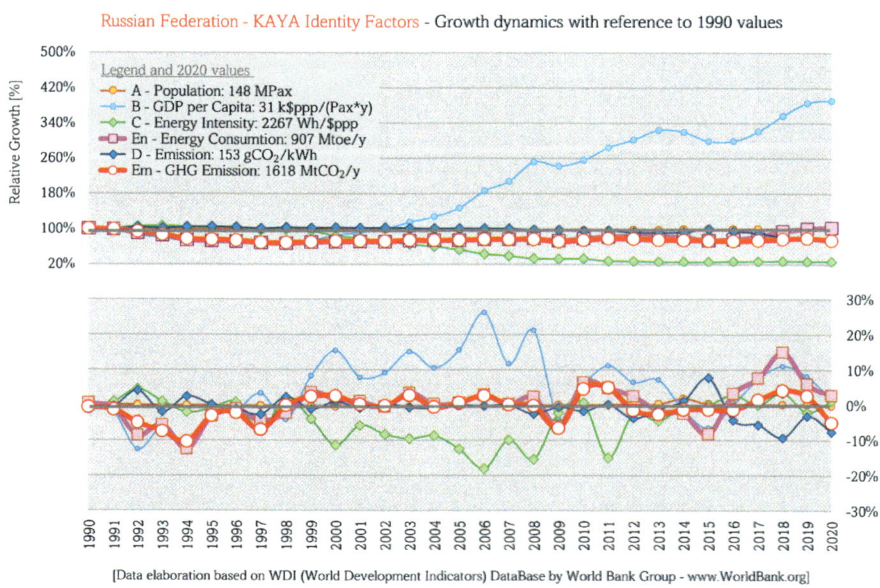

FIGURE I.7.17 Russian dynamics for all the Kaya factors: curves in the upper part are normalized by their initial values in 1990 and therefore depict variable relative growth; the bottom part, in the same colours, reports yearly growth rates of the single curves above.

Unfortunately, from a political point of view, the less you emit, the less your voice tends to be important in the definition of world decarbonization policies.

- Lastly, Figure I.1.17 shows how the Russian economy shows an extended period of depression during the 1990s and then started to grow again. This recovery was mainly due to a series of new economic relations Russia had with the EU, especially based on methane import/export. For a limited period, these brought about a new development phase for both sides, however this was stopped due to the Russia–Ukraine crisis in 2009 which degenerated into the Russia–Ukraine war in 2020.

I.8 Decarbonization Pathways

In Chapter I.6, we explored how, since 1988, the IPCC (on behalf of the UN Environment Program) has provided scientifically robust reports to assist world decision-makers in formulating policies aimed at achieving more sustainable global development. Unfortunately, the implementation of these policies cannot be local, because a single country implementing them would have great difficulty in terms of competitiveness with non-implementing countries, especially in the near and mid-term.

Decarbonization policies, therefore, should and must be coordinated on a global level, with the highest possible involvement of CO_2-emitting nations. To this aim, the UN, since the 1980s, within its Framework Convention on Climate Change (UNFCCC), has promoted effective dialogue among the parties.

In particular, in 1992 a milestone summit was organized in Rio De Janeiro, known as the "Earth Summit", where the UNFCCC was adopted as the first global treaty to address climate change, recognizing that the Earth's climate is a common concern of humankind.

The UNFCCC then decided to organize periodic Conferences of Parties (known as COPs). The first of them, COP1, took place in Berlin in 1995, setting the stage for what would become a regular, annual gathering of world leaders, negotiators, and experts. The COP was envisioned as a forum where countries could discuss the implementation of the UN Framework Convention on Climate Changes, review its progress, and make decisions on future climate actions.

In the early years, the COPs focused on developing mechanisms to reduce greenhouse gas emissions and set specific targets for industrialized nations, who were seen as the primary contributors to climate change. A defining moment came in 1997 at COP3 in Kyoto, Japan, where the so-called Kyoto Protocol was adopted. This agreement established legally binding commitments for developed countries to reduce their emissions of six key greenhouse gases (carbon dioxide, methane, nitrous oxide, and three fluorinated gases) by an average of 5.2% below 1990 levels, to be achieved in a five-year implementation period.

The Kyoto Protocol marked a milestone in international climate diplomacy, but it also revealed the challenges in reaching global consensus. In fact, while the

DOI: 10.1201/9781003631125-9

agreement was a significant achievement, it was met with criticism for not requiring developing countries, such as China and India, to take binding commitments. The United States, under President George W. Bush, ultimately withdrew from the Protocol, citing concerns about its economic impact and the lack of commitments from developing nations. This decision highlighted the ongoing tension between developed and developing countries, a division that would continue to shape climate negotiations in the years to come.

The Kyoto Protocol was not automatically effective for the signing countries and had a very long gestation, coming into actuation only in the period 2008–2012, more than 10 years after COP3 in Kyoto was held.

The early 2000s, in fact, saw the COPs continuing to evolve, but also facing difficulties in achieving substantial progress on climate action. COP6, for example, which was held in The Hague (Netherlands) in 2000, was a crucial moment for the implementation of the Kyoto Protocol. The talks ended in failure, as disagreements over the details of the Protocol's mechanisms, such as emissions trading and carbon offsets, prevented consensus.

By COP15 in Copenhagen in 2009 (while the Kyoto Protocol was finally ongoing), the urgency of the climate crisis had reached a tipping point. Expectations for a legally binding, universal climate agreement were high. That agreement should have defined "post Kyoto protocol" rules, but the summit ended with a highly criticized and non-binding "Copenhagen Accord", which should have at least recognized the need to limit global temperature rise to 2°C above pre-industrial levels.

Another 5 years (and five COPs) passed by without any substantial decisions, until COP21 was held in Paris, just days after a large Paris terrorist attacks. After nearly two decades of negotiations, the international community achieved, at least formally, a second historic breakthrough with the adoption of the so-called "Paris Agreement". For the first time, all 195 countries agreed to limit global warming to "well below 2°C" above pre-industrial levels, and to pursue efforts to limit the temperature increase to "1.5°C".

The Paris Agreement changed the approach with respect to the Kyoto Protocol, introducing a bottom-up approach to climate action, where each country submits its own Nationally Determined Contribution (NDC) for emission reductions, with the understanding that these targets would be reviewed and progressively strengthened every 5 years.

The Paris Agreement also established mechanisms for "climate finance", to help developing countries mitigate and adapt to the effects of climate change. The agreement was therefore seen as a historic moment of "global solidarity", with countries acknowledging that climate change is a shared responsibility that requires collective action.

COP21 also marked a shift in the global climate conversation, moving beyond the traditional divide between developed and developing countries. For the first time, developing nations were included in the global effort to reduce emissions, and there was an explicit commitment to supporting their adaptation needs through financial resources and technology transfer.

After the preliminary success of Paris, the subsequent COPs have focused on trying to ensure the full implementation of the Paris Agreement and ramping up ambition in response to the growing urgency of the climate crisis.

COP26 in Glasgow (2021) was one of the most significant COPs since Paris, as countries came together to reaffirm their commitments to the 1.5°C goal, with pledges to phase out coal, end deforestation, and scale up climate finance. However, critics pointed out that the promises made still fell short of the necessary actions to prevent the worst impacts of climate change.

Unfortunately, the COPs that followed Glasgow have largely been considered failures, including the most recent COP29 in Baku, where the absence of key players like the US, the Netherlands, France, and European Commission Chairwoman Ursula von der Leyen underscored a lack of global leadership on the issue.

History has shown that global political attention to climate issues tends to peak in cycles, with significant focus every 5–7 years. However, the fight against global warming demands sustained, year-on-year action and a continuous, coordinated effort. The intermittent nature of political engagement severely hampers the consistent progress needed to address the climate crisis effectively.

However, even though Paris Climate Agreement is formally on-going, as we will see in more detail, most of the world's countries are implementing it in a greatly insufficient way to reach its objectives, so that currently it is already almost certain that the +1.5°C GW goal will not be reached, and also reaching the +2°C GW is significantly at risk.

Let us now look in more detail at the contents of that Agreement and discuss its implementation state.

1.8.1 PARIS AGREEMENT

As briefly introduced, the primary goal of the Paris Agreement is to limit global warming to well below 2°C above pre-industrial levels, with an aspirational target of 1.5°C. The Agreement acknowledges that a rise in temperature beyond this threshold would have devastating impacts on ecosystems, human societies, and economies.

To this aim, the Paris Agreement establishes four key goals:

- **Mitigation**
  Reducing global greenhouse gas (GHG) emissions to curb the rate of climate change.
- **Adaptation**
  Enhancing resilience and adapting to climate change impacts that are already unavoidable, especially in vulnerable regions.
- **Climate Finance**
  Supporting developing countries in their mitigation and adaptation efforts through financial mechanisms.
- **Transition**
  Encouraging the shift towards a low-carbon economy, leveraging technological innovation, and promoting sustainable development.

A key feature of the Agreement is its much higher flexibility with respect to the previous Kyoto Protocol. In fact, it allows countries to set their own policies, named Nationally Determined Contributions (NDCs), to emissions reduction, based on their own capacities and circumstances.

Each country is required to submit its NDCs and update them regularly. These NDCs outline each country's targets for reducing emissions, and actions related to climate adaptation.

Emission reduction targets are non-binding, which means that countries are not legally compelled to meet specific emission cuts. Each country instead is held accountable and reliable through a global transparency framework that tracks progress and reports on climate actions.

One of the most crucial elements of the Paris Agreement is its emphasis on climate finance, especially to assist developing countries in adapting to the inevitable impacts of climate change.

The Agreement acknowledges the historical responsibility of developed nations, who have contributed the most to global greenhouse gas emissions, in helping to support developing countries through financial assistance, technology transfers, and capacity-building.

The commitment was made by developed countries to mobilize $100 billion per year by 2020 in climate finance for developing nations, with the goal of supporting their climate adaptation and mitigation efforts. This funding is intended to enhance resilience, especially in countries that are most vulnerable to climate impacts, such as small island nations and low-lying coastal regions.

The Paris Agreement underlined the ambition to maintain GW under the 1.5°C target, reflecting great urgency for decarbonization, since, according to the prevision made by the IPCC, this is necessary to avoid catastrophic impacts such as extreme weather events, droughts, food insecurity, and biodiversity loss.

The NDCs approach, granting flexibility to single nations, was proposed by the US and China to overcome some problems with the Kyoto Protocol. However, this could result in a drawback, since the determination of the single nation to perceive the objectives becomes crucial, and if some of the more relevant emitting countries show a lack of political motivation towards the Paris Agreement, the whole objective goal is put at risk.

As an example, in recent years, the US explicitly withdrew from the Paris Agreement under the Trump administration, and came back in for 4 years during the Biden presidency. Now, US adherence to the Paris Agreement is newly in doubt. This obviously puts the whole Agreement at serious risk.

According to the Paris Agreement, parties shall periodically take stock of their implementation to assess the collective progress towards achieving the purpose of the Agreement and its long-term goals. It enables countries and other stakeholders to take inventory, to see where they are collectively making progress towards meeting the goals of the Paris Agreement – and where they are not.

The first global stocktake was in 2023 and was concluded at COP28 in Dubai. Its outcome recognized that progress has been made towards the Paris Agreement goals, reaching near-universal climate action.

The outcome also tells us that these efforts are, however, insufficient, and that the world is not on track to meet the long-term goals of the Paris Agreement. The world has a narrowing window of opportunity to get back on track.

Despite their prudence, the official report by the UN about the global stocktake is clear in stating that the current state of implementation of the Agreement is substantially insufficient, remarking that:

- While the Agreement has spurred global action on climate change and reduced future warming forecasts, there is a glaring gap in emissions reductions, with current national commitments falling short by 20.3 to 23.9 $GtCO_{2e}$ compared to the levels required to limit warming to 1.5°C by 2030.
- To achieve the Paris Agreement's targets, GHG emissions need to be cut by around 43% by 2030 and 60% by 2035 from 2019 levels, aiming for net-zero CO_2 emissions by 2050.
- Many countries recognized that adaptation was no longer a matter of choice, but a matter of urgency in the face of intensifying climate impacts. These impacts escalate with each fraction of a degree of global warming, with some effects becoming irreversible if temperatures rise beyond 1.5°C.
- The role of finance, technology, and capacity-building are critical in enabling effective climate action. The international financial system should be transformed to support low-emission, climate-resilient development, and it is important to mobilize support for climate action in developing countries.

1.8.2 EMISSION SCENARIOS DISCUSSION

As already explained, Paris Agreement implementation will probably not reach its goals, but it still is producing some positive effects. Most of the world's countries, in fact, are currently making and depositing their own NDCs. Since the Paris Agreement was signed, its application makes it clearer and more foreseeable how the world will behave in the coming years, and probably decades.

Summarizing, minimal (let us say "optimistic") perspective world CO_2 emissions are now substantially known (while defined by the sum of NDCs) up to 2035. A few states, including the European Union, the UK, and Canada, have also deposited their NDC policies for a net Zero Carbon Policy up to 2050 in coherence with the 1.5°C pledge, but most did not.

Since the physical bases of the GW phenomenon are now known and agreed, they can be applied in stimulating NDC implementation, leading to previsions whose probability of happening is very high in the coming decades.

This is how most institutions studying GW and climate change (including the IPCC) produce a so-called BAU ("Business As Usual") scenario. This scenario is the most likely to happen, but things could evolve either better or worse with respect to that, depending on the "momentum" of the world (and especially that of the most CO_2-emitting countries) towards GW issues.

taking into account these considerations, the IPCC developed several scenarios, comparing them and evaluating their probability of happening and, in its recent AR6 report (released in 2023) stated that:

- Global warming will continue to increase in the near term (2021–2040) mainly due to increased cumulative CO_2 emissions in nearly all considered scenarios and modelled pathways. In the near term, global warming is more likely than not to reach 1.5°C even under the very low GHG emission scenario and likely or very likely to exceed 1.5°C under higher emissions scenarios. In the considered scenarios and modelled pathways, the best estimates of the time when the level of global warming of 1.5°C is reached lie in the near term".
- "Global warming declines back to below 1.5°C by the end of the 21st century in some scenarios and modelled pathways. The assessed climate response to GHG emissions scenarios results in a best estimate of warming for 2081–2100 that spans a range from 1.4°C for a very low GHG emissions scenario to 2.7°C for an intermediate GHG emissions scenario and 4.4°C for a very high GHG emissions scenario".
- "Continued emissions will further affect all major climate system components. With every additional increment of global warming, changes in extremes continue to become larger. Continued global warming is projected to further intensify the global water cycle, including its variability, global monsoon precipitation, and very wet and very dry weather and climate events and seasons (high confidence). In scenarios with increasing CO_2 emissions, natural land and ocean carbon sinks are projected to take up a decreasing proportion of these emissions (high confidence)."
- "With further warming, every region is projected to increasingly experience concurrent and multiple changes in climatic impact-drivers. Compound heatwaves and droughts are projected to become more frequent, including concurrent events across multiple locations (high confidence). Due to relative sea level rise, current 1-in-100 year extreme sea level events are projected to occur at least annually in more than half of all tide gauge locations by 2100 under all considered scenarios (high confidence). Other projected regional changes include intensification of tropical cyclones and/or extratropical storms (medium confidence) and increases in aridity and fire weather.

1.9 Small Space-Time Scale Environmental Impacts

Let us now discuss a further very critical issue which is normally generated by energy systems on a local and temporary basis.

As is well known, all those energy processes involving the combustion of chemical energy sources and/or carriers determine temporary and local air pollution which may cause, due to the emissions in the atmosphere, small or large modifications in air quality characteristics.

It is obvious, but still noteworthy, to observe that these problems are generated by the chemical nature of the energy carrier used and are substantially unmodified by the origin of this carrier, which could be green or not (fossil, biological, electro-synthetic).

These modifications may produce critical near-/mid-/long-term effects on the natural ecosystems and on human and animal health, depending on the amplitude in time and space of the exposure to unhealthy air. However, air quality fortunately can be restored in a relatively short time (hours or days at most) if the emission sources are removed in a nearby area (within a few kilometres) of the affected site.

Therefore, even though it is very critical, this problem is not a sustainability problem.

However, it is clear that the solution to local pollution problems calls for utilization of greener final-use energy technologies whose introduction into energy systems must be decided and coordinated with all the sustainable policies available to the energy sector.

Also, from an administrative point of view, all decisions concerning air quality are within the competences of local administrations (especially large metropolitan areas), but their decisions have a huge effect at a national level, with most of inhabitants nowadays living (and consuming energy) within urban contexts.

All energy policies should therefore be decided, shared and implemented through effective and close coordination at all administrative levels.

1.9.1 PRIMARY AND SECONDARY AIR POLLUTANTS

All things considered, it is noteworthy here to provide some brief details about the main chemical and physical phenomena that may determine locally alarming conditions about air quality.

DOI: 10.1201/9781003631125-10

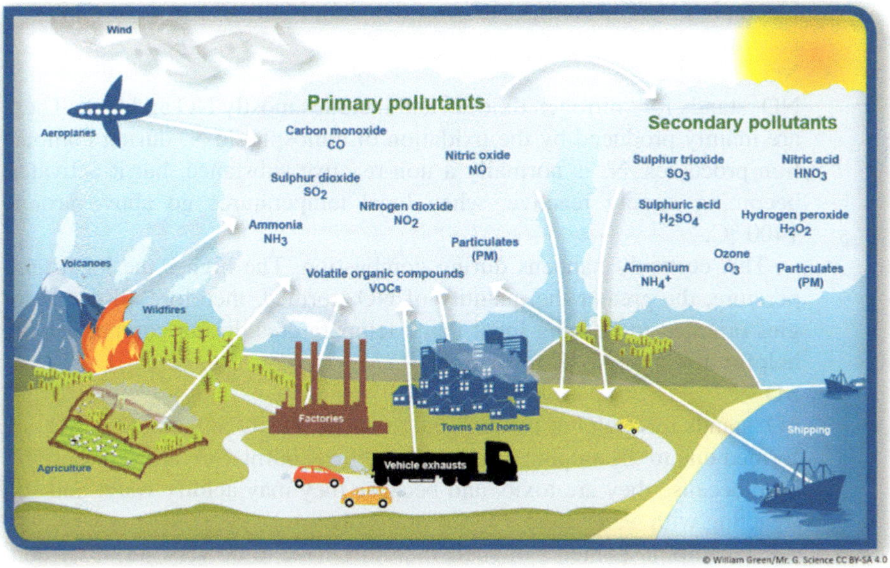

FIGURE I.9.1 Graphical sketch identifying the main primary and secondary air pollutants and explaining their differences in originating mechanisms.

From this point of view, we firstly introduce an important categorization among air pollutants as either primary or secondary pollutants (see Figure I.9.1):

- **Primary pollutants**
  Primary pollutants are those which are emitted by anthropic or non-anthropic sources (e.g. natural fires). We will limit our discussion to pollutants produced by energy processes (and therefore mainly from fuel combustion). These are those which are creating the biggest problems in terms of air quality because these processes normally exist in urban environments.
  Obviously, many other industrial and non-industrial activities may produce significant impacts on air quality (including on water and soil), however these impacts are beyond the objectives of this book. Limiting our attention to energy processes, the main pollutant categories to be considered are these five:
  - **CO**
    Carbon monoxide is mainly a product of incomplete oxidation (to CO_2) of fuels during combustion. Although it is less abundant than CO_2, CO poses significant health risks due to its toxicity, as it can impair oxygen delivery to the body.
    Combustion processes have evolved to minimize CO emissions, but its presence remains a concern in poorly ventilated or high-traffic urban areas.
  - **HC**
    HC stands for "unburnt hydrocarbons" and identifies a wide category of pollutants which are measured and normed together. They are a product of

incomplete combustion processes (like CO). Many of them are toxic and harmful.

- **NO_x**
NO_x stands for "nitrogen oxides" and includes mostly NO and NO_2. These are mainly produced by the oxidation of atmospheric N_2 during combustion processes. N_2 is normally a non-reactive substance, but it activates, becoming largely reactive, when local temperatures go above around 1400 °C.
  This certainly happens during combustion. The higher the local temperature, the greater the quantity of NO_x formed, therefore all technologies oriented to limiting local flame temperatures during combustion are helpful in limiting NO_x formation.
- **SO_x**
SO_x stands for "sulphur oxides" and includes mostly SO_2 and SO_3. SO_x are formed due to the sulphur content of the fuels burnt. SO_x are very harmful both because they are toxic, and because they may acidify water and soil (reacting with water vapour in the atmosphere).
  The sulphur contents of fuels therefore have to be very seriously controlled. NG does not contain sulphur. Fuel derivates for the automotive sector (gasoline and diesel), also have very low sulphur contents (especially in OECD countries), but other oil derivatives and especially coal fuels may have a very relevant sulphur content, therefore potentially producing very high SO_x emissions.
  Huge problems are therefore encountered if coal thermo-electric plants are too close to residential areas. This may happen if urbanization processes are too fast and at least partially uncontrolled (such as those that have occurred in many Chinese and Indian megalopolis in the last few decades).
  In these cases, the "easiest" solution is to purify gas exhausts from SO_x, but these processes, as already introduced, are very costly, both economically and energetically, and lead to very high CO_2 emissions for unit net electricity produced by the plants (up to 5–6 times more emitting, in terms of CO_2, than methane-fed thermoelectric power plants)
- **PM**
PM stands for "particular matter" and identifies solid particles emitted into the atmosphere. They are produced from burning liquid and/or solid fuels. The main urban-persisting PM sources are internal combustion engines for automotive applications (mainly diesel), and some oil- and solid fuel-fed power and heating plants (especially in non-OECD countries that are still using those energy carriers in urban contexts).
  PM's harmfulness strictly depends both on their permanence time in atmosphere and/or their "breathability" and interaction with human organs. Regarding the first aspect, permanence time depends on the equilibrium between volume (weight) and surface (viscosity) forces acting on the particle, and define its buoyancy attitude. This force balance depends on a characteristic geometric dimension of the particle: the most commonly

used particle characterization is that depending on their so-called Sauter mean diameter (SMD).

Normally, emitted PMs have many different SMDs, so that their population must be characterized in terms of typical SMD statistical distributions to define their harmfulness. However, also based on the foreseeable standard distribution of PMs coming from various types of emitting technologies, normally emitting limits on PMs are defined based on a limiting maximal SMD.

As an example, PM10 defines all these emitted particles having an SMD below 10 μm. As already mentioned, the smaller the PM is, the more harmful it is considered. This is because, when breathed in it can overcome most of the "filter devices" the human body has (nose, larynx, pharynx, trachea, bronchi, lungs), while also being able to reach the blood and hence transferred to other human organs.

PM chemical composition is not really defined, as particles are formed by two parts: an internal mainly carbonic core coming from unburnt material, usually named "soot"; and an external part made by the product which is later deposited on the particle that depends on the fuel's original composition but also on the other pollutants that the particle will encounter in air.

This external part of the particle is named the SOF (soluble organic fraction) and is usually the most harmful part. PM is therefore particularly dangerous because it is the "vehicle" transporting pollutants within the human body.

- **Secondary pollutants**

  Secondary pollutants are not emitted directly but are formed in the atmosphere through chemical reactions involving primary pollutants. Therefore, the distinction between primary and secondary pollutant interactions says nothing about the pollutant's harmfulness: a secondary pollutant may be more harmful than a primary one, with the additional problem that it was not emitted directly, so we may have difficulties in knowing how to limit its presence in the atmosphere (if we don't exactly know the physics and chemistry governing its dynamical formation).

  One of best secondary pollutants is ozone (O_3). This should not be confused with stratospheric ozone, whose presence is limited to very high altitudes in the atmosphere and is largely positive for human beings because it filtrates UV solar wavelengths.

The distinction between primary and secondary pollutants is important also from a regulatory point of view, because homologation and installation limits may be imposed in urban contexts only on their emitting potentialities in terms of primary pollutants.

Instead, air quality characteristics must be measured and controlled for all harmful pollutants (both primary and secondary). In this regard, normally air quality not only depends on the pollutant concentration in breathable air, but also on mean exposure time.

Directives also must consider that most secondary pollutants are created through photo-activated reaction cycles (whose start depends on the presence of high-frequency solar wavelengths), and therefore typically have significant hourly variations during the day.

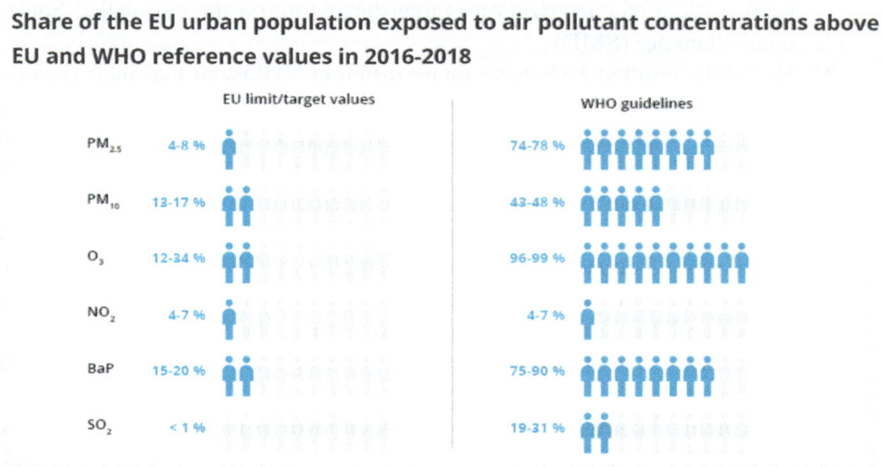

FIGURE I.9.2 Percentage of EU residents exposed to harmful pollutants exceeding EU and WHO air quality standards limits.

Therefore, air quality standards (AQS) locally define various thresholds for harmful pollutants in an attempt to safeguard people's health, to inform the population (so that people can decide to limit their exposure risk to pollutants) and finally to actively limit the presence of pollutant sources in the urban context (for example by limiting car use). The different thresholds defined for each pollutant are normally set to at least three limits of "attention", "alert", and "alarm".

The question is, how frequently are these thresholds overcome, and consequently which parts of the population are exposed to potentially harmful concentrations of pollutants?

The World Health Organization (WHO) normally studies these aspects, mostly based on measurements taken in EU nations, while also comparing WHO non-regulatory AQS guidelines with regulatory EU AQS. The results of one of these studies are reported in Figure I.9.2, showing that a significant part of the population is affected by pollution phenomena in a non-negligible way.

I.9.2 MAIN POLLUTANT ATMOSPHERIC DIFFUSION MECHANISMS

To what extent is air pollution a "local" issue? The answer to this question depends mainly on the way pollutants are diffused in air, making them harmless at a sufficient distance and after a sufficient time from the emission point and time. A full discussion about diffusion mechanisms is beyond the objectives of this book, and is normally the subject of full academic courses in atmospheric physics and/or computational fluid dynamics curricula.

Here, only a simple discussion of the effects of the diffusion mechanism is provided for its didactic implications, usually referred as the Gaussian plume

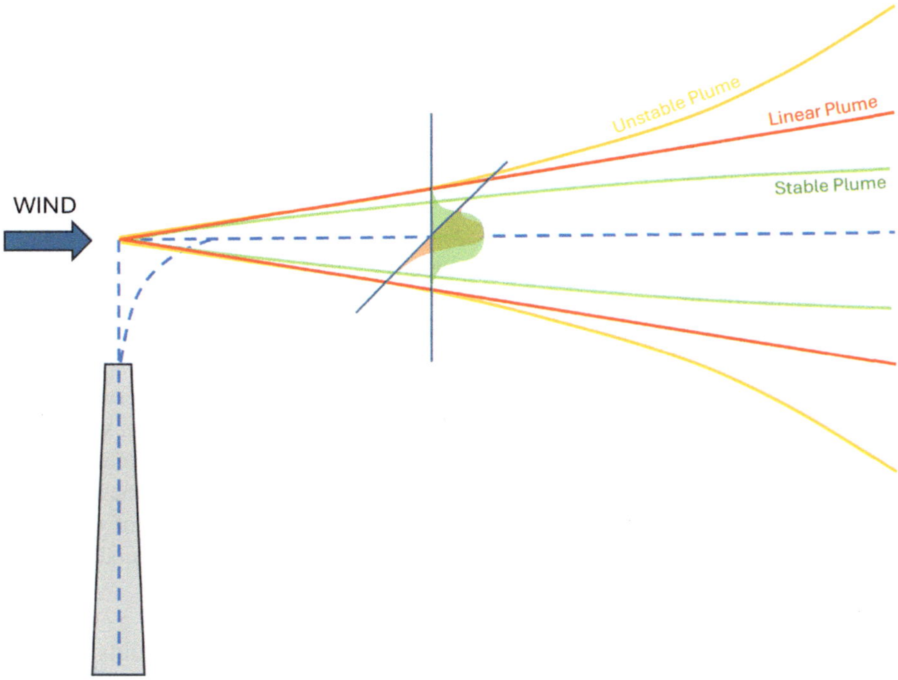

FIGURE I.9.3 Schematic description of Gaussian plume.

(graphically depicted in Figure I.9.3). This model provides a foundational framework for understanding pollutant dispersion in the atmosphere and is often used for educational purposes and preliminary simulations.

As already mentioned, this "plume" illustrates a simpler possible way (though sufficiently accurate for some purposes) in which pollutant diffusion in air may be described. It briefly assumes that the diffusion volume for the pollutant is a "Gaussian cone" defined by the statistical dispersion of the pollutant with respect to a primary diffusion path defined by the vectorial speed of the air flow transporting the pollutant. the key characteristics of a Gaussian plume include:

- Its Gaussian Distribution
  The concentration of pollutants is assumed to follow a Gaussian (normal) distribution perpendicular to the primary diffusion path. The plume's width expands as the pollutant disperses further from the source.
- Its Standard Deviation (σ)
  The width of the plume is quantified by the σ of the Gaussian distribution, which increases over time and with distance from the source. The rate of this increase depends on factors such as wind speed and atmospheric turbulence.

Therefore, the amplitude of this cone may be defined by the value of σ of the Gaussian distribution which increases with time, and therefore with the distance from the source. This increase in width of the plume may be less or more than linear, depending on the wind speed and corresponding air turbulence and instability levels.

The Gaussian plume is a simplification and assumes steady-state conditions, which makes it suitable for scenarios where pollutant dispersion can be modelled as a relatively constant and uniform process. Obviously, this is a very simplified model, which may need to be corrected considering several other aspects.

Just as examples, consider the following:

- The temperature of the emitting source, for example, may be considered by moving the emitting point in a virtual and higher position with respect to the real emitting point, considering the local buoyant effect given by the lower density of exhaust gas with respect to surrounding air.
- Pollutant diffusion occurs in the lowest atmospheric layers, where air temperature and density can vary significantly with altitude. These variations may be rapid and non-monotonic, potentially causing temporary temperature gradient inversions that strongly influence pollutant dispersion. These atmospheric conditions, which are normally created by a succession of meteorological phenomena (more frequent in certain areas), may be modelled by non-symmetric but still Gaussian plumes.

Despite its simplicity, the Gaussian plume model is widely used in air quality studies, especially for scenarios involving multiple emission sources within a confined area. Its computational efficiency makes it suitable for preliminary assessments, regulatory compliance checks, and scenarios where high-fidelity CFD models are impractical. However, the model's limitations should be acknowledged:

- It assumes steady-state conditions and uniform wind fields, which may not represent real-world variability.
- It does not account for complex terrain, buildings, or vegetation, which can significantly alter pollutant dispersion patterns.

Even though Gaussian plumes are based on the observation of effects more than on physically consistent CFD, models are still useful to imagine and reconstruct reality and are used in many simulations, especially when and where the number of sources in a restricted dominion increases and makes it difficult to use more physically consistent CFD approaches.

1.9.3 CONVENTIONAL URBAN ENERGY TECHNOLOGIES

To conclude this brief discussion about small space–time scale environmental impacts and corresponding air quality issues, it is noteworthy to recall some important aspects:

- These problems usually become relevant only where the combination of two conditions is present: a high concentration of polluting sources and a high concentration of potentially damaged population.

- It is therefore obvious that urban contexts are those where we should concentrate our aims to avoid the presence of combustion technologies.
- In normal situations, pollutant emission sources insisting on urban contexts are mainly: conventional ICE-based cars; conventional residential heaters based on NG or oil burners. In fortunately limited (but not-at-all-negligible) cases, very rapid and uncontrolled urbanization processes have placed residential areas in close proximity to both industrial sites and thermoelectric plants (including coal-fed). Those cases must be highlighted and treated separately.
- Thanks to technological advances which were mainly due to sustainability issues, nowadays very effective technologies are available to solve local pollution problems. In fact:
    - Conventional ICE-based urban cars can/must be urgently substituted by electric vehicles. The same urgency is not required for long-distance mobility, especially considering that the benefit of electric mobility is not obvious if the weight of the vehicle (due to the high mileage required for long-distance mobility, with vehicle consumption and therefore CO_2 emissions being proportional to the vehicle's weight).
    - Conventional residential heaters (based on NG or oil burners) may be replaced with heat pumps driven by electricity produced by RESs (especially photovoltaic [PV] installed on building roofs).

Some more details on these technological aspects are given in Chapters I.11– I.13 about key energy technologies for energy transition implementation.

I.10 Energy Transition to Sustainable Energy Systems

As we discussed earlier in Chapter I.2, an energy system could be defined as: "the set of technological, political, economic and social choices which are made to define a territory/population energy demand and the ways to fulfil it".

In essence, an energy system encompasses all interconnected components and processes involved in the production, conversion, storage, distribution, and consumption of energy.

Narrowing the focus to the technological domain, we identified key components of an energy system, including:

- Energy sources and their conversion technologies.
- Energy storage systems (ESSs) and energy carriers (also referred to as energy vectors).
- Final energy use technologies.

In earlier chapters we defined and discussed the main sustainability and local short-term issues that energy systems define and are called upon to solve. Among those are:

- Energy dependency (Chapter I.3)
- Energy sources exhaustibility (Chapter I.4)
- Global warming and climate change (Chapters I.5–I.8)
- Small space–time scale environmental impacts and clean air issues (Chapter I.9).

With this foundation, we can now revisit the evolution of energy systems in greater detail, addressing:

- The structure of national energy systems developed during the industrial revolution and persisting into the 20th century.
- The key improvements already made towards sustainability, resulting in the configuration of today's energy systems.
- The necessary transformations required over the next two decades to achieve a full transition to sustainable energy systems.

Starting from the first point, referring to Figure I.10.1, we can state that up to the end of the 20th century the main characteristics of all national energy systems were the following:

DOI: 10.1201/9781003631125-11

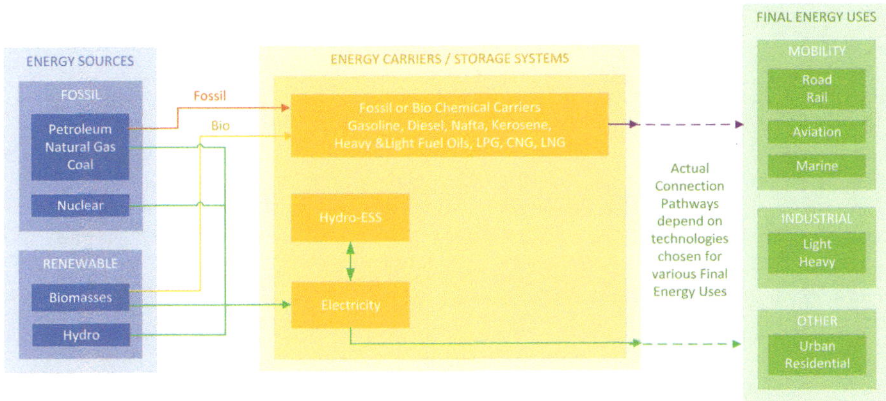

FIGURE I.10.1 A graphical sketch of an energy system of the 20th century.

- About energy sources: what we now call "conventional sources" were:
  - Fossil sources (all hydro-carbon sources [oil, NG, and coal]), and nuclear energy (only in some countries, starting from the 1970s).
  - Some RESs: mainly conventional "basin" hydro-energy and biomasses (especially in low-income nations) and some limited geothermal use of (in selected nations, such as Italy, Iceland, the US).
- About energy carriers and ESSs: what we now call "conventional carriers and ESSs" were:
  - Electricity: mainly grey electricity (coming from fossil sources) and limited green electricity (coming from hydro, biomasses, and geothermal energy sources). The energy mix used to produce electricity mainly privileged fossil potential energies so that the need for electricity storage was limited and due to cost optimization aims.

    This need was normally satisfied through hydro-ESS technologies (pumping water to an upper basin during the night, storing electricity, and making it flow back through a turbine mostly during central daylight hours to regain electricity).
  - Several types of chemical energy: Carriers for the various final use sectors include: mainly gasoline and diesel fuels for road transport; mainly NG-, LPG- and oil-based carriers for heating and other industrial uses. All these carriers were largely "grey", that is, produced by fossil energy sources.
- About final energy use technologies:
  - Most transportation systems use chemical energy carriers (e.g., gasoline, diesel), whereas rail-guided mobility solutions and trolleybus systems primarily operate on electricity.
  - Residential heating is mainly chemical based, and uses mostly NG in Europe, and NG and various oil derivatives in most other countries.
  - Most of the other final energy uses (including a significant portion of industrial ones) are fed by electricity.

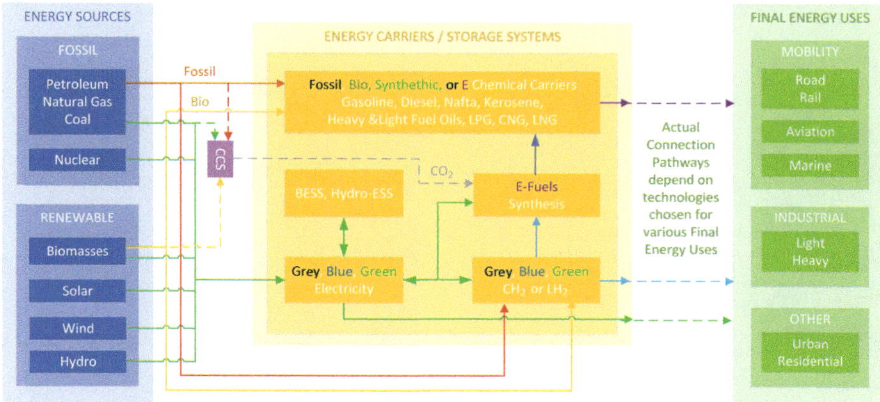

FIGURE I.10.2 A graphical sketch of an energy system in its actual configuration.

From this starting point, many things have already changed in most world energy systems, so that nowadays energy systems are much more complex than previously and are effectively represented by the sketch in Figure I.10.2.

The main characteristics which have been newly introduced in the energy systems are the following:

- About energy sources: new RESs are now in use, and some new uses are made of conventional ones:
  - Wind energy has been introduced since the early 2000s, both in-shore and off-shore, to produce electricity.
  - Solar energy has been introduced, initially for thermal application and then mainly for electricity production through PV applications. Concentrated solar thermodynamic plants are also an option for electricity production.
  - Hydro energy use was increased, installing new basin and flowing-water power plants. Some demonstrational plants have been realized using wave energy and tidal waves.
  - Biomasses are now more extensively used to produce bio-chemical energy carriers.
  - Nuclear energy utilization remains substantially unchanged.
  - Some demonstrational carbon capture and sequestration (CCS) technologies have been installed in fossil thermoelectric plants, but many issues remain to be solved in this sector.
- About energy carriers and ESSs, many more possibilities are being experimented with, in particular:
  - The increased penetration of non-programmable RESs (especially wind and solar) in the electricity sector have produced a much higher need for ESS installation. In most countries, hydro-ESSs are not an option anymore because their environmental impact is generally judged too high locally. In these cases, the only two available options seem to be BESSs

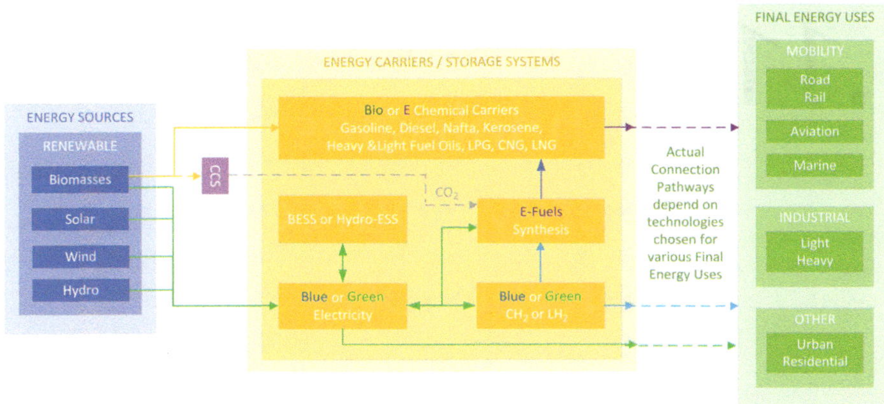

FIGURE I.10.3 A graphical sketch of a fully sustainable energy system.

(electrochemical batteries, mainly Li-based) and/or H_2 production through electrolysis.
  - Synthetic biofuels and/or e-fuels are being increasingly experimented with, with the future aim of feeding those hard-to-abate final use sectors that are difficult to supply by electricity.
- About final energy use technologies:
  - Urban road mobility is slowly becoming electric. Most of the other mobility uses still use conventional technologies.
  - In newly built residences, low-impact and high-sustainability concepts are being diffused, using electric heating/conditioning and cooking appliances, with electricity mostly being locally produced through PV plants. Attention is growing also about building energy consumption.

All the processes activated in the last two decades must proceed as fast as possible with the aim of finally having only fully sustainable energy systems like those presented in Figure I.10.3.

This implementation path is what we normally call energy transition. The European Union is trying to open the road in this direction and is planning to reach this final objective by 2050, and so has implemented a fully decarbonized energy system, possibly implementing only RES, while still accepting nuclear energy use as an option.

In Chapters I.11–I.13 we discuss the key enabling technologies (KETs) which must be diffused and installed to reach this objective. We describe them individually, giving some very brief technical details. In particular:

- Chapter I.11 focuses on KETs for energy sources conversion.
- Chapter I.12 focuses on KETs for energy carriers production and ESSs.
- Chapter I.13 focuses on KETs for final energy use.

I.11 Key Enabling Technologies for Energy Sources Conversion

As previously discussed in this book, the choice of energy sources plays a pivotal role in determining the sustainability of energy systems. These, in fact, should be as much as possible:

- Unconventional (at least for the energy system under consideration), to increase energy independency.
- Renewable, to increase their expected availability during time.
- CO_2 free (or at least low on an LCA [life cycle assessment] basis) to decrease the energy system carbon footprint.
- Convertible in proper energy carriers which could:
  - Feed the most possible final uses, using available technologies with a sufficiently high technology readiness level (TRL)
  - Be easily distributed to final energy use sites, without significant energy losses.
  - Be easily storable to temporally match energy sources availability with final energy demand.

All these considerations make it evident that the only way to make an energy system fully sustainable in the near future is to rapidly move towards a fully RESs energy mix.

Obviously, each national energy system has its own starting conditions and constraints to implement this transition, defined by existing infrastructure, economic considerations, and geopolitical realities.

To evaluate these conditions, the data reported in Figure I.11.1 depict 15 of the world's main consuming nations' actual energy mixes, both in absolute energy values (left side of figure, in EJ), and as percentages of the total local energy sources consumption (right part of figure). Those nations together consume about 75% of the total world primary energy.

RESs' utilization is depicted in the upper part of the figure, while fossil use is in the lower part. In this regard, fossil consumption is disaggregated as oil, NG, coal, and nuclear energy contributions, while RESs are disaggregated as hydro, wind, solar, and other RESs (including geo-thermal, biomasses, and other possibly used RESs).

DOI: 10.1201/9781003631125-12

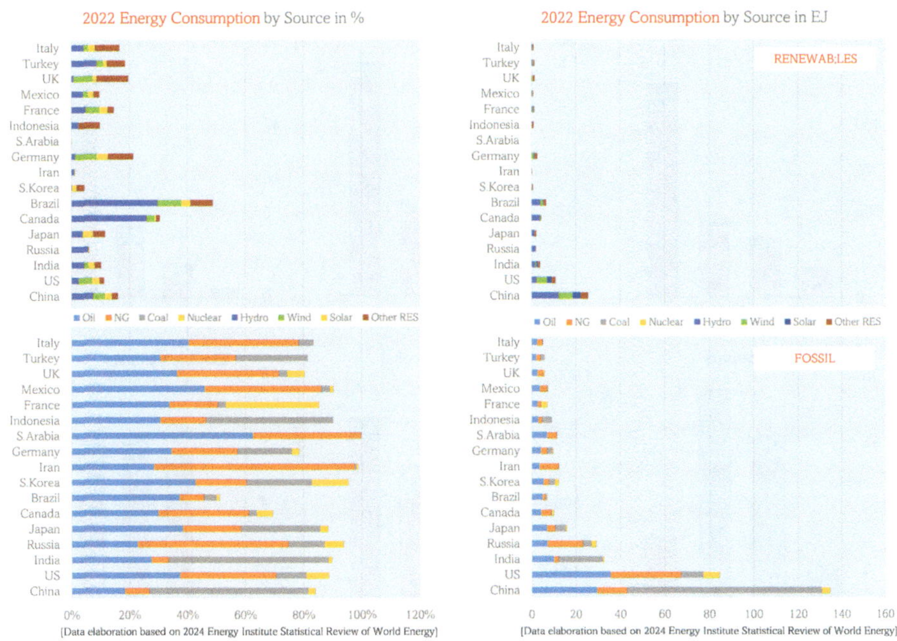

FIGURE I.11.1 Energy sources mix for the 15 most energy-consuming nations.

Data show that, at a global level, China is nowadays the most consuming country (26% of total world consumption). Its energy mix is mostly fossil based (84%) with a strong prevalence of coal among fossil sources (55% of its total consumption), followed by oil (18%), NG (8%), and nuclear (2%).

As already said, this coal dependency defines a particularly unsustainable energy mix: it grants sufficient energy independency but leads to a very high emission intensity and overall CO_2 emissions, and also contributes to significant local pollution problems.

During recent years, China has finally been moving towards RESs. Even though RES penetration is still poor in its energy mix (16%), overall figures in terms of installed power on an annual basis make China the most important market for all RES typologies.

Paradoxically, this evidence is now somehow decreasing the attitude towards RESs of other countries and economic groups (European Union, US, and all other OECD members) who up to some years ago were fostering RESs. This is because they now fear losing a previously dominant industrial position within this sector, so that their investment in new RES plant installations has become progressively less economically affordable.

These considerations have been confirmed by the International Energy Agency and published, for example, in their Energy Technology Perspectives Report 2024.

Figure I.11.2, in particular, shows some of its conclusions both about the RES market evolution (solar PV panels and wind aerogenerators) in recent years, and that

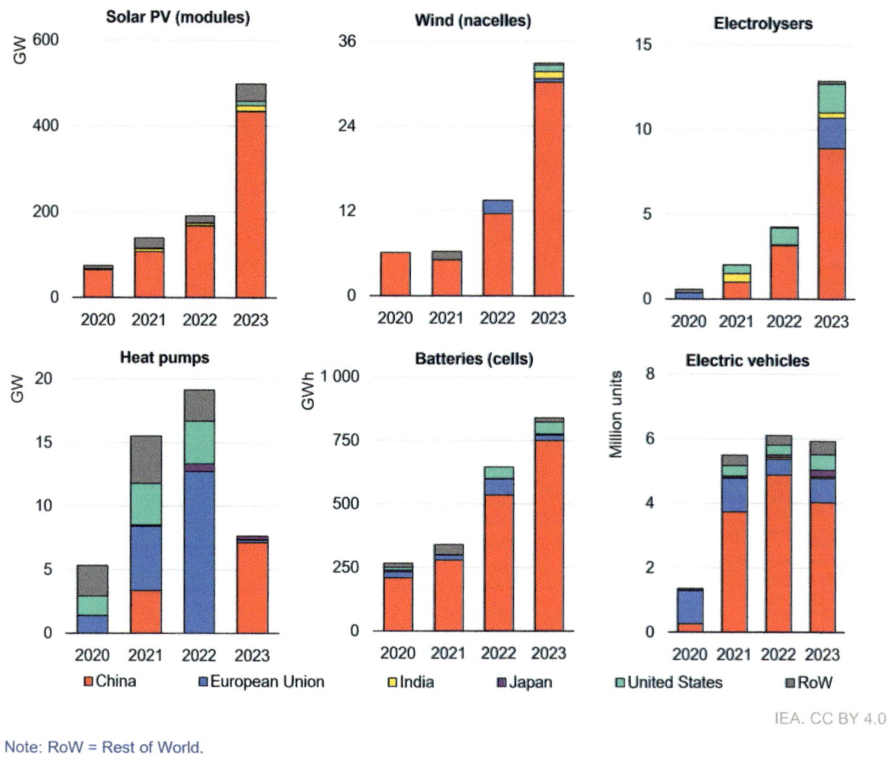

FIGURE I.11.2 Net manufacturing capacities for selected clean energy technologies (IEA – Energy Technology Perspectives Report 2024).

of other KETs in the energy sectors (which will be treated in following chapters of this book). In all these markets, China is the most important world sub-market regarding both technologies production and installation sites.

Going back to the Figure I.11.1 data discussion, the second-highest global energy-consuming nation is the US (15% of total world consumption). Its overall economy (GDP values) is now very similar to that of China, but it consumes much less energy due to a much lower energy intensity (35% less than China).

The US is still particularly fossil (89%) and especially oil (37%) dependent, even though in recent years it has been progressively moving towards NG produced by fracking to increase its independence.

This approach, unfortunately, may determine a relevant decrease in its attitude towards RESs in the near and mid-term future, mainly due to progressively increasing political tensions on the global scale between the US and OECD countries on one side, and BRICs countries on the other, which is strongly affecting all possible pathways towards effective decarbonization policies.

Some other additional considerations may be obtained through the analysis of Figure I.11.1 data on the world's nations' actual energy mix:

- Among the 15 highest consuming nations, nuclear energy use as a fossil alternative primary source (with respect to hydrocarbons) is limited (12% in S. Korea; 8% in the US, 6% in Russia, the UK, and Canada; under 3% in all the other cases), with the only exception of France (32%).

  Most of the world's nuclear plants were installed in the 1970s and 1980s, apart from some new plants installed mostly in France and China over the last two decades.

  In Germany, nuclear energy played a significant role in electricity generation since 1962, accounting for about 18% of the country's electricity production in 2011. However, in 2023, Germany made the decision to decommission all its nuclear plants, a move that has led to both increased energy costs and higher CO_2 emissions.
- Canada and France are the two countries, amongst these 15, with the lowest use of hydrocarbon sources in their energy mix (respectively, 50% and 53%), followed by the UK (74%) and Germany (76%).
- Brazil, amongst these 15 nations is has a more RESs-based energy mix (49%), based mainly on hydro sources (30%) but also on wind (8%) and biomass (8%). For RESs' overall penetration in the energy mix, Brazil is followed by Canada (31%, mostly due to hydro energy), Germany (21%, using no hydro energy), UK (20%); Türkiye (18%), Italy (17%), and China (16%).
- As already said, Brazil and Canada are also the two nations having the highest penetration of hydro energy in their energy mix (respectively, 30 and 26%), due to favourable territorial conditions. They are followed by Türkiye (9%), China (8%), and Russia (6%).
- Germany has the highest penetration of wind sources in its energy mix (8%), followed by the UK (7%), France and US (5%), and China (4%).
- Germany (4%), Japan, Brazil, and France (3%) are the nations having the highest penetration of solar energy in their energy mix. All these figures about use of solar energy are expected to increase a lot in the near-middle term, as will be explained later.

Let us now have a closer look also at the dynamic data about nuclear energy and RESs to understand whether things are improving or worsening in these sectors. In the following sections, when needed, we will also very briefly introduce the main KETs that are currently available for RES energy sources conversion and consequent utilization.

I.11.1 NUCLEAR ENERGY

Figure I.11.3 depicts data about overall world energy production by nuclear sources, disaggregated by nations.

These data reveal that the United States has consistently been the world's largest producer of nuclear energy. However, its production has remained relatively stable since the 1990s, reflecting a plateau in nuclear expansion.

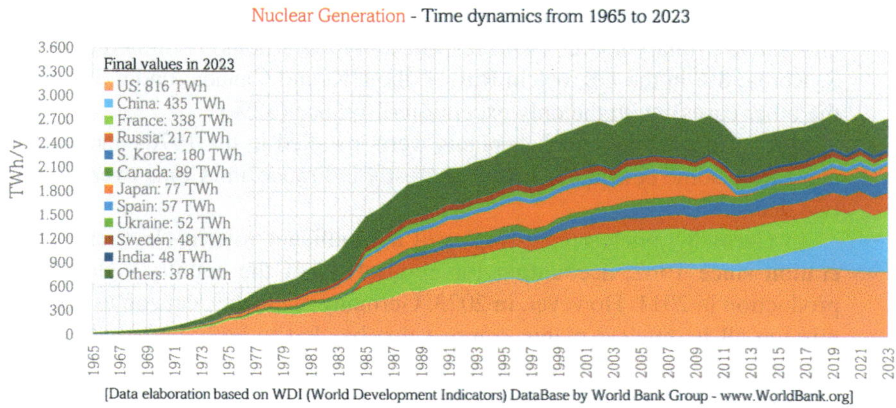

FIGURE I.11.3 World energy production by nuclear sources since 1965 disaggregated by nations.

Over the past three decades, two key developments have significantly shaped the nuclear energy sector:

- Japan's shift away from nuclear energy: Following the catastrophic earthquake and the Fukushima nuclear disaster in 2011, Japan drastically reduced its reliance on nuclear power, effectively halting most of its nuclear energy production.
- China's rapid nuclear growth: In contrast, China began incorporating nuclear energy into its energy mix at the start of the 21st century. Over the last decade, China has substantially increased its nuclear production, emerging as a key player in the global nuclear energy landscape. This rapid expansion underscores China's efforts to diversify its energy portfolio.

Nuclear energy is not a renewable nor a fully sustainable energy source. Its utilization can increase energy system sustainability in the mid-term, so that it was the first and more widely used among alternative energy sources in the 1970s. However, most of the nuclear plants around the world were built before or in the early 1980s, as clearly shown by the data depicted in Figure I.11.4.

I.11.2 HYDRO ENERGY

Figure I.11.5 depicts data about overall world energy production by hydro sources, disaggregated by nations. The Figure I.11.5 data clearly show that hydro energy can be considered, by all effects, to be a conventional RES.

Most of its utilization has basically been stable since the 1970s with some exceptions:

- Brazil only started using its huge hydro energy potentialities in the 1970s and kept increasing them during the following decades.
- Some huge hydro powerplants were built in Russia in the 1980s.

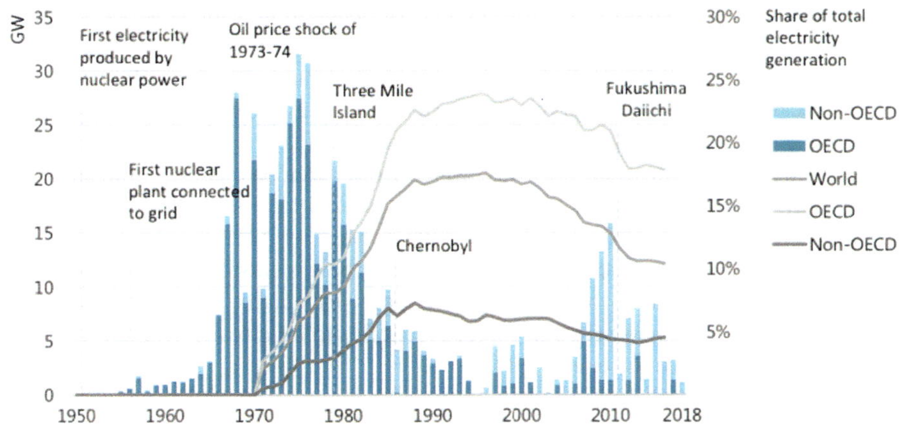

FIGURE I.11.4 Trend in installed nuclear capacity since 1950.

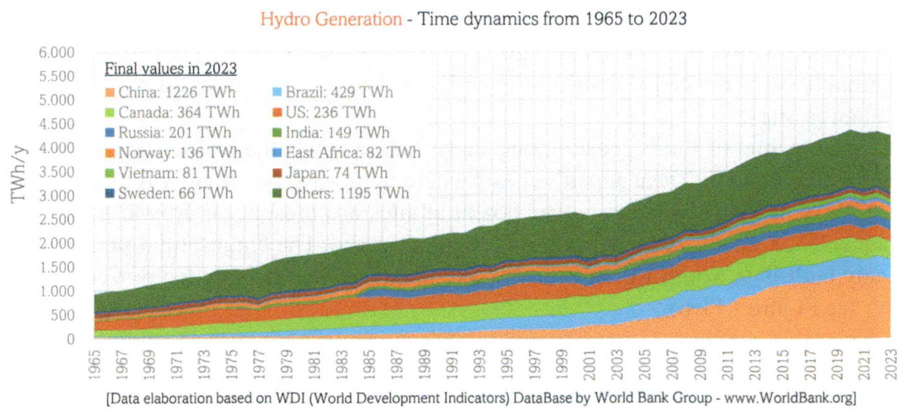

FIGURE I.11.5 World energy production by hydro sources since 1965 disaggregated by nations.

- China has been increasing more and more its use of hydro energy in the last two decades, building the biggest ever hydro power plants. As an example of this, the so-called "Three Gorges Dam" powerplant was built about 20 years ago and has an installed power of more than 22.5 GW and produces more than 100 TWh of hydro energy each year. Approximately 3% of China's energy consumption is attributed to this source.

However, its installation and operation have resulted in significant challenges, including the displacement of local populations previously inhabiting the area and the disruption of upstream and downstream ecosystems, causing adverse impacts on local wildlife due to alterations in the natural environment.

This example underscores the importance of using renewable energy sources (RESs) thoughtfully, highlighting that no matter how "clean" a technological solution may appear, true sustainability requires evaluating it comprehensively.

A holistic perspective must be applied, considering not only the broader environmental benefits but also the localized impacts and consequences of individual applications.

As already said, hydro energy using water from river basins is a conventional method. Other hydro energy sources are however available: this is the case for both wind-induced and gravity-induced marine water waves.

In both cases, the energy content in these water movements could be sufficiently dense to become of interest for energy applications, especially when this energy is made available in near-coast sites in direct proximity to an already-existing electric grid which could receive the energy produced.

Kinetic power (P) included in a wave for a metre length of its wavefront is proportional to the square of its height (H) and to its oscillation period (T) ($P = k * H^2\ T$). Therefore, high-frequency wind-induced waves only become interesting if their height is sufficient, and in those places where they are regularly present.

Tidal-wave energy is intrinsically more powerful and interesting but, in order to be technologically used, requires the realization of a water basin at the shore to be filled and emptied as the tide comes and go. The combination of high tidal waves and a potential basin at the coast identifies several world sites as potential tidal-powerplant installation sites.

Certainly, one of the biggest among those is in Kamchatka (a Russian peninsula which is "sufficiently" near Japan's coasts where tidal wave potentiality for more than 100 GW is present) but many other potential sites are present all around the globe (France, UK, Canada, Australia, New Zealand, Canada). The biggest tidal-powerplant already built is in South Korea (Shiwa Lake plant, having an installed power of 260 MW; see Figure I.11.6).

FIGURE I.11.6 Shiwa-lake tidal wave power plant.

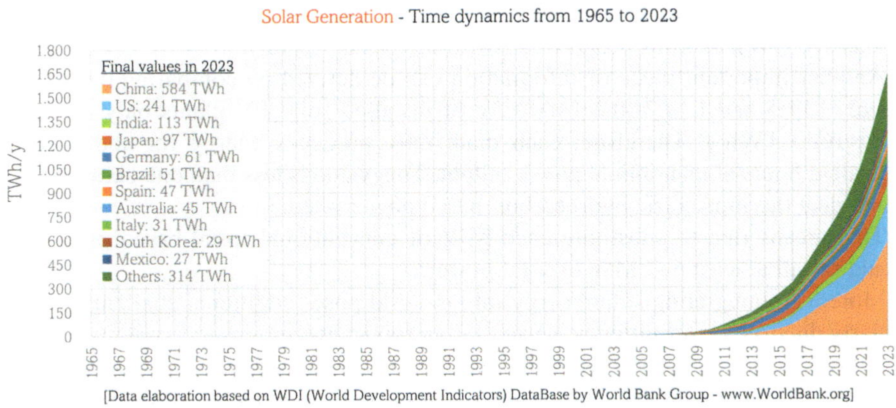

FIGURE I.11.7 World electric energy production by solar sources since 1965 disaggregated by nations.

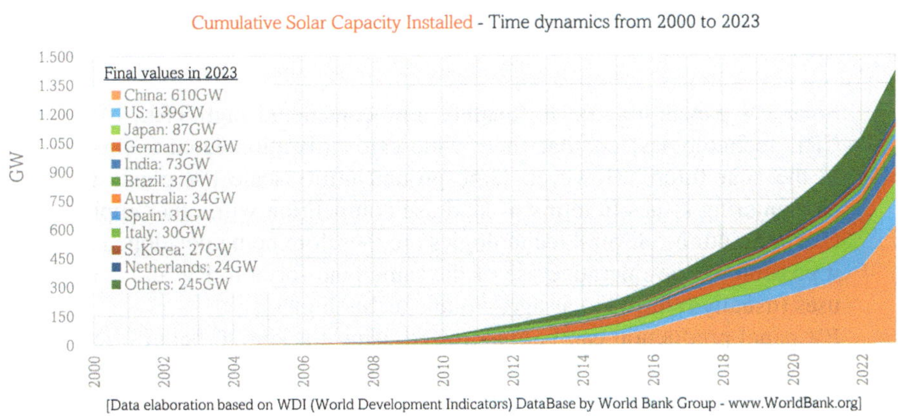

FIGURE I.11.8 World electric energy total capacity installed by solar PV power plants since 2000 disaggregated by nations.

I.11.3 SOLAR ENERGY

Figure I.11.7 depicts data about overall world electric energy production by solar source, disaggregated by nations.

Data show clearly that dynamic evolution of the solar energy sector is very young and that its increase is now almost "exponential". This was expected and is the result of the increase in the technologic, manufacturing, and commercialization readiness levels (respectively known as TRL, MRL, and CRL) of the KETs which are progressively expected to be introduced in this sector.

As a direct consequence of that, Figure I.11.8 shows the increment in total capacity of the power plants installed all over the world, disaggregated by main installing nations. These data clearly show that, even though OECD countries were amongst the first movers in this sector (especially Italy and Germany paying a role in the early years), currently almost half of installed capacity globally (around 1400 GW) is in China (610 GW), and this proportion is still increasing rapidly.

Another important piece of information can be derived by comparing data from Figures I.11.5 and I.11.7: the 1400 GW of installed solar PV plants only produce about 1650 TWh/y. Therefore, each plant only produces energy for about 1200 equivalent hours at its nominal power output. This value is less than one-sixth of that of a typical thermoelectric hydrocarbon-based power plant.

This means that to substitute each MW of conventional fossil-based plant, we should install at least 6–7 MW of solar power.

Moreover, integrating solar energy into the grid poses additional challenges, particularly due to its intermittency, necessitating robust energy storage systems (ESSs) and grid management solutions, which will be explored in subsequent chapters.

Figure I.11.9 highlights current solar market technologies and potential future directions. It categorizes existing solar PV applications, from rooftop and utility-scale installations to emerging innovations like agro-voltaic systems and PV-integrated building materials.

A full discussion and description of all the technologies depicted in Figure I.11.9 is beyond the objectives of this book. Therefore, here only some brief considerations are reported:

- Solar PV panels may be fortunately now considered high TRL, MRL, and CRL technologies, so that their widescale utilization is now foreseeable in the near future. However, large ground utility-scale PV plants are now encountering issues in terms of land use competition with other applications (i.e. agriculture). Several technologies are therefore being developed to make it possible for co-operative use of the same land for both PV and agronomic uses (usually referred to as agro-voltaic technologies).
- PV panel production can be maximized using trackers to better orient them towards solar beams. Normally, cost considerations do not make these applications preferrable with respect to fixed PV mounting. Instead, PV panels are usually mounted on civil buildings, especially on rooftops, where plant complexity must be limited to a minimum.
- The PV technologies market is evolving in at least two directions:
  - An increase of coverage of civil structure surfaces, especially in newly built buildings: in these cases new low-cost (and low-efficiency) PV materials (especially based on so-called thin-film PV technologies) are being developed with the intention of covering every possible construction material at negligible costs, including wall surfaces and window glass, with it also being possible to actively change their transparency depending on the amount of solar radiation.
  - An increase in PV panel efficiency and specific performance (for unit use of land and extension of the technology installed). In this case, the best available PV cells are selected (based on multiple PV junctions). These cells, which can have an electric efficiency of 0.35 (much higher than the maximum 0.25 of mono-crystalline silicon PV normally used for other PV

FIGURE I.11.9 Most used technologies in the current solar market and the main directions in which the solar plants market for electricity generation could evolve in the next decade.

technologies), due to their cost, are very limited in surface (about 1 cm^2 each cell) and are reached by solar energy using a very high concentration ratio (up to 1000 in some cases), obtained through precisely manufactured solar-concentrating lenses and very efficient solar-tracking devices.

- The bottom part of Figure I.11.9 is dedicated to solar thermodynamic plants (also referred to as concentrated solar plants). In these technologies solar energy is used to heat a thermo-vector fluid up to high temperatures which can feed a thermodynamic cycle producing mechanical energy, then converted into electricity by an electric generator. If the temperature reached is sufficiently high, these plants can have an electric efficiency which is much higher than corresponding PV technologies (up to 0.45).
  Due to the low energy density of solar energy, which may reach 1 kW/m^2 at most, solar beams must be highly concentrated before reaching an absorbing material and transferring thermal energy to the thermo-vector fluid. Depending on the technology used for concentrating purposes, the main CSP plants currently developed are distinguished as follows:
  - Dish/engine CSP plants. These are small-scale plants (some kW output at most) using glasses mounted on a parabolic concentrating device and a gas fed Stirling thermodynamic cycle. These technologies are very compact, oriented especially to various civil/urban uses and may be particularly suitable for co-generating purposes.
  - Solar tower CSP plants. The concept behind these plants is the same as the above, but the glass surfaces are all plane and mounted on the ground. Each of them has a separate tracker and is oriented to focus the sun beams on a restricted area placed on the top of tower. There, a heat-absorbing surface is placed which heats up and continuously transfers energy to a thermo-vector fluid.

    Both of these applications and in the following linear-parabolic CSP, the fluid normally is chosen to maintain its liquid state up to high temperatures (even above 400°C). Both diathermic fluid and molten salts can be used for this. The choice is made to enable the use of this fluid as an ESS by incorporating two tanks, one before and one after the solar heat exchanger. These tanks are dimensioned so that, independently from solar radiation, thermal energy is steadily taken from the tanks to be exchanged with a motor-fluid (normally water) which is vaporized and feds thermal energy to a thermodynamic high-efficiency vapour cycle (up to 0.4–0.45).
  - Parabolic through CSP plants. This technological concept is very similar to that of tower CSP plants, but the concentrator only moves using one degree of freedom, so that energy is linearly concentrated on a tube running near and just in front of the concentrating parabolic glasses. A glass evacuated tube must be used to limit convective energy losses towards the external environment, while radiative losses are limited using the greenhouse effect. Fluid flowing in the inner part of the tube accumulates energy, reaching a very high operational temperature. Fluid comes from different concentrating strings (each of which is approximatively 100 m long) which are connected in parallel, increasing the fluid flow rate, and later it is sent to a storage tank. This is used to feed a centralized vapour cycle-based thermoelectric plant just as in the previous example.

I.11.4 WIND ENERGY

Figure I.11.10 depicts data about overall world energy production by hydro sources, disaggregated by nations. The data show clearly that dynamic evolution of the wind energy sector is not as young as the solar sector. It started in the 1990s and followed a similar evolutionary pattern, though increasing less rapidly.

Data about cumulative installed capacity (Figure I.11.10) also here show that some OECD countries acted as the first movers in the 1990s and 2000s (e.g. Germany and the US), but later in the 2010s China started installing aerogenerators at a rapid pace, so that installed capacity in China (442GW) now represents almost half the world's installed capacity.

As previously said for the solar energy sector, another important piece of information can be derived by comparing data from Figures I.11.10 and I.11.11: 1000 GW of installed wind generators only produce about 2350 TWh/y. Therefore, each plant only produces energy for about 2350 equivalent hours at its nominal power output. This value is less than one-third that of a typical thermoelectric hydrocarbon-based power plant.

This means that to substitute each MW of conventional fossil-based plant, we should install at least 3–4 MW of wind power.

In the same way, as stated before for solar power, wind energy production is largely non-programmable and has additional issues to be addressed for RES-based fully sustainable energy systems.

Electric wind-energy generators started being used in the early 2000s so that their technologies can now be considered almost conventional, although they are still not consolidated. In particular, Figure I.11.12 presents the current most used technologies in the wind energy sector.

This figure clearly shows that, though in the very first developing stage it seemed that small-scale generators would have a significant market share, nowadays the wind

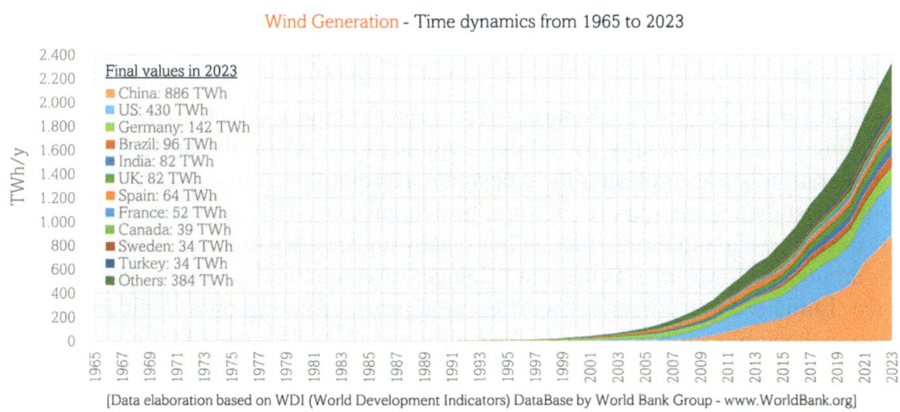

FIGURE I.11.10 World energy production by wind sources since 1965 disaggregated by nations.

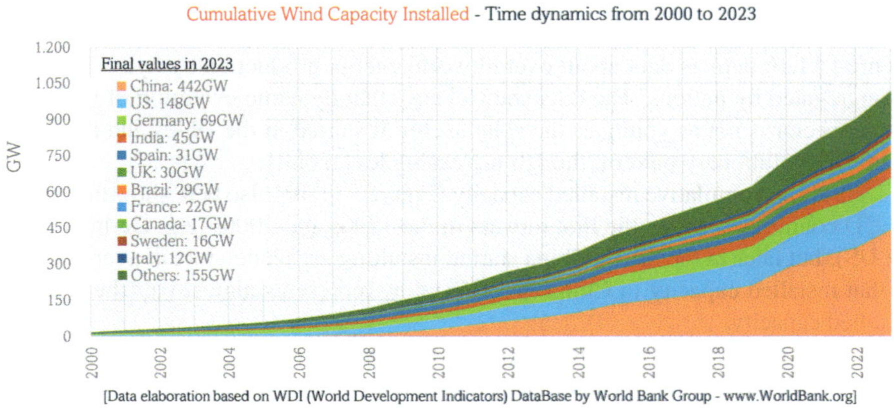

FIGURE I.11.11 World energy total capacity installed by wind sources since 2000 disaggregated by nations.

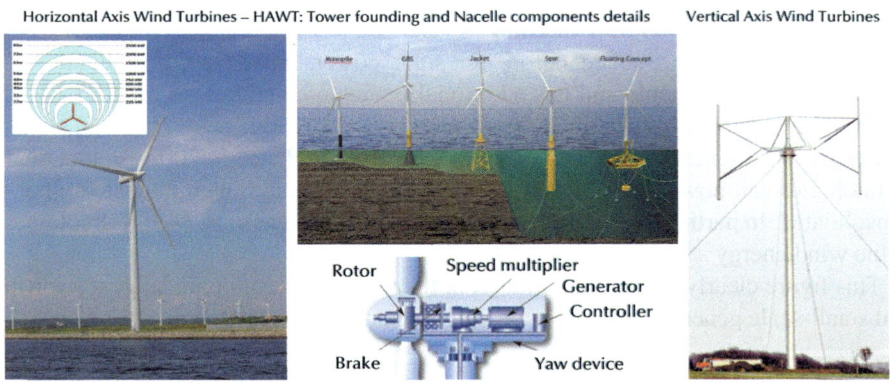

FIGURE I.11.12 Most used technologies in the current wind market.

energy sector is mainly manufacturing very large structures, which are progressively increasing in size, especially for offshore applications, where the noise induced by their operation is a less relevant local design constraint.

Most installed generators rotate on a horizontal axis. This approach makes their CFD design easier and more efficient but makes it necessary to move the nacelle and rotor around the tower axis as the wind changes direction, using a proper yaw control device. Limited applications use a different approach rotating on a vertical axis, as shown in the application depicted in the right-hand part of Figure I.11.11.

I.11.5 GEOTHERMAL ENERGY, BIOMASSES, AND OTHER RESS

Figure I.11.13 depicts data about overall world energy production by all other aggregated RESs (geothermal, biomass, and other minor RESs), highlighting their

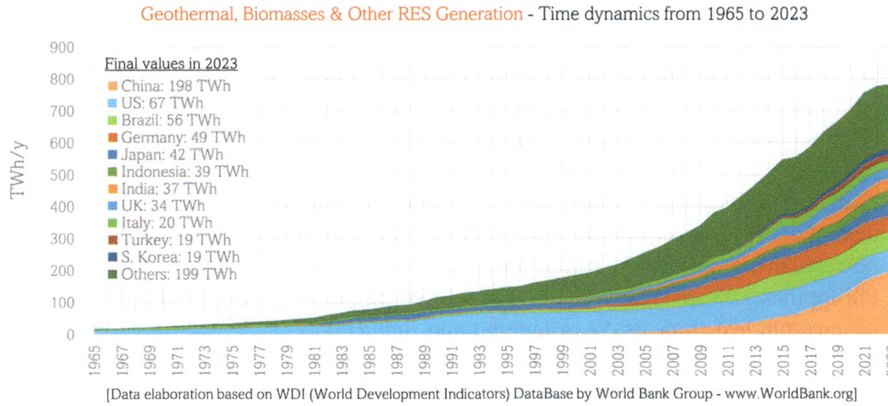

FIGURE I.11.13 World energy production by geothermal biomass and other minorly used RESs since 1965 disaggregated by nations.

aggregated contribution over time. As a general comment to data shown in Figure I.11.13 we can say that:

- Use of these RESs has increased in recent decades, especially due to biofuel production for automotive applications: bio-gasoline and bio-diesel. These biofuels are respectively produced through sugar-containing and vegetable oil substances mixed with fossil-based gasoline and diesel fuels and used in normal cars without engine modifications, since, up to a certain bio-origin fraction they can still be defined as “drop-in” fuels.

  These processes will probably proceed again even though higher percentages of fuel mixtures are becoming increasingly difficult to tolerate by vehicles (especially older ones). Relevant issues are also arising in terms of land usage distortion from agricultural to energy markets (e.g. oil cultivation in Mexico now producing oil for the US energy market and not for internal agricultural needs).
- The term geothermal energy use refers to those technologies oriented towards the utilization of the heat flowing from the Earth’s nucleus through to its surface. This thermal flow generally is very poor (much less than 1 W/m^2) on Earth’s surface. Its presence becomes of value only in those places where the flow is much more intense, due to the ascent near the surface (up to 10 km or less) of magma reservoirs. Also in these conditions, energy flow is too low to configure any possible energetic use. The presence of geothermal fluid must also be ensured in the soil. Ideally, it should migrate deep underground near the magma reservoir, where it is heated and then convectively driven back to the surface. This whole mechanism (referred to as the “geothermal system”) may bring up to the Earth surface significant energy that can of interest for energy utilization.

  The geothermal system may be fully natural and/or partially artificial (obtained by modifying local permeability of the terrain and/or adding a certain amount of fluid – normally simply water) and is normally classified depending on the enthalpy of the produced fluid. Only the most intense of these systems may be used to produce electricity through a thermodynamic cycle. Moreover, it also must be considered that the geothermal fluid can be toxic and/or corrosive, determining additional issues to its utilization.

With all this considered, geothermal electricity production remains a sufficiently interesting energy source for some countries. Currently, about 15 GW of these electric powerplants are installed around the world (about 4 in the US, 2 in Indonesia and the Philippines, 1.5 in Türkiye, and about 1 GW in New Zealand, Italy, and Kenya, with somewhat less than 1 GW in Iceland and Japan).

I.12 Sustainable Choice of Energy Carriers

As already explained, RES-based energy systems are increasing the use of electricity as an energy carrier.

This effect is already largely visible, as clearly show by Figure I.12.1, and is expected to increase more rapidly in the coming years.

However, some final uses will always be very hard to be fed by electricity. These are normally referred to as hard-to-abate final energy sectors. Among these are the following:

- “Unguided” long-distance flexible mobility (i.e. freight and passenger services using road infrastructures, as well as air and water). These are difficult to store on board the amount of energy needed to fulfil their mission without the intrinsically high storage capabilities of chemical carriers.
- Industrial processes making use of high-temperature fluids that are difficult to produce using electricity-fed technologies.

Some chemical energy carriers will therefore still be needed in fully RESs-based energy systems.

To this aim, the most obvious hypothesis is to use biomass in place of fossil fuels to derive them. Most of the thermochemical processes acting on fossil fuels (i.e. all pyrolysis and gasification technologies) also work on biomass and may be used to produce so-called biofuels for many final users’ applications including ICEs. So-called bio-diesel and bio-gasoline can be used for these applications, but also bio-naphtha, bio-NG, bio-LPG, and any other bio-oil are already produced (even though mostly only in demonstration plants).

An alternative would be industrially synthetizing chemical energy carriers without using biomass but using only electricity from RESs and other carbon and other non-fossil and widely available hydrogen- and carbon-containing molecules. This is the idea behind the concept known as electro-synthesis, producing what we refer to as e-fuels: e-diesel, e-gasoline, e-NG, e-LPG, e-oils and so on. The most obvious primary resources for C and H would be CO_2 and H_2O, leading to very clean processes.

Most of the e-synthesis technologies, however, still are characterized by low TRL, MRL, and CRL, so that in the near-/mid-term future the bio-fuels path remains probably the most promising alternative to produce chemical energy carriers for hard-to-abate final uses.

A second big issue is being increasingly faced as far as energy systems are moving towards RESs. This is due to the intrinsic non-programmability of most of the

DOI: 10.1201/9781003631125-13

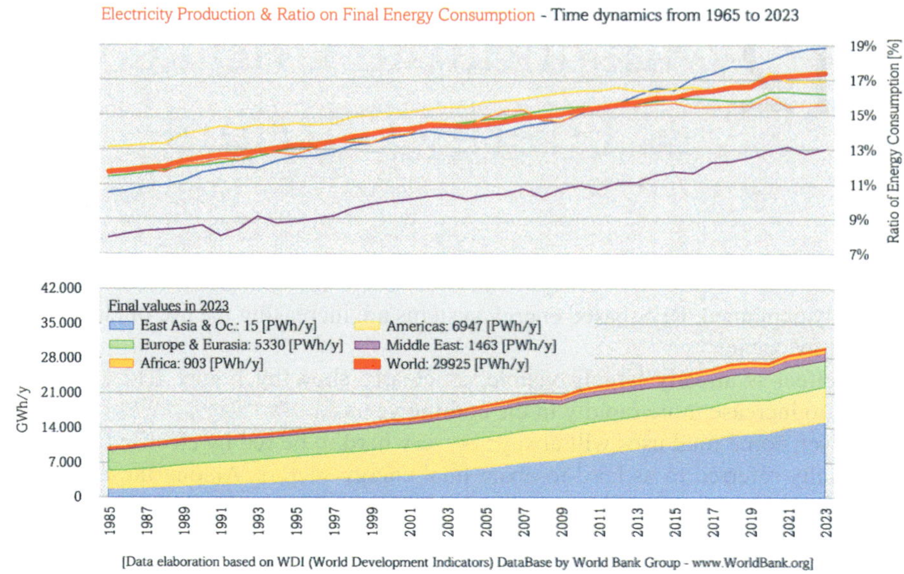

FIGURE I.12.1 Electricity production and ratio of final energy consumption (1965–2023).

available RESs (especially solar and wind energy, but also partially hydro energy, which are non-programmable).

Therefore, the electricity produced has to be stored somewhere (at energy source conversion sites, within distribution infrastructures, and/or in final uses technologies) to make them available for final use only when and where needed. Use of electricity ESSs should therefore grow in the coming decades by at least two orders of magnitude with respect to electric ESSs currently manufactured and implanted on a global level.

The most obvious option is using electro-chemical batteries, i.e. BESS technologies. These are technologically similar to BESS technologies installed on board electric vehicles for mobility applications, but they are placed inside containers and/or dedicated buildings and connected to the grid. This means that their weight and volume for unit energy stored are not a particularly binding design constraint, so that different electro-chemistries and technologies can be designed and developed for the two sectors.

Figure I.12.2 shows how rapidly BESS installed capacity, even though still limited, grew in the last decade, especially in East Asia (since China actually holds most of the technologies and prime materials used in this sector).

An alternative could be using other fluid-fed electrochemical processes to store electricity, these are called invertible fuel cells (FCs). The best known among these are alkaline, polymer-membrane, and solid-oxide fuel cells, which may, very basically, produce H_2 and O_2 from H_2O (and therefore "storing" electricity), and then realize the inverse reaction when and where needed producing electricity (consuming H_2 and O_2 and producing H_2O).

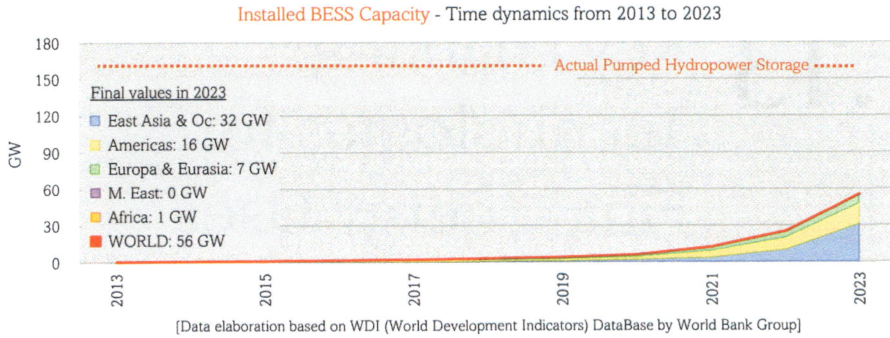

FIGURE I.12.2 Recent trend of installed BESS capacity worldwide disaggregated by world regions.

Up to now the cost, efficiency, and transient performance of these components are unfortunately still not competitive with BESSs, but their TRL, MRL, and CRL will hopefully increase in the near-/mid-future.

Wide-scale stationary use of these technologies would mean also having the possibility to use this synthesized E-H_2 as an energy carrier, making it possible to use it also for energy distribution purposes as an alternative to electricity transmission networks.

Moreover, the E-H_2 could also be used as a prime material for further e-synthesis processes to produce light liquid fuels for hard-to-abate applications: E-NH_3 (ammonia) and E-CH_3OH (methanol) seem to be two of the currently most promising options in this regard.

In summary, achieving a fully RES-based energy system requires a strategic combination of advanced energy carriers and storage solutions. While biofuels and BESSs represent the most immediate and scalable options, continued investment in emerging technologies like e-fuels and invertible FCs will be essential for addressing the diverse and evolving demands of a sustainable energy future.

I.13 Key Enabling Technologies for Final Energy Uses

I.13.1 TRANSPORTATION SECTOR

Worldwide policies on energy transition and sustainable development, aimed at mitigating GW, are very ambitious for the transportation sector, since most of the energy consumption in the transportation sector is nowadays due to road transport systems and these are mainly based on internal combustion engines (ICEs) powered almost exclusively by fossil-based chemical energy carriers (mostly gasoline and diesel fuels).

It is therefore better to distinguish what the objectives are that can be reached in shorter timescales from those that are more reasonable to be imposed over a longer space–time horizon. In the near term, it is convenient to address and solve mainly small space–time scale issues (such as local pollution) introducing new technologies that, in the mid-term can help us also in addressing and solving wider scale problems (especially energy dependency and decarbonization issues).

Transport (land, air, and maritime) is among the most difficult sectors to decarbonize (a so-called hard-to-abate sector) and is currently responsible for 37% of direct CO_2 emissions worldwide, with annual emissions which, in 2020, reached 7.2 Gt_{CO2}.

Furthermore, transport also causes emissions of air pollutants (NO_x, CO, HC, PM) in urban environments, which, in turn, have been estimated by the World Health Organization (WHO) to be responsible for at least 4 million premature deaths worldwide.

The European Union, according to the most ambitious +1.5° C GW objective of the Paris Agreement, already stated internal directives (which are binding for all its Member Nations) towards a full decarbonization of its transport sector, setting the long-term objective to eliminate greenhouse gas emissions (“net zero emissions” – NZE) by 2050.

This is envisaged in the European Green Deal (EGD), which is the 1000-billion-euro investment plan presented on 11 December 2019 by the European Commission. Exactly one year later, the EU also proposed an increase in the short/medium-term objectives of the EGD, increasing the percentage of reduction of total GHG emissions from 40% to 55% (compared to 1990) by 2030.

To better delineate quantitatively and temporally the impact of these objectives on the consequent actions that need to be prepared in EU Members’ national transport

DOI: 10.1201/9781003631125-14

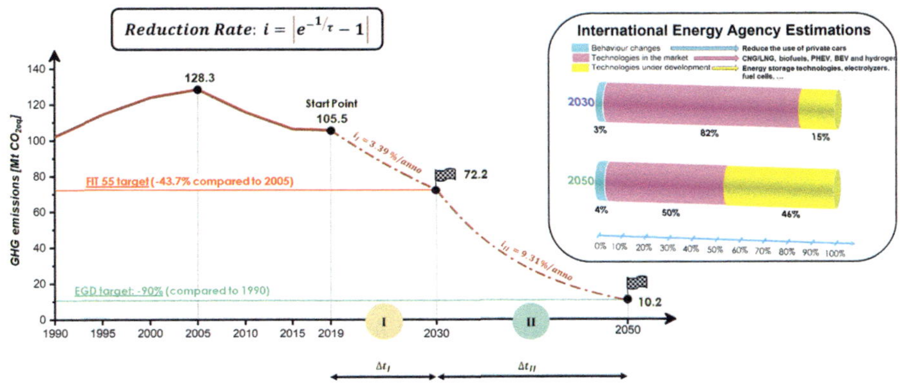

FIGURE I.13.1 Expected effect of the updated objective stated by EU Green Deal Policies on the overall Italian GHG emissions by the transportation sector.

sectors, as an example, we here present the translation of updated EU Green Deal objectives on the Italian transport sectors. The foreseen effects of these policies in terms of reduction of GHG emissions are sketched in Figure I.13.1.

Figure I.13.1 clearly shows that EU Green Deal policies will be implemented in two stages, having an increasing level of ambition, and therefore foreseeable difficulty, in their implementation. Figure I.13.1, in fact, permits three phases in the GHG emission trend to be identified:

- **Recent Past: 1990–2019**
  In 2019, the Italian transport sector was responsible for 30% of total greenhouse gas emissions, of which 92.6% were attributable to road transport. In that year, emissions amounted to approximately 105.51 Mt CO_2eq, the result of a historical trend characterized, starting from 1990, by an average increase in emissions of 3.2%, with a peak of approximately 128.3 Mt CO_2eq recorded in 2005. In fact, although vehicular unit emissions (grCO_2eq/km) have tended to decrease over the years, the global value of emissions has increased following the increase in the rate of motorization.
- **Phase I: short/medium term (2020–2030)**
  The objectives aim to reverse the historical trend of the previous phase by reducing emissions by 2030 to approximately 72.2 Mt CO_2eq, corresponding to a decrease of 43.7% in line with what is foreseen by the EU Green Deal. Achieving this goal requires a reduction rate of 3.39% each year compared to the previous year, starting in 2019.
- **Phase II: medium/long term (2031–2050)**
  Given that the most ambitious objective for this phase would be the achievement of net zero emissions (NZE), the minimum objectives envisaged by the EGD aim, however, to reduce GHG emissions to approximately 10.22 Mt CO_2eq, corresponding to a 90% decrease compared to 1990.

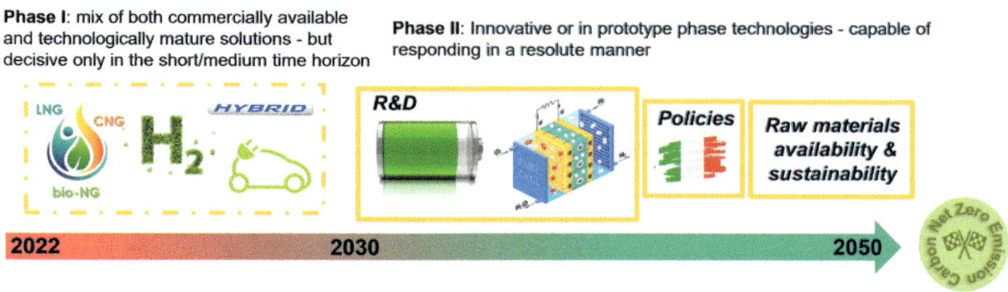

FIGURE I.13.2 Schematic representation of a possible roadmap for the introduction of new technologies into the transportation sector towards net zero 2050 GHG emissions objective.

Although the scenario shown in Figure I.13.1 was elaborated for the Italian transportation sector, it is certainly representative of most of the OECD countries and of most world, highlighting the need to take sudden actions to mitigate GHG emissions in transportation, if Paris Agreement objectives are to be maintained.

A possible roadmap towards this objective is schematically presented in Figure I.13.2.

In the short term (2022–2030), efforts must focus on deploying existing technologies and improving transportation efficiency. By contrast, the long-term phase (2031–2050) will necessitate the adoption of advanced and innovative solutions currently in early development stages.

One of the main drivers regarding the mitigation of GHG emissions in the sector lies in the limitation of the impact per unit of transport service satisfied, which needs, among other things, to be achieved with the following actions:

- Significant changes in users' behaviour, i.e. resorting to a more efficient and effective modal split with actions that discourage travel by individual means, especially in urban contexts; at the same time, it is necessary to strengthen and structure collective transport systems in such a way as to make them competitive, in terms of generalized costs, compared to private transport systems.
- Adoption of technological solutions already available on the market capable of increasing the efficiency of current propulsion systems, but also of reducing consumption based on the optimization of aerodynamics and masses. Although already largely mature, for these technologies it will, however, be essential to continue research and development activities with the aim of increasing their efficiency and making them increasingly compatible with the use of energy carriers that can be produced through renewable sources (biofuels). These are only partially conclusive solutions, but, considering the high level of maturity, necessary to guide the transition towards NZE.
- Promotion and funding of R&D activities towards maturity increase of those technologies which are still in the prototype phase and, in any case, are not yet widely present on the market (characterized by a low TRL), which will most likely be able to respond in a decisive manner to the needs of emission mitigation.

I.13.1.1 Phase I: Short/medium Term (2020–2030)

As said, in the next 5–10 years (Phase I) it will be only possible to promote and foster market introduction of those technologies that, although not decisive, still allow the overall environmental impact to be reduced. Those listed below are some examples of those consolidated technologies, with a high level of TRL (technology readiness level) and therefore are ready to be implemented helping transportation decarbonization and/or lowering its local pollution:

- **Compressed natural gas (CNG) vehicles.**
  This option allows for a good cost/benefit ratio. CNG is an intrinsically cleaner and more sustainable fuel than gasoline and diesel due to less carbon atoms in its molecular structure and gaseous state. For this, compared to the latter, it implies a potential reduction in CO_2 emissions of up to 20% (over 40% in the case of biomass gas) and almost total abatement of solid particulate matter (PM) emissions.

  The mainly used technologies are industrially mature and widely available on the market: vehicles, both light and heavy duty (LD and HD), with spark-ignition engines powered by natural gas are already commercially widespread. However, the major issues associated with the use of natural gas are related to GHG emissions, due to methane's global warming potential (GWP), which is 28 times higher than for CO_2. Indeed, even if CNG can potentially reduce CO_2 releases, fugitive emissions of methane during extraction, transport, and distribution phases and methane slip after the on-board combustion process significantly limit the environmental benefits.
- **Dual-fuel vehicles (mainly for heavy-duty vehicles)**
  They are based on dual diesel and natural gas fuelling, which can be obtained through after-market conversion kits, while ensuring adequate autonomy and flexibility of use typical of diesel vehicles, representing an interesting alternative for replacing HD diesel vehicles normally used in the transportation sector for long distances.

  Dual-fuel technology is particularly interesting when the fuel is introduced into the tank in liquid form (LNG – liquefied natural gas), allowing a large quantity of gas to be stored on board without the need for high storage pressures as for CNG (higher energy density).

  Dual-fuel vehicles lead to lower NO_x emissions compared to diesel vehicles (not considering abatement devices), due to a lower end combustion temperature and, depending on the quantity of diesel used, there is a more or less marked reduction in solid particulate emissions.
- **Hydro-methane vehicles**
  These are based on the use of a mixture of hydrogen and methane (up to 20% in energy) which can be used without major changes to current propulsion technologies, guaranteeing improvements in terms of reductions in climate-altering gases up to at around 20%. This option is still in the testing phase, and is not completely mature (low TRL).

- **Hybrid vehicles**

  They allow to achieve a good cost/benefit ratio, especially in the urban context characterized by continuously variable energy demands and difficult to manage by conventional engines without significant increases in emissions. In hybrid vehicles, propulsion is ensured by multiple energy sources available on board (typically one chemical, one fuel, and one electric).

  The restoration of the electrical energy necessary for traction can take place on board (using the chemical source to produce electrical energy and/or recovering the inertial and potential energy of the vehicle itself, where possible) or through suitable charging stations, in the so-called vehicles plug-ins.

  There are many different types of hybrid vehicles, differing in degree of hybridization, level of electrification, type (series, parallel, mixed), and energy storage system (batteries, super-capacitors, flywheels). They all entail significant advantages related to the under-sizing of the ICE and to its operation in much less transient and pollutant-emitting conditions per unit of energy supplied to the wheels.

  Even in non-plug-in solutions (which therefore still make exclusive use of chemical fuels) significant advantages are obtained in terms of reducing consumption (even over 20–30%) and therefore polluting emissions. Hybrid technologies are also suitable for use in vehicles with constrained (trams) and semi-constrained driving (trolleybuses): in these cases, the use of on-board accumulation systems and any additional thermal engines allows for the effective recovery of braking energy thereby increasing the flexibility of the transport system.
- **Electric vehicles**

  This option represents the future solution for the complete reduction of emissions of primary pollutants at the exhaust, once complete technological maturity has been reached and once the charging infrastructure network has been adequately implemented. In these vehicles, propulsion is purely electric with electrochemical storage on board.

  These vehicles are normally referred to as battery electric vehicles (BEVs). BEVs have a series of advantages compared to vehicles with endothermic propulsion systems, including the absence of pollutants and climate-altering gases in the exhaust, reduced noise, and low maintenance costs. Furthermore, electricity can be produced more easily through renewable sources than chemical fuels.

  However, the diffusion of BEVs is still limited mainly to short-distance applications due to some economic, technological, and environmental barriers. Firstly, a strong interaction of vehicles with the national electricity transmission and distribution network is required, with the consequent need for an adequate infrastructure plan (Smart Grid and Vehicle to Grid – V2G).

  Another limiting factor for the diffusion of BEVs lies in the limited availability of the raw materials necessary for the creation of electrochemical storage systems (lithium and other rare metals) and the definition of suitable procedures for their end of life and their total recovery from a circular economy perspective. In the specific calculation of GHG emissions, it is also necessary to consider the environmental impact associated with the entire life cycle (LCA – "Life Cycle Assessment") of the electrochemical storage systems.

Recent LCA studies conducted on different types of lithium-ion batteries used in electric vehicles highlight consistent GHG emission values varying between 70–175 $kgCO_2eq$ per kWh of accumulable energy. Assuming correct recycling of materials or reuse of the same batteries for other applications, it is possible to achieve a reduction in emissions for an average value of 20 $kgCO_2eq/kWh$. Furthermore, BEVs are characterized by high investment costs.

Currently, the battery pack can represent up to 50% of the cost of the entire vehicle and must, as a rule, be supported more than once during the useful life of the vehicle. In addition to the initial investment for the vehicle, it is also necessary to consider the costs related to the creation of an adequate charging infrastructure.

Finally, another issue with BEVs concerns the low energy and power density of the batteries which causes significant problems in terms of vehicle autonomy and the times required for recharging. Battery charging times can be very long (up to 8 hours for a "slow" charge) and the need for a widespread and adequate network of public and private charging infrastructure, currently under construction, must also be assessed.

In the current transition phase towards the electrification of the transport sector, the penetration rates of electric mobility are rapidly increasing worldwide. There are over 10 million vehicles powered by electric motors – including both BEVs and plug-in hybrid vehicles (PHEVs).

Figure I.13.3 shows that in a Stated Policies Scenario (STEPS), i.e. a situation that reflects all existing policies and objectives that have been legislated

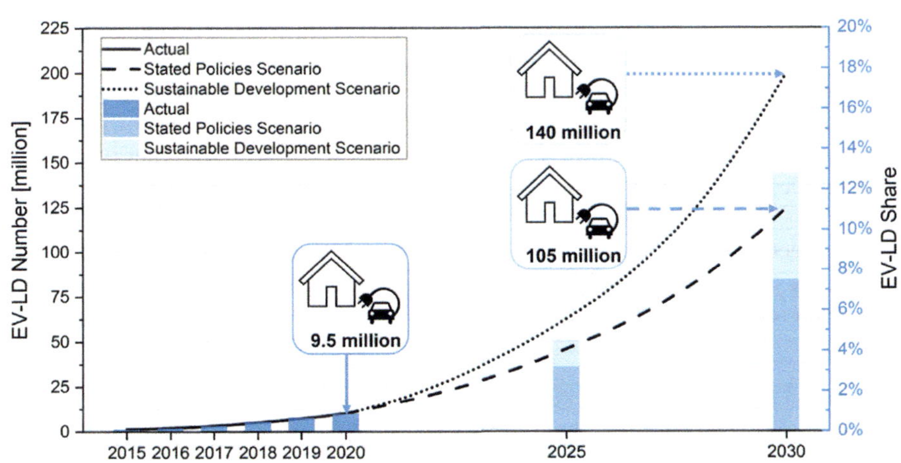

FIGURE I.13.3 Scenarios about perspective shares of EV penetration in the passenger light-duty car market and corresponding forecasts of vehicle production and sales.

or announced by governments around the world, the average annual growth rate of the stock of electric vehicles is estimated to be 30%, reaching almost 140 million light commercial vehicles sold by 2030.

In a Sustainable Development Scenario (SDS), it is expected that over 220 million electric LDVs will be in circulation by 2030, corresponding to almost 12.8%. SDS is a more ambitious scenario, in line with the signatories of EV30@30, a campaign launched at the 8th Clean Energy Ministerial (CEM) in June 2017 and supported by 15 member countries and several international companies and organizations, whose objective is to achieve a 30% share of sales of light electric vehicles in 2030.

- **Hydrogen vehicles**
  This option is based on the use of the hydrogen energy carrier. The first option of using hydrogen as a transportation energy source relies on fuel cell electric vehicles (FCEVs) in which the propulsion system is still purely electric but is powered by energy produced on board via electrochemical devices known as fuel cells which transform the binding energy of the hydrogen carrier combined with atmospheric oxygen directly into energy electric. The hydrogen is stored on board the vehicle in pressurized tanks (from 350 to 700 bar).

  FCEVs benefit from all the typical advantages of BEVs (zero local emissions, low noise, and reduction of vibrations on board the vehicle), while allowing, at the same time, greater autonomy, representing a valid solution especially for heavy-duty vehicles and buses.

The refuelling times of a hydrogen vehicle are around 10 minutes, remaining similar to those of a conventional vehicle. Depending on the primary energy source and the process through which it is produced, hydrogen can have a more or less significant environmental impact.

FCEV buses are currently deployed in several European cities for demonstration purposes. The main critical issues that limit the diffusion of FCEVs concern the absence of a widespread distribution network and the low TRL of FCEVs.

This requires a need for high investment. Operation and maintenance costs and limited spread of H_2 sources and the consequent high costs especially for so-called green hydrogen (hydrogen from water electrolysis with electrical energy needed for the process produced by renewable sources), estimated at between 2.5 and 5.5 euros/kg.

It is expected, however, that the cost of green hydrogen will fall below 2 euro/kg by 2030, thus approaching the current cost of grey hydrogen (hydrogen from carbon fossil steam reforming).

I.13.1.2 Phase II: Medium/long Term (2031–2050)

As already said, after all these technologies have been introduced and experimented with, a further decarbonization phase (2035–2050) will be started, aiming for full decarbonization to be hopefully obtained by 2050. These objectives, in fact, will require the use of a wide range of technology options which are only partly available in current market. Up to now, the most significant options (which are not alternatives but concurrent) for achieving the Phase II objectives are:

- **Hydrogen use in transport applications**.
  To this aim, at least in an initial introductory phase, it will probably be useful to diversify H_2 production processes. These should be obviously as much as possible oriented towards green hydrogen (through the development of increasingly efficient and cheaper electrolysers), but blue hydrogen (produced by biomass using CO_2 capture and sequestration systems) will also be an option.
  By 2050, renewable hydrogen technologies will hopefully reach the required high TRL so that they can be deployed on a large scale. At a European level, total investments are estimated at between 180 and 470 billion euros, with the aim of producing about 500 GW of green hydrogen.
  Further efforts will have to be directed towards the development of high-efficiency fuel cells (powers from 50–125 kW) to be used on board vehicles, to increase their duration and efficiency and reduce costs.
  The long-term goal set by the US Department of Energy (DOE) concerns the development of an FC system for motor vehicles with 65% efficiency and duration (defined as the number of hours with performance losses of less than 10%) of over 8000 hours, mass-produced at a cost of approximately 25 €/kW.
  It will also require the development of on-board vehicle storage systems capable of increasing the energy density of the carrier in complete safety, including gaseous storage at high pressures (over 700 bar), in liquefied form (in cryogenic tanks) and technological alternatives regarding the so-called "slush hydrogen" (solid and liquid mixture at the temperature of the triple-state transition point).
  The development of storage forms within liquids or solids (metal hydrides, organic liquids with high absorption capacity and surface adsorption using organic/metal materials) are also interesting.
  Finally, it will be necessary for the development of territorial infrastructures for the distribution, storage, and supply of hydrogen in large volumes and over long distances, and the resolution of legislative and regulatory issues and the implementation of internationally recognized safety standards. The expected global transport consumption by fuel type in the net-zero emissions scenario (2050) is shown in Figure I.13.4. As depicted, fossil fuels represent just 10% of transport energy, while electricity and hydrogen-based fuels account for more than 70% of transport energy demand..
- **Development of innovative ESSs for self-moving electric applications**.
  As explained in previous chapters, the development and diffusion of highly efficient ESSs will be increasingly crucial in the coming years. Here attention is limited to those ESSs which are oriented to self-moving applications, needing mechanical energy for the final transportation purposes. This sector is developing and experimenting with new technologies, such as compressed air equipment (CAES), inertial electromechanical equipment (FESS – Flywheel Energy Storage System), and several other possible electricity ESSs mainly based on batteries and other electro-chemical devices (BESSs).

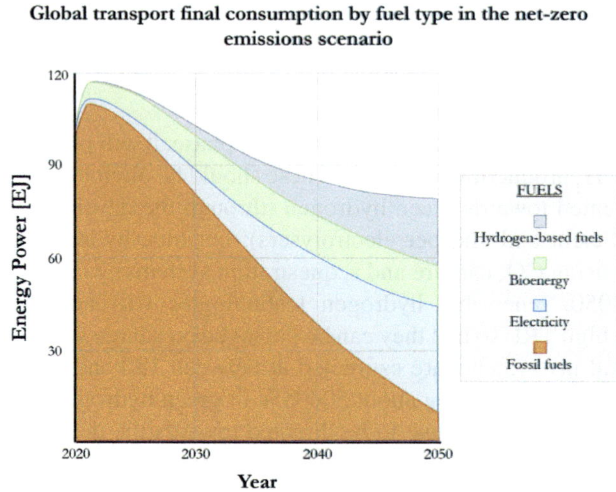

FIGURE I.13.4 Expected global final transport consumption by fuel type in the net-zero emissions scenario.

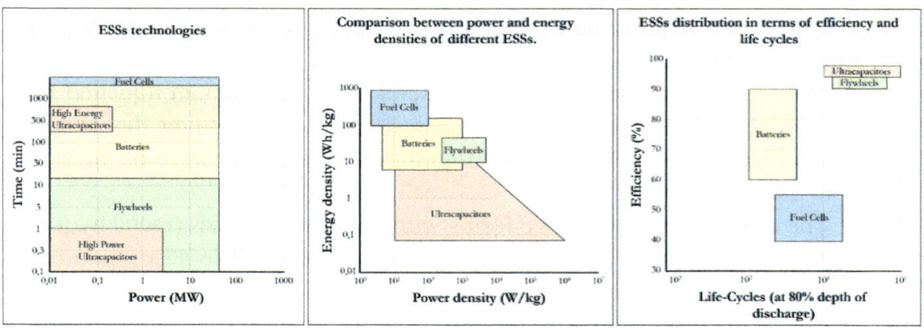

FIGURE I.13.5 Actual and expected performances of vehicular ESSs under development. Performances of actual and developing ESSs.

All these systems are competing to meet the very challenging required performances in terms of specific energy and density (for unit mass and volume). Avoiding as much as possible mass (and volume) in transportation is crucial so as not to increase significantly their energy consumption and therefore lose a relevant part of the advantages in terms of GHG emissions that the introduction of these technologies in vehicles is expected to produce.

Moreover, it is also important to verify that the utilization of these new technologies on a wide scale (millions of new vehicles) will not create critical problems from other viewpoints. Figure I.13.5 compares various energy storage systems (ESSs) based on technology type, power vs. energy density, and efficiency vs. life cycles.

- **Development of new ICT technologies for transportation applications.** As an example, several new ICT technologies have been designed in recent years, mainly (but not only) using artificial intelligence (AI) and/or machine learning techniques to the current level of vehicle automation and prospectively to introduce autonomous driving vehicles on a progressively wider scale.

  Moreover, please consider that the whole mobility sector, especially (but not only) in urban contexts, will probably be subject to a revolutionary transformation which will include a number of new development platforms including flexible high-frequency mass transport platforms (normally referred as Mobility-as-a-Service Platforms: MaaS), intelligent transport systems (ITS), V2V (vehicle-to-vehicle), V2G (vehicle-to-grid), and V2X (vehicle-to-everything) communication protocols and technologies.

  All these novel approaches to the mobility sector are here only briefly described, because explaining them in full detail would go far beyond the objectives of this book, however they still give a sufficiently clear idea of the disruptive evolution that will affect the mobility sector in the next 15–20 years.

I.13.2 RESIDENTIAL SECTOR AND SMART CITIES

Depending on the local climate, residential buildings typically require thermal energy for heating during winter and may require cooling during summer. They also need thermal energy for cooking and electricity for lighting. These are the basic energy requirements of a household.

In conventional energy systems, thermal energy requirements (for both cooking and heating) are usually met through chemical carriers, an option that is likely unsustainable in the near future. As explained in Chapter 12, sustainable energy systems based on RESs will provide significantly more electricity and fewer chemical carriers compared to conventional systems. Consequently, it will become easier and more cost-effective to meet thermal energy needs using electric technologies. Furthermore, electricity can be at least partially generated locally using photovoltaic (PV) panels installed on rooftops.

These factors are driving a significant transformation in the technical systems used for energy management in buildings.

Additionally, increasing attention on building energy performance has led to the establishment of new standards. For example, the European Union has defined minimum energy performance requirements for all new buildings under construction, which set a maximum allowable fossil energy use per unit of floor area. To meet these requirements, designers must incorporate electricity self-generation (and local storage in BESSs) and enhance local self-consumption through a range of technical solutions. To this aim, building technical plants commonly include:

- **A properly designed and dimensioned PV system**: The building's roof is typically optimized for orientation and inclination to maximize solar power generation.

- **A battery energy storage system (BESS)**: This optimizes the local consumption of electricity produced by the PV system.
- **A heat pump**: Powered by electricity from the PV system, BESS, or the grid, the heat pump provides both heating and cooling.
- **A low-temperature heating system**: Designed to operate with fluid temperatures between 30–40°C, these systems maximize the efficiency of heat pumps. They typically use floor-embedded heat exchangers, which also enhance indoor comfort.
- **A cooling system**: Split units installed in individual rooms are powered directly by the heat pump.
- **An induction cooking appliance**: These high-efficiency devices replace traditional gas stoves.

Adopting this approach on a large scale implies that newly constructed urban neighbourhoods and entire cities (especially in developing regions) could be built without the need for fuel distribution infrastructure. This marks a significant departure from traditional practices.

Moreover, the elimination of combustion within buildings would reduce the emission of local pollutants, improving urban air quality.

However, this shift is not the only transformation expected in cities from an energy perspective. Electric mobility will increasingly permeate urban areas, necessitating widespread charging infrastructure. Charging points should primarily be installed in residential buildings, workplaces, and commercial parking lots. If this infrastructure is implemented widely, most vehicles will remain connected to the grid when not in use.

In this scenario, vehicle batteries could theoretically provide grid services. They would charge when renewable electricity is being produced and discharge to supply local appliances when renewable electricity is unavailable. Achieving this requires a significantly upgraded electric grid (a so-called "Smart Grid") as well as smart, bidirectional vehicle chargers capable of supporting vehicle-to-grid (V2G) technologies, which are already under development.

These developments represent just two examples of technological advancements and smart applications that will transform urban environments as part of the "Smart City" paradigm.

Smart Cities are not merely a futuristic concept; they are already taking shape through pioneering projects. For instance, Singapore employs AI-driven energy management systems, while Barcelona has implemented an extensive network of smart lighting and waste management technologies. These initiatives highlight the potential of integrated, technology-driven solutions to reshape urban living for a more sustainable future.

Further Readings

1. Sustainable Development

Boulding, K. E., The Economics of the Coming Spaceship Earth, in *Environmental Quality in a Growing Economy*. John Hopkins University Press, pp. 3–14, Baltimore, 1966.

Chancel, L., Piketty, T., Saez, E., Zucman, G. et al. – *World Inequality Report 2022*, World Inequality Lab., 2022.

Energy Institute, 2024 Energy Institute Statistical Review of World Energy. Energy Institute, 2024.

Georgescu-Roegen, N., Energy and Economic Myths. Southern Economic Journal, *41*(3), 347–381, (Jan., 1975).

Meadows, D. H, Meadows, D. L., Randers, & J. Beherens III, W. W., *The Limits to Growth*. Potomac Associates – Universe Books, 1972.

Sachs, J. D., *The Age of Sustainable Development*. Columbia University Press, 2015.

Smith, A., *The Wealth of Nations: A Translation into Modern English.* Industrial Systems Research, 2015.

World Bank, *World Development Indicators*. The World Bank Group, 2024.

World Commission on Environment and Development, *Our Common Future*. Oxford University Press, 1987.

UN General Assembly, *Transforming our World: the 2030 Agenda for Sustainable Development*, A/RES/70/1. UN, 2015.

United Nations, *The Sustainable Development Goals Report 2022*. UN, 2022.

United Nations, *Human Development Report 2023-24*, UN Development Programme. UN, 2024.

2. Energy Systems

Energy Institute, *2024 Energy Institute Statistical Review of World Energy*. Energy Institute, 2024.

Grubler, A., *Energy Transitions: History, Requirements, Prospects*. Cambridge University Press, 2012.

International Energy Agency., *World Energy Outlook 2022*. IEA, 2022.

Smil, V. *Energy Transitions: History, Requirements, Prospects*. Santa Barbara: Praeger/ABC CLIO, 2010.

Smil, V., *Energy and Civilization: A History*. MIT Press, 2017.

World Energy Outlook 2022. IEA, 2022.

3. Energy Dependency

Goldthau, A., & Sovacool, B. K., *The Uniqueness of the Energy Security, Justice, and Governance Problem*. Energy Policy, 2012.

International Energy Agency, *Energy Security – Reliable, Affordable Access to All Fuels and Energy Sources*. IEA, 2021.

Security of Clean Energy Transitions. IEA, 2021.

Society for Petroleum Engineers, *Guidelines for Application of the Petroleum, Resources Management System*. SPE Oil and Gas Reserves Committee (OGRC), 2011.

Yergin, D., *The Prize: The Epic Quest for Oil, Money & Power*. Free Press, 1991.

4. Energy Sources' Exhaustibility

Ayres, R. U., & Warr, B., *The Economic Growth Engine: How Energy and Work Drive Material Prosperity*. Edward Elgar Publishing, 2009.

Energy Institute, 2024 Energy Institute Statistical Review of World Energy, Energy Institute, 2024.

Heinberg, R., *The Party's Over: Oil, War and the Fate of Industrial Societies*. New Society Publishers, 2003.

Höök, M., & Tang, X., *Depletion of Fossil Fuels and Anthropogenic Climate Change – A Review*. Energy Policy, 2013.

Miller, R. G., *Future Oil Supply: The Changing Stance of the International Energy Agency*. Energy Policy, 2011.

5. Greenhouse Effect and Global Warming

Archer, D., *The Long Thaw: How Humans Are Changing the Next 100,000 Years of Earth's Climate*. Princeton University Press, 2009.

Arrhenius, S., On the Influence of Carbonic Acid in the Air upon the Temperature of the Ground. *Philosophical Magazine and Journal of Science Series* 5, *41*, 237–276, April 1896.

Climate Change 2021: The Physical Science Basis. IPCC, 2021.

Dessler, A. E., & Parson, E. A., *The Science and Politics of Global Climate Change: A Guide to the Debate*. Cambridge University Press, 2019.

Masson-Delmotte, V., P. Zhai, A. Pirani, S. L. Connors, C. Péan, S. Berger, N. Caud, Y. Chen, L. Goldfarb, M. I. Gomis, M. Huang, K. Leitzell, E. Lonnoy, J. B. R. Matthews, T. K. Maycock, T. Waterfield, O. Yelekçi, R. Yu, and B. Zhou (eds.) *IPCC – Climate Change 2021: The Physical Science Basis*. Cambridge University Press, 2021.

IPCC, *Climate Change 2023: Synthesis Report, Contribution of Working Groups I, II and III to the Sixth Assessment Report of the Intergovernmental Panel on Climate Change*. IPCC, 2023.

Ripple, W. J., et al., *World Scientists' Warning of a Climate Emergency*. BioScience, 2020.

6. Global Warming Effects and Related Issues

IPCC. *Climate Change 2022: Impacts, Adaptation, and Vulnerability*. Cambridge University Press, 2022.

Kolbert, E., *The Sixth Extinction: An Unnatural History*. Henry Holt and Company, 2014.

Lenton, T. M., et al., *Climate Tipping Points—Too Risky to Bet Against*. Nature, 2019.

Mann, M. E., *The New Climate War: The Fight to Take Back Our Planet*. PublicAffairs, 2021.

Steffen, W., et al., *Trajectories of the Earth System in the Anthropocene*. Proceedings of the National Academy of Sciences, 2018.

Wallace-Wells, D., *The Uninhabitable Earth: Life After Warming*. Tim Duggan Books, 2019.

7. Decarbonization Policies Design

Davis, S. J., et al., *Net-Zero Emissions Energy Systems*. Science, 2018.

Energy Institute, 2024 Energy Institute Statistical Review of World Energy. Energy Institute, 2024.

Grubb, M., Hourcade, J-C., & Neuhoff, K., *Planetary Economics: Energy, Climate Change and the Three Domains of Sustainable Development*. Routledge, 2014.

Pacala, S., & Socolow, R., Stabilization Wedges: Solving the Climate Problem for the Next 50 Years with Current Technologies. *Science*, *305*(S686), 968-972, 2004.

Rogelj, J., et al., *Mitigation Pathways Compatible with 1.5°C in the Context of Sustainable Development*. IPCC Special Report on Global Warming of 1.5°C. IPCC, 2018.

United Nations, *World Population Prospects 2024*. UN Dept of Economic and Social Affairs, 2024.

van Vuuren, D. P., et al., *An Energy Vision: The Transformation Towards Sustainability—Interconnected Challenges and Solutions*. Current Opinion in Environmental Sustainability, 2012.

8. Decarbonization Pathways: The Kaya Identity

Helm, D., *The Carbon Crunch: How We're Getting Climate Change Wrong – and How to Fix It*. Yale University Press, 2012.

International Energy Agency, *Net Zero by 2050: A Roadmap for the Global Energy Sector*. IEA, 2021.

IPCC, *Climate Change 2023: Synthesis Report, Contribution of Working Groups I, II and III to the Sixth Assessment Report of the Intergovernmental Panel on Climate Change*. IPCC, 2024.

Masson-Delmotte, V., P. Zhai, H. -O. Pörtner, D. Roberts, J. Skea, P. R. Shukla, A. Pirani, W. Moufouma-Okia, C. Péan, R. Pidcock, S. Connors, J. B. R. Matthews, Y. Chen, X. Zhou, M. I. Gomis, E. Lonnoy, T. Maycock, M. Tignor, and T. Waterfield (eds.) *IPCC – Global Warming of 1.5°C*. Cambridge University Press, 2018.

IRENA, *World Energy Transitions Outlook 2023*. IRENA, 2023.

Markard, J., et al, *Sustainability Transitions: An Emerging Field of Research and Its Prospects*. Research Policy, 2021.

Smil, V., *How the World Really Works: The Science Behind How We Got Here and Where We're Going*. Viking, 2022.

UNFCCC, *Kyoto Protocol to the United Nations Framework Convention on Climate Change*, 2303/U.N.T.S./162. UNFCCC, 1997.

UNFCCC, *The Paris Agreement*, FCCC/CP/2015/10/Add.1. UNFCCC, 2015.

UNFCCC, *Outcome of the first global stocktake*, FCCC/PA/CMA/2023/16/Add.1 Decision 1/CMA.5. UNFCCC, 2024.

9. Small Space–Time Scale Environmental Impacts

Fiore, A. M., Naik, V., & Leibensperger, E. M., Air Quality and Climate Connections. *Journal of the Air & Waste Management Association*, *65*(6), 645-685, 2015.

Jacobson, M. Z., *Air Pollution and Global Warming: History, Science, and Solutions*. Cambridge University Press, 2012.

Lelieveld, J., et al., *The Contribution of Outdoor Air Pollution Sources to Premature Mortality on a Global Scale*. Nature, 2015.

Seinfeld, J. H., & Pandis, S. N., *Atmospheric Chemistry and Physics: From Air Pollution to Climate Change*. Wiley, 2016.

10. Energy Transition to Sustainable Energy Systems

Cherp, A., et al., *Integrating Techno-Economic, Socio-Technical and Political Perspectives on National Energy Transitions: A Meta-Theoretical Framework*. Energy Research & Social Science, 2018.

Geels, F. W., et al., *The Socio-Technical Dynamics of Low-Carbon Transitions*. Joule, 2017.Sovacool, B. K., *Energy Transitions: Accelerating a Low-Carbon Future*. Oxford University Press, 2016.

Sovacool, B. K., The Political Economy of Clean Energy Transitions. Oxford University Press, 2017.

11. Key Enabling Technologies for Energy Sources Conversion

Energy Institute, 2024 Energy Institute Statistical Review of World Energy. Energy Institute, 2024.

International Energy Agency, Energy Technology Perspectives 2024. IEA, 2024.

Larminie, J., & Lowry, J., *Electric Vehicle Technology Explained.* Wiley, 2015.

Peake, S., *Renewable Energy: Power for a Sustainable Future*. Oxford University Press, 2017.

Sorensen, B., *Hydrogen and Fuel Cells: Emerging Technologies and Applications.* Academic Press, 2017.

Ter-Gazarian, A. G., *Energy Storage: Systems and Components.* Institution of Engineering and Technology, 2015.

Tester, J. W., Drake, E. M., Driscoll, M. J., Golay, M. W., & Peters, W. A., *Sustainable Energy: Choosing Among Options*. MIT Press, 2016.

12. Sustainable Choice of Energy Carriers

Jacobson, M. Z., *100% Clean, Renewable Energy and Storage for Everything*. Cambridge University Press, 2020.

Vazquez, S., Lukic, S. M., Galvan, E., Franquelo, L. G., & Carrasco, J. M., Energy Storage Systems for Transport and Grid Applications. *IEEE Transactions on Industrial Electronics*, *57*(12), 3881–3895, 2015.

13. Key Enabling Technologies for Final Energy Use

Hannan, M. A., et Al., *Review of energy storage systems for electric vehicle applications: Issues and challenges,* Renewable and Sustainable Energy Reviews, Elsevier, 2017.

Hensley, R., Knupfer, S. M., & Pinner, D., *Electrifying Insights: How Automakers Can Drive Electrification*. McKinsey & Company, 2018.

Liu, K., Song, H., & Zhang, X., *Electric Vehicle Integration into Modern Power Networks.* Springer, 2019.

Natarajan, J., et al. *The Future of Road Transportation Electrification and Automation,* CRC Press, 2024.

Net Zero by 2050 A Roadmap for the Global Energy Sector, IEA, 2021.

Ruilong, D.,et Al., *A Survey on Electric Buses—Energy Storage, Power Management, and Charging Scheduling*, IEEE Transactions on Intelligent Transportation Systems, IEEE, 2021.

Valle, R. G., & Pecas Lopes J. A., *Electric Vehicle Integration into Modern Power* Networks. Springer. 2019.

Detailed Contents

PART II Sustainable Energy Sources: Characteristics, Key Conversion Technologies

Part II

Sustainable Energy Sources

Characteristics, Key Conversion Technologies

INTRODUCTION

Part II of this book explores energy conversion technologies, with a detailed analysis of key renewable sources such as solar, wind, geothermal, hydro, and biomass. Each chapter delves into the technical characteristics, conversion technologies, and integration challenges of these energy sources. Readers will also learn how advancements in energy storage and grid technologies are transforming the energy landscape, paving the way for a low-carbon future.

- **Chapter II.1: Sustainable Energy Sources' Characteristics**
  This chapter introduces the characteristics of renewable energy sources, including their availability, intermittency, and environmental benefits. It discusses the challenges of integrating renewables into energy systems and the importance of technological advancements in overcoming these barriers.
- **Chapter II.2: Solar Energy**
  The chapter explores solar energy technologies, including photovoltaic systems, solar thermal plants, and concentrated solar power. It discusses their efficiency, cost trends, and integration challenges, highlighting their potential to meet global energy needs sustainably.

DOI: 10.1201/9781003631125-15

- **Chapter II.3: Wind Energy**
  This chapter examines wind energy potential, focusing on turbine technologies, site selection, and environmental impacts. It discusses the role of wind energy in decarbonizing power systems and its integration with other renewables.
- **Chapter II.4: Hydro Energy**
  The chapter covers hydro energy technologies, including river, wave, and tidal systems. It discusses their environmental and social implications, emphasizing sustainable practices in harnessing hydropower.
- **Chapter II.5: Geothermal Energy**
  This chapter explores geothermal energy systems, including their resource potential, extraction technologies, and applications. It highlights geothermal energy's role in providing stable and low-carbon energy.
- **Chapter II.6: Biomass, Biofuels, and e-Fuels**
  This chapter discusses the role of biomass and synthetic fuels in sustainable energy systems. It examines their production processes, environmental impacts, and applications in sectors like transportation and industry.

11.1 Sustainable Energy Sources' Characteristics

When we think about sustainable energy sources compared to fossil hydrocarbons and nuclear energy, we generally refer to what are commonly called renewable energy sources (RESs).

In Part I of this book, we extensively discussed energy systems' sustainability challenges. With that foundation, we now have the conceptual tools to define sustainable energy sources more precisely and effectively.

To qualify as a sustainable energy source, an energy form should possibly be:

- **Available and Concentrated**
  An energy source must exist in sufficient quantity within a specific territory to meet a significant portion of energy demand. The greater the portion of demand it can fulfil, the more strategic the source becomes. Additionally, the higher the energy concentration, the smaller and more cost-effective the technologies required for its conversion.
- **Alternative**
  Sustainable energy sources should reduce energy dependency by providing unconventional options. They should diversify and expand the energy supply paths of nations, helping to address energy security concerns.
- **Renewable**
  A renewable energy source is one that can naturally replenish or regenerate over time. However, renewability depends on maintaining the rate of usage at or below the resource's natural regeneration rate. Overexploitation risks depleting the resource, undermining its sustainability.
- **Low Carbon Emitting (Ideally Net-Zero)**
  A sustainable energy source should minimize greenhouse gas (GHG) emissions, such as CO_2, CH_4, and N_2O, to combat climate change. This involves some further considerations:
  - **Life Cycle Assessment (LCA)**
    Emissions evaluations must account for the entire life cycle of the energy source, from resource extraction to end use. While unconventional sources reduce dependency on fossil fuels, their adoption often requires additional energy (particularly during infrastructure development) potentially increasing temporary emissions due to reliance on conventional systems.
  - **Context-Specific Impact**
    Emissions are strongly influenced by the carbon intensity of the local energy system. As the energy transition progresses, emissions typically decrease.

DOI: 10.1201/9781003631125-16

- **Avoided LCA Emissions**
  Even if unconventional sources generate measurable emissions, they must be compared with the significantly higher LCA emissions of the conventional fossil fuels they replace. For example, conventional fuels incur emissions not only from combustion but also from mining, refining, distribution, and other industrial processes. These processes often more than double the total emissions compared to combustion alone.
- **Global Impact**
  Emissions tied to fossil fuels often occur outside national boundaries and are excluded from national inventories. In contrast, unconventional sources consistently show more favourable LCA carbon balances.

- **Environmentally Friendly ("Green")**
  Sustainable energy sources should minimize or eliminate local pollutant emissions, especially by avoiding the combustion of chemical fuels. This is particularly critical in residential areas where air quality impacts are most pronounced.
- **Economically Affordable**
  The cost of utilizing sustainable energy should be as low as possible to ensure economic viability. For end users, the energy transition should not result in a significant increase in energy costs. Assessing the total cost of unconventional energy sources involves many factors, including infrastructure, maintenance, and operational expenses. Among these, the most important are:
  - **Conversion Technology CAPEX and Capacity Factor**
    Powerplant investment costs are not directly comparable on an installed power basis because different plants may have quite different expected "capacity factors" (C_f, ratio between yearly annual production and installed power, normally expressed in % or in h/y), and therefore would produce quite different energy quantities within their lifetime. These capital expenditures (referred as CAPEX) are normally expressed for unit installed power in M$/MW.

    These considerations become particularly relevant in the electricity sector since electricity is difficult to store and is therefore more conveniently produced on-demand.

    Conventional thermoelectric powerplant capacity factors mainly depend on economic issues, being connected with the need to fulfil a very oscillating and partly unpredictable energy demand.

    Some "high-capacity" powerplants are always on and therefore have a very high capacity factor (>85%: mainly nuclear and high efficiency combined-cycle powerplants), while others are more flexible and have lower capacity factors (<40%: mainly small methane or oil-fed gas turbine powerplants).

    The capacity factor of electric powerplants using unconventional and/or renewable sources, instead, is normally also strongly influenced by the time-availability of the energy source. This may be pulsating and unpredictable (refer, as an example, to sun, water, or wind availability). Corresponding capacity factors are therefore normally much lower than those of more conventional powerplants.

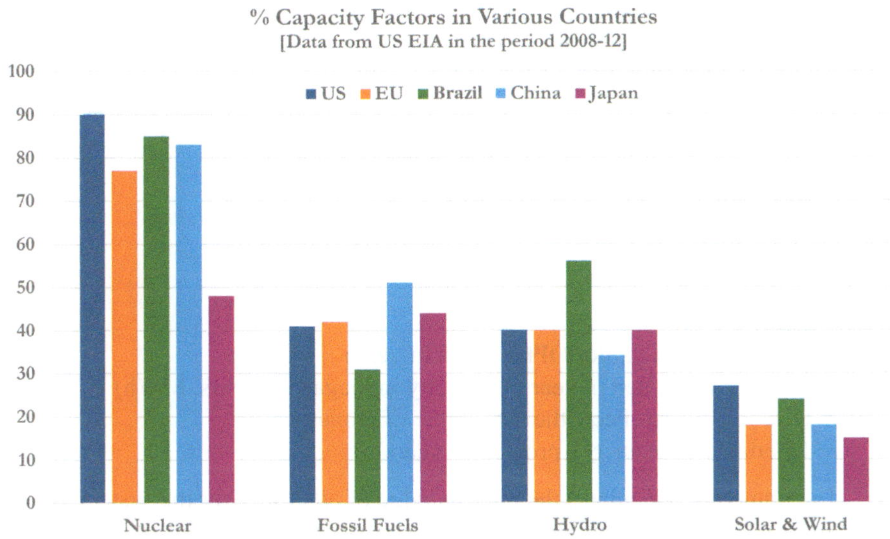

FIGURE II.1.1 Electric powerplant mean capacity factors (2008–12 average) for various sources and regions. (Data from the US Information Energy Agency).

As already explained, powerplant capacity factors also depend on many local issues connected with RES penetration factors, and grid management rules. Some reference values for the single sources are reported in Figure II.1.1.

- **Conversion Technology OPEX**
  Powerplant exercise costs may also vary depending on the source, and are normally much lower than those of conventional plants. This is due to the territorial availability of most of the sustainable sources (sun, wind, water, biomasses). These operational expenditures (referred to as OPEX) are normally expressed for a unit of installed power in \$/MWh.
- **Conversion Technology LCOE**
  The previous two points are normally gathered by a conventional technique used (mainly in the electricity sector) to derive an indicator comparing different possible powerplants, known as the Levelized Cost of Electricity (LCOE).

  In fact, LCOE is a metric used to calculate the average cost of generating one unit of electricity (usually measured in \$/MWh) over the lifetime of an energy-producing asset. It incorporates all costs, including CAPEX, OPEX (together with fuel costs, if and when applicable, and other exercise maintenance costs) divided by the total energy output over the system's operational life.

  Therefore, the LCOE enables comparisons of different energy technologies, such as solar, wind, coal, and nuclear, by providing a standardized measure of cost affordability.

A lower LCOE indicates a more cost-effective energy source, making it a critical tool for investment and policy decisions in the energy sector. LCOE can be expressed with the following formula:

$$LCOE = \frac{\sum_{t=1}^{n} \frac{C_t}{(1+r)^t}}{\sum_{t=1}^{n} \frac{E_t}{(1+r)^t}}$$

where:
C_t: Total CAPEX and/or OPEX costs in year t
E_t: Energy generated in year t (typically in megawatt-hours, MWh)
r: Discount rate (reflects the time value of money)
n: Lifetime of the energy project (in years)

The numerator calculates the present value of all costs over the system's lifetime, while the denominator calculates the present value of all electricity generated over the system's lifetime.

LCOE values for unconventional powerplants are expected to decrease over time due to advancements in technology, leading to higher technology readiness levels (TRLs), manufacturing readiness levels (MRLs), and commercial readiness levels (CRLs). Conversely, LCOE values for conventional plants are projected to increase due to rising fuel costs.

Several political, environmental, and economic institutions engage in calculating the evolution of LCOE values to use them in policies oriented towards an energy transition implementation.

Figure II.1.2 reports a graphical representation of data from Lazard about LCOE trends from different powerplants in the period 2009–24.

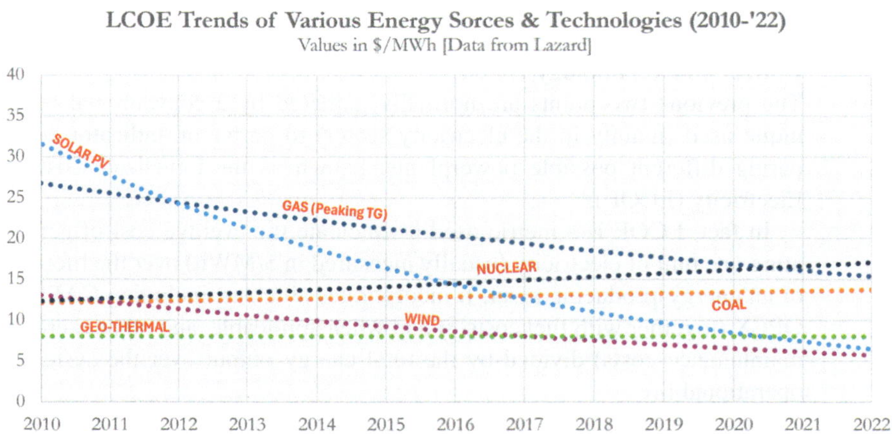

FIGURE II.1.2 LCOE trends in the period 2009–24 for different technologies. (Data from the Lazard report "Levelized Cost of Energy+" – June 2024).

- **Avoided Externalities**
While LCOE provides a standardized cost metric, it does not explicitly account for the environmental and societal costs of energy production, often referred to as "externalities".
This omission disadvantages sustainable energy adoption by:
- **Unaccounted Costs of Damage**
Fossil fuels contribute to global warming and climate change through significant CO_2 emissions, which are not reflected in their direct costs. Similarly, the health impacts caused by air pollution in urban areas, such as respiratory diseases, impose heavy burdens on healthcare systems, yet these costs are not included in traditional LCOE calculations.
- **Inadequate Compensation via Taxes**
In some cases, fossil fuels are subject to "Pigouvian taxes", aimed at internalizing environmental damage. However, these taxes are often far below the actual estimated costs of environmental and health externalities.
LCOE evaluations and comparisons do not explicitly consider the cost of environmental externalities. This lack disfavours the introduction of sustainable energies.
The very costs of the damage produced by CO_2 emissions and consequently by global warming and climate change issues are not considered.
In the same way, the cost produced on the health system by the increased occurrence of many illnesses directly connected to urban air quality issues is not at all considered.
All these costs only indirectly affect fossil fuel purchasing price, and/or some plant management production costs, which, in a very site-dependent way, may be affected by some "Pigouvian" damage compensation taxes. Even when and where present, these compensation costs are far below the estimated costs of the cited externalities.
- **Energy System Charges**
On the other hand, LCOE evaluations and comparisons also do not explicitly consider the other energy system modifications which are expected to be needed due to sustainable energies utilization.
These modifications (as fully explained in Part I) will be required both in the energy carriers and ESS sector, and in many of the final-use sectors (especially in the transportation sector). All these costs must be sustained for a complete energy transition and should therefore be considered as additional contributions to the LCOE of sustainable sources.
Renewable energy sources offer significant benefits, including reduced emissions and long-term availability. However, challenges such as intermittency and integration require ongoing technological improvements.

II.2 Solar Energy

II.2.1 SOURCE CHARACTERIZATION

Solar energy reaches the Earth as electromagnetic radiation emitted by the Sun. This radiation is approximatively like that of a black body at a temperature of about 5250 °C. Its spectrum consists of ultraviolet rays in the range of 200–400 nm, visible light in the range 390 nm (violet) to 740 nm (red), and infra-red in the range 700 nm to 2500 nm (see Figure II.2.1).

This radiation delivers an enormous cumulative power of about 173 petawatts (PW) to the Earth. Annually, this translates to 1515 exawatt-hours (EWh) of solar energy hitting the planet, which is roughly 60,000 times greater than the world's current annual energy consumption of 25 petawatt-hours (PWh). This underscores solar energy's immense strategic importance among sustainable energy sources.

Distributing this energy over the whole Earth surface, we may calculate the so-called solar constant. This, in fact, is formally defined as the amount of solar energy that reaches a unit area of a surface perpendicular to sunbeams, located at the outer edge of Earth's atmosphere.

Therefore, the solar constant is a measure of the total solar radiation received per square metre: its mean value is approximately 1361 W/m². It is not perfectly stable but varies a little due to the elliptical characteristics of the Earth's orbit (about 3.3% amplitude) and to variations in solar activity (about 0.2% amplitude), which has its own 11-year periodic cycle.

Only a part of this energy directly or indirectly reaches the Earth's surface, due to atmospheric interactions.

II.2.1.1 Solar Energy Interaction with the Atmosphere

In Chapter I.5, we described the typical interactions which solar radiation has with the atmosphere, determining its energetic equilibrium. these are mainly the following:

- About 30% of incoming solar energy is optically reflected into space. This phenomenon is called planetary albedo. Each astronomical body has its own albedo coefficient.
- About 18% is scattered in the atmosphere, especially due to the water vapour particles which it contains, depending on season, latitude, as well as climatic

DOI: 10.1201/9781003631125-17

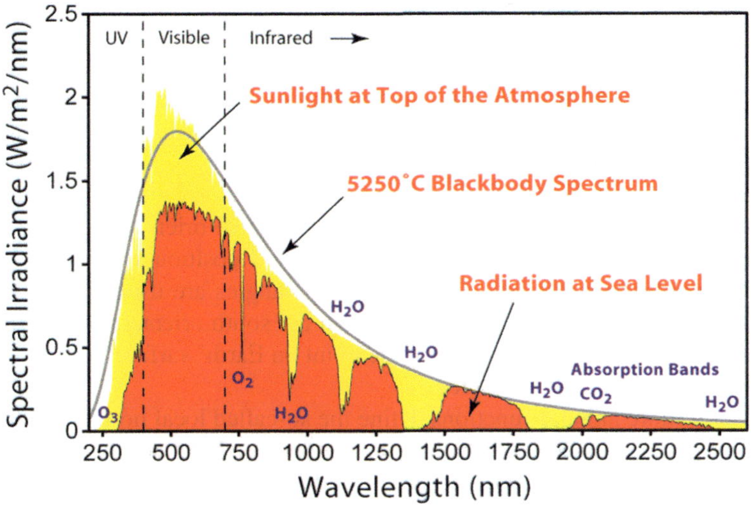

FIGURE II.2.1 Solar Radiation Spectrum. Source: figure by Robert A. Rohde – Global Warming Art project. Licensed by CC BY-SA 3.0 – https://creativecommons.org/licenses/by-sa/3.0/deed.en

and meteorological phenomena. This scattering causes sunbeams to spread out in different directions, but energy still reaches the ground.

- About 25% is absorbed by the atmosphere, due to its composition in different gases and particles. Absorption converts solar energy into heat, warming the atmosphere. Absorbed energy is re-emitted by the atmosphere in all directions, with a different wavelength spectrum.
- The remaining part, about 27%, directly reaches the Earth's surface, having no interactions with the atmosphere.

On a cloudless day, scattering and absorption phenomena are minimal, while planetary albedo remains active, so that the maximum radiation reaching the surface is around 1000 W/m².

Therefore, the value of 1 kW/m², also referred as **1 Sun**, is commonly used to define nominal operative conditions of solar technologies. These are normally measured in lab environments using properly designed and certified "solar lamps", which mimic the solar spectrum. Solar lamps must be correctly oriented and posed at the correct distance from the tested equipment to make the test valid.

According to the above-discussed phenomena, the amount of energy which instantaneously reaches a certain surface at ground depends on several aspects. Among these, the most influential are:

- **Astronomical considerations**
  These define the relative instantaneous positions of the lighted surface relative to incoming sunbeams. To this aim, four angles have to be defined:

    - Two of them, latitude and longitude, are the two angular coordinates commonly used to locate a point on the Earth's surface.
    - The other two angles (which can be substituted by knowledge of the UTC – Coordinated Universal Time) are needed to define the position of the Earth in its two astronomical movements (rotation on its axis and revolution on its elliptical orbit around the sun).
- **Geometric considerations**
  Another two angles are used to define the relative orientation of the surface with respect to the normal vector in the geoid position where it is located. Commonly, azimuth and elevation (or tilt) angles are used. These, respectively, refer to the angular distances from the south orientation, and from the perpendicular vector (connecting the point on Earth' surface with its centre).
- **Presence of optical obstacles**
  One or more angular functions define the so-called local horizon on the 360° (commonly referred to north direction clockwise). It defines the angular height of the obstacles which would not permit direct sunbeams to reach the surface.
- **Instantaneous local atmospheric conditions**
  Atmospheric events, including wind speed, relative humidity, and temperature, affect the number of particles and concentration of gas that the sun beams must overcome to reach the receiving surface.
  These last factors determine the presence and amplitude of atmospheric scattering and absorption phenomena and therefore, as will be shown later in more detail, not only affect the quantity of radiation but also its foreseen wavelength distribution and incoming direction. These characteristics are important, in turn, because they may influence and affect the functioning of energy conversion technologies.

Nowadays, several online apps are available which make it possible to fully view the solar irradiance on a surface during a standard meteorological year. Many of these apps and databases are publicly available and free. The meteorological conditions of the location are statistically characterized using climatic geo-referenced databases.

One of the most used in Europe is PVGIS, designed by the Joint Research Centre of the European Commission, and accessible at the site https://re.jrc.ec.europa.eu/pvg_tools/en/.

These apps are normally oriented to photovoltaic applications design (which will be discussed in detail in the following paragraphs) but may be used also for simple calculation of the characteristics of the incoming radiation.

Calculations can be made both about the global energy amount hitting the surface (as a sum of all the possible contributions), and about the contribution of directly incoming sunbeams only (reaching a surface being unaffected by scattering and/or absorption and re-emission phenomena). In more detail, the calculated values about irradiance which are most commonly used in technological evaluations are:

- **DNI (direct normal irradiance)**
  DNI represents the amount of solar radiation received per unit area by a surface that is kept perpendicular to incoming radiation at its current position in the sky.

In fact, to maximize the amount of irradiance received annually, it would be ideal to move the surface, instantaneously keeping it perpendicular to the incoming solar radiation.

DNI is particularly important for concentrating solar thermal systems and solar installations that track the Sun's movement.

- **DHI (diffused horizontal irradiance)**
  DHI refers to solar radiation that does not come directly from the Sun, but instead has been scattered by clouds, particles, and other components in the atmosphere. This radiation arrives from all directions, being equally distributed across the sky. It may also have a different wavelength spectrum with respect to incoming sun radiation.
- **GHI (global horizontal irradiance)**
  GHI represents the total amount of solar radiation received by a horizontal surface (parallel to the ground) from the sky above. It combines both direct and diffused radiation.

DNI, DHI, and GHI are not independent but are correlated as follows:

$$\text{GHI} = \text{DHI} + \text{DNI} \times \cos(\theta)$$

where θ is the so-called zenith angle, defined as the angular distance between sun beams and the verticality.

II.2.1.2 Irradiance Maps

Calculations about solar irradiance can be repeated by varying the latitude and longitude. Mean irradiance values can be calculated during single days, months, or the whole mean meteorological year.

Correspondent average maps can be drawn, permitting to identify the most promising locations, and showing the variability of energy availability in a territory.

Figures II.2.4 show public data from the Italian Solar Atlas, published by ENEA (the Italian National Agency for New Technologies, Energy and Sustainable Economic Development):

- Data in Figure II.2.2 are relative to DNI values on an optimally oriented fixed surface (not tracking the sunbeam direction, but oriented to catch the highest possible energy amount over the year).

  Values in January (left part of figure) vary from 1 to 2.5 kWh/m^2 day. This means that daylight overall irradiance is equivalent to 1 to 2.5 hours of full irradiance at 1 Sun, with 1 Sun, as already described, being the nominal irradiance condition, any possibly installed conversion technology can only have a maximum capacity factor ranging from 1/24 (≈4%) to 2.5/24 (≈10%) in January.

  However, in July these values increase a lot, ranging from about 5/24 (≈20%) to almost 8/24 (≈33%).

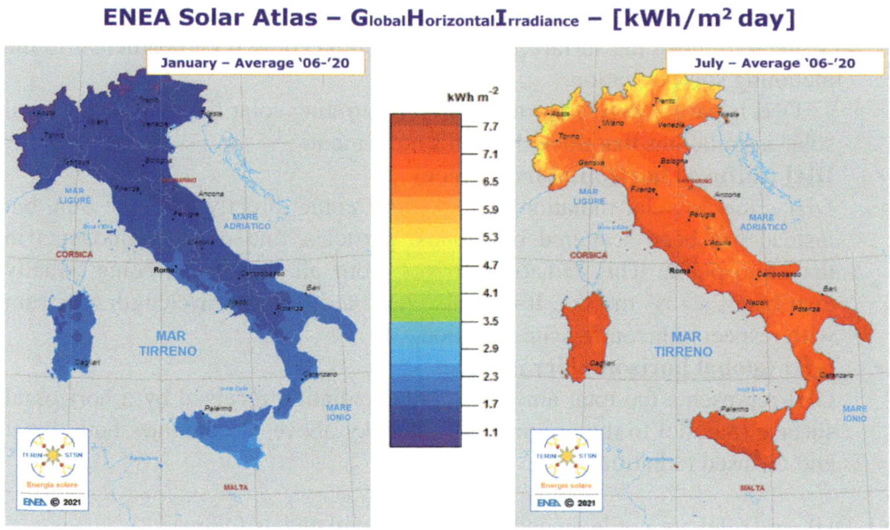

FIGURE II.2.2 ENEA Solar Atlas – Mean DNI Values in Italy in January and July (2006–2020).

- Data in Figure II.2.3 are relative to global horizontal irradiance (GHI), and show similar patterns.
- Data in Figure II.2.4 are relative to mean yearly irradiance. These are obtained by summing throughout the whole mean meteorological year the data calculated at every moment of the day.

  Values in the bottom left corners show the overall values reached in Rome and in the Italian cities where the highest and lowest values are obtained.

  The highest values of both GHI and DNI are obtained in Agrigento in Sicily (1717 kWh/m^2/year for GHI and 1798 for DNI, on a fixed optimally oriented surface).

  The lowest values are obtained in Sondrio in the Alps (1288 kWh/m^2 y for GHI and 1302 for DNI on an optimally oriented surface).

  This means the same solar technology would produce approximately 25–27% less energy in Sondrio than in Agrigento.

  The data also show that the maximum yearly average expected capacity factors of fixed solar technologies in Italy are expected to be around 1798 h/year, or equivalently 1798/(24*365.25)≈0.20 (20%).

II.2.1.3 Energy Density

As just described, the expected capacity factor for solar technologies in Italy is expected to be between 15% and 20%, referring to a nominal energy input of 1 Sun (1 kW/m^2). Similar calculations can be repeated in different countries.

Therefore, if we imagine converting this energy into its final use with a reference efficiency of 0.25, we could feed a final use of around 400 kWh/year for each

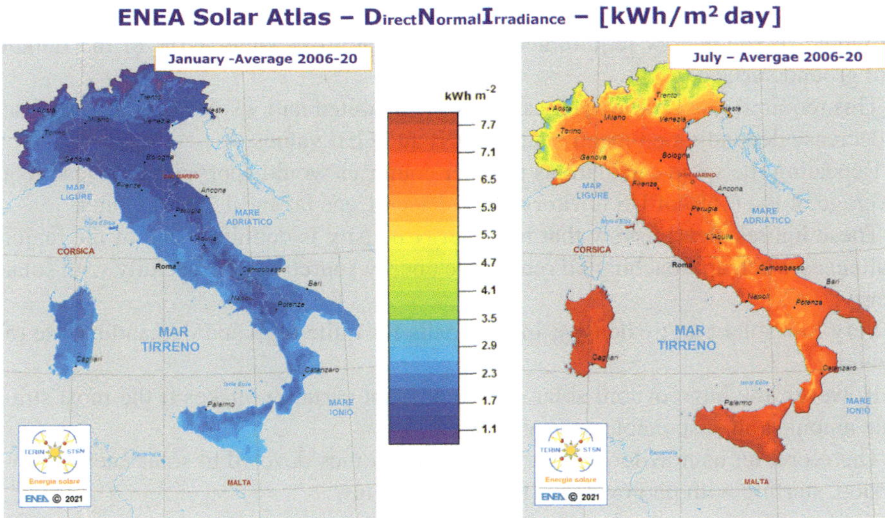

FIGURE II.2.3 ENEA Solar Atlas – Mean GHI values in Italy in January and July (2006–2020).

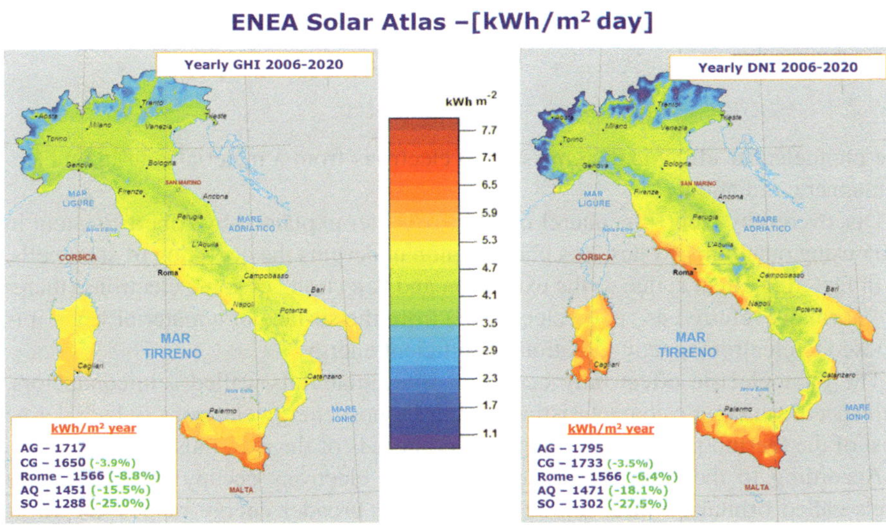

FIGURE II.2.4 ENEA Solar Atlas – Yearly GHI and DNI values in Italy (2006–2020).

installed m^2 of solar energy conversion technologies (i.e. 1 kW nominal power having 1600 h/year, V18% capacity factor, and 0.25 efficiency).

Now, consider that Italian energy consumption for all its final uses is nowadays approximately 5.6 EJ/y, which means about 1.6×10^{12} kWh/y.

This would mean that we should install ≈4 × 10^9 m^2 of solar technologies to cover the whole Italian energy requirements, which is approximately 1.1% of the Italian total ground surface.

This required ratio is lower in less densely inhabited nations. It may also increase or decrease depending on mean country DNI and GHI values.

Repeating the same calculations in other countries would approximately lead to a coverage of the 0.6% of US territory, the 0.1% of Canada, and 1.1% of China.

These figures clearly show that low energy density is definitely one of the major challenges when approaching all renewable energy sources, but especially with solar energy.

Very low solar energy density, in fact, calls for quite high and demanding use of ground.

However, the huge overall solar energy amount certainly makes it the most strategic amongst all sustainable energy sources.

Therefore, let us now look at the main technologies devoted to solar energy conversion, starting with photovoltaic (PV) technologies.

II.2.2 PHOTOVOLTAIC TECHNOLOGIES

The goal of photovoltaic technologies is the direct conversion of radiant energy into electrical energy. To achieve this, solar cells are used to convert solar energy into electrical energy directly using the so-called photoelectric effect.

II.2.2.1 Photoelectric Effect and Solar Cells

The photoelectric effect is the emission of electrons from a material caused by electromagnetic radiation.

Discovered by Henri Becquerel in 1839 and later explained by Albert Einstein in 1905 using quantum physics, this phenomenon underpins the operation of solar cells.

In fact, this effect is due to the excitation and "ejection" of some electrons (therefore normally referred as "photoelectrons") from the surface of a material when it is exposed to electromagnetic radiation of sufficient energy.

The photoelectric effect only occurs in some materials, called semiconductors, with specific energetic characteristics of their highest energy electrons (which define most of the thermo-physical and chemical properties of the materials).

According to the Bohr model of substances, electrons occupy specific energy levels called orbitals, which can be grouped into broader energy ranges known as bands. The two key bands in solid materials are the valence band and the conduction band, and the relative positions of these bands determine the material's electrical properties.

- **Valence band:**
  - This band comprises energy levels associated with electrons in valence orbitals, which allow electrons to orbit individual nuclei and contribute to molecular bonds.

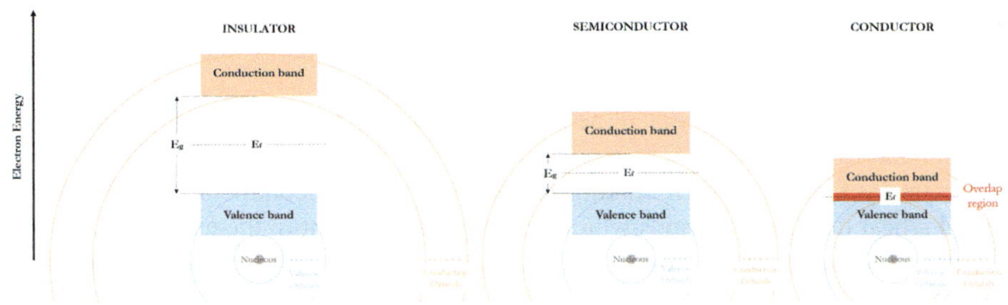

FIGURE II.2.5 Bandgap and materials' behaviour.

- Electrons in this band are tightly bound to atoms and do not move freely within the material, conferring it an electrically insulating behaviour.
- **Conduction band:**
  - This band consists of energy levels associated with conduction orbitals, where electrons can move freely throughout the material.
  - Electrons in the conduction band are responsible for electrical conductivity.

The electric properties of each material are therefore defined by the reciprocal energetic position of the valence band and the conduction band and by the potential presence of electrons occupying these orbitals.

The mean energy level of the valance band is always lower than that of the conduction band. This means that the valence and conduction bands may be partially overlapped, slightly separated (by a small energy gap also called the "band gap"), or clearly separated (by a sufficiently relevant energy gap), according to the sketch in Figure II.2.5

In these three distinct cases, the material, respectively, is named:

- Conductor (metal)
- Semiconductor
- Insulator.

These different band configurations confer the material with a different electric behaviour. In fact, the highest energy electrons statistically occupy the more stable energetic positions amongst those available. Therefore, they first fill the lowest energy orbitals.

In conductors, some of these electrons occupy conduction orbitals, while both in semiconductors and insulators no electrons are present in the conduction band, conferring insulating properties to both of them.

However, the amplitude of the bandgap in these latter two cases makes the difference between conductors and semiconductors, making it possible to explain the photoelectric effect.

This bandgap, in fact, must be compared with the quantum energy contributions that the electro ns staying in valence orbitals may receive from the exciting interactions with the surrounding environment. Depending on the energy they may receive, their excited state may justify or not their "jumping" in the conduction band and the corresponding change in the electric behaviour of the whole material (from an insulator to a conductor).

Clearly, this may only happen if the band gap is sufficiently small, so distinguishing semiconductors from conductors.

The energy threshold discriminating conductors from semiconductors is determined by the highest quantum energies which are normally present in the external environment and may be therefore transferred to the highest energy electrons orbiting in the valence band.

The highest available environmental quantum energy, in turn, is given by the highest energy photons coming from the Sun (having a violet wavelength).

In fact, Planck's law ($E = hv = h/\lambda$) states that the quantum energy of a photon travelling in an electromagnetic wave is inversely proportional to its wavelength λ by the Planck's constant h. Therefore, the lowest wavelength sunbeams (violet wavelength) are those transferring the highest unit quantum energy.

As a single quantum of energy is very small, it is convenient to express it referring to the voltage difference it may produce if transferred to a single electron. Using this approach, energy quantities are expressed in electron-volts, eV, with the electron-volt being defined as the amount of energy gained or lost by the unit charge of a single electron ($1.602\ 10^{-19}$ Coulomb) when it moves through an electric potential difference of 1 volt. Therefore, numerically: $1\text{ eV} = 1.602 \times 10^{-19}$ J.

Planck's constant is $h = 1.2398$ eV/μm, and the solar spectrum ranges from 0.2 to 2.5 μm (ultraviolet rays, 200–390 nm; visible light, 390–740 nm from violet to red; infra-red rays 700–2500 mm).

Therefore, according to Planck's law, photon energy ranges from about 0.5 to 5–6 eV.

For this reason, a reference value of 5 eV is commonly used for the band gap to discriminate semiconductors from conductors.

Summarizing, the photo-electric effect may be explained as follows: if an electron of a semiconductor material is hit by a sufficiently energetic photon, it "jumps" from a valence orbital into a conduction one, conferring the material with a conductive behaviour. This presence of electrons in an excited state also evokes a voltage difference between parts of the material which are differently hit by the sun beams (the so-called photovoltaic PV effect).

II.2.2.2 Silicon-Based PV Cells

One of the best-known and most widely available semiconductors is silicon, which has a bandgap of 1.12 eV.

Solar cells are devices made of semiconductive materials that use the photoelectric effect for various electric applications. The development of solar cells has progressed from niche uses to scalable energy production.

The first silicon solar cell was created by Chapin, Pearson, and Fuller in 1954, achieving a 6% conversion efficiency. These solar cells were initially used in satellite applications starting in 1958.

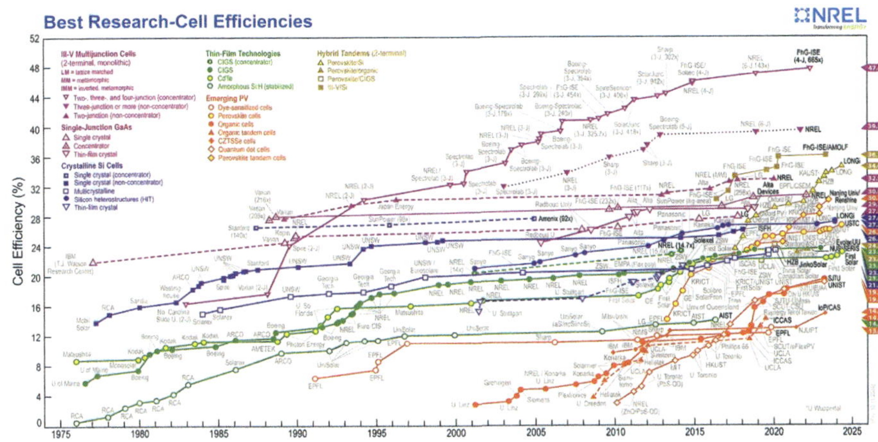

FIGURE II.2.6 Evolution of PV technologies performances in the period 1975–2024.

In the decades since, the efficiency of solar cells has improved significantly, and both solar cells and solar power plants have become cheaper and more competitive with other power generation methods. This has enabled the widespread adoption of solar technology across the globe.

There are various methods for producing silicon solar cells, including single-crystal, polycrystalline, and amorphous forms of silicon. Additionally, other semiconductors, such as indium and cadmium, are also used in certain types of solar cells.

Figure II.2.6 presents the results of a comprehensive research analysis conducted by the US National Renewable Energy Laboratory (NREL), showing the evolution of PV technology performance over time. According to technical and economic considerations, the majority of solar cells currently used in power plants are still silicon-based cells (represented by the blue curves in the figure).

These plants normally reach a maximum conversion efficiency of around 20-25%, but may have also a much lower conversion efficiency (down to 5–10%) when making use of cheaper materials and technologies (i.e. amorphous silicon and others).

To obtain a silicon-based PV cell, it is essential that the silicon used is extremely pure. Silicon is a fourth-order material that has a fully populated valence band, with many electrons that can jump into the conduction band when excited.

Typically, pure silicon's structure is chemically modified to enhance the photoelectric effect and improve the overall conversion efficiency. This chemical modification is normally referred to as the "doping" process.

The doping process can be achieved by high-temperature diffusion of the dopant material into the pure silicon structure, with the dopant introduced as a vapour. The process basically involves introducing small amounts of specific elements (dopants) into pure silicon. These dopants, such as phosphorus (for n-type doping) or boron (for p-type doping), alter the balance of electrons in the silicon crystal.

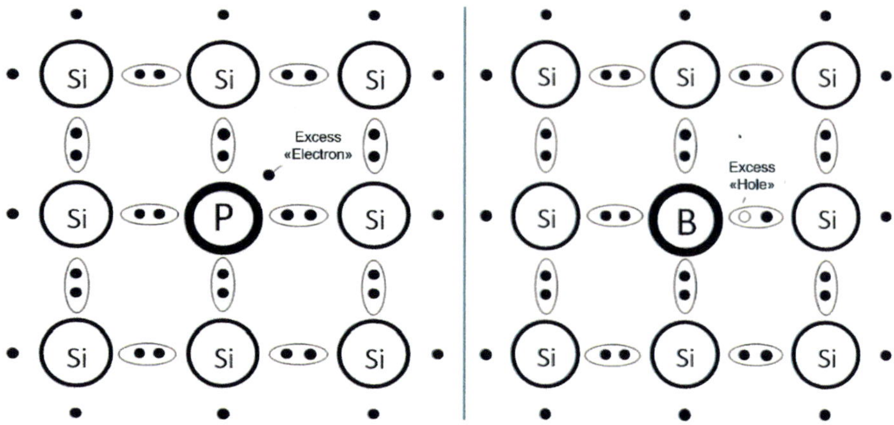

FIGURE II.2.7 Sketch of the typical bonds that the dopant elements create within the pre-existing silicon layers.

In n-type doping, phosphorus atoms, which have an extra electron compared to silicon, are added. This extra electron enhances the material's conductivity by providing more free electrons that can move through silicon.

On the other hand, in p-type doping, boron atoms, which have fewer electrons than silicon, create "holes" in the crystal, allowing the movement of positive charge carriers, or "holes", through the material.

Figure II.2.7 shows a sketch of the typical bonds that the dopant elements create within the pre-existing silicon layers.

A p–n junction may then be realized joining two types of semiconductor material: p-type and n-type silicon. When the two regions are brought together, electrons from the n-type region begin to diffuse into the p-type region, where they recombine with holes. This creates a depletion region, where no free charge carriers exist. As a result, an electric field is formed across the junction, creating a potential barrier that prevents further movement of charge carriers.

The p–n junction has unique electrical properties that are crucial for many electronic devices. When a voltage is applied, the junction allows current to flow in one direction but blocks it in the opposite direction. This behaviour is what makes the p–n junction essential for the operation of diodes, transistors, and solar cells. In solar cells, for example, the p–n junction allows sunlight to generate electron–hole pairs, creating a flow of electricity when the external circuit is connected.

Figure II.2.8 illustrates two different sketches realized, respectively, from the US EIA (Energy Information Administration) and the US Chemistry Association. Both show how the p–n junction is realized and how it is expected to behave when it is hit by a certain sun flux and connected to an external utilization circuit.

When the n-region is exposed to sunlight, high-energy photons transfer their energy to valence electrons exciting them. These "photoelectrons" are moved in conductive orbitals, generating electrostatic potential energy. To collect this energy, an

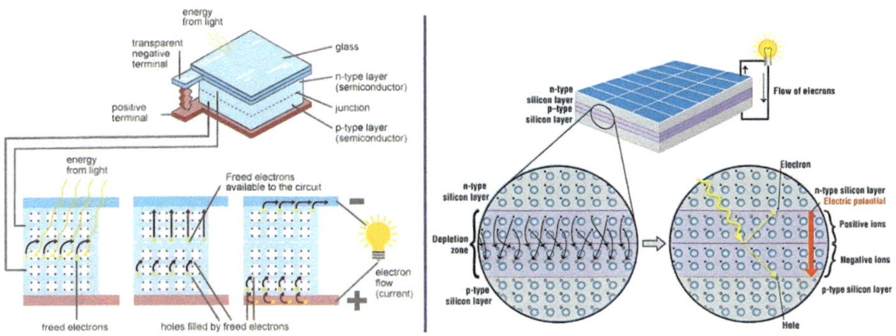

FIGURE II.2.8 Two different sketches realized respectively from the US EIA (Energy Information Administration) (left) and from US Chemistry Association (right).

external circuit is created to harness the electron movement via electrical contacts (e.g., copper wires).

Specifically, a front contact grid is used to help the extraction of the photoelectrons going as near as possible to them, so limiting resistive Joule-effect losses. This grid must be designed to avoid too much interference with incoming sunlight photons, ensuring it is neither too spread out, nor made from conductors that are too thin.

The extracted electrons then flow through the external circuit, providing electricity, before returning to the p–n junction via a rear electrical contact connected to the p-type region. This rear contact is usually a simple metallic layer that also serves to mechanically support the device.

Due to the electrostatic potential difference (measured in volts) between the n and p zones, electrons circulate through the external load.

However, they cannot transfer all their potential energy to the external load. A residual potential difference is consumed inside the p–n junction to enable the electrons to return to their original position in the valence orbital. This is necessary so that these electrons can be re-excited by incoming photons, close the circuit, and be available for continuous electricity generation.

This residual potential difference is an energy loss for the device.

Mathematically, the component mostly behaves linearly with electron flow intensity and therefore with circuital current I. Its behaviour is therefore described through the definition of an equivalent internal resistance r, which accounts for all the energy losses concentrated or distributed within the junction due to the electrons' flow.

These losses, in turn, strictly depend on the position of the silicon, which can be determined or random, depending on the chemical structure of the silicon used in the junction production.

In fact, silicon solar cells are typically made starting from thin slices (wafers) of pure solid silicon, having a different chemical structure (see also Figure II.2.9 for reference). The structure obtained for silicon substrate depends on the production processes used to obtain solid silicon, starting from purified liquid silicon:

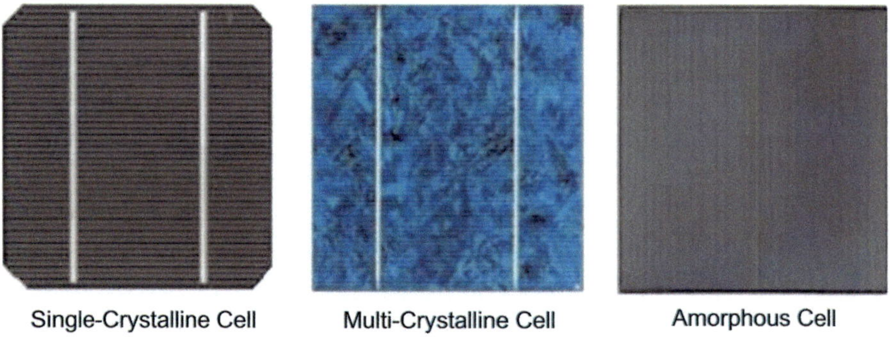

FIGURE II.2.9 Different silicon PV cell types.

- **Single-crystalline wafers**
  To produce single-crystalline silicon wafers, the liquid silicon must be extremely pure and solidified very slowly to achieve a perfect single-crystalline structure. The silicon is shaped into a large cylindrical ingot at around 1400°C through a complex and expensive manufacturing process. This ingot is then sliced into wafers with thicknesses ranging from 0.3 to 0.5 mm.

  The p–n junction formed in single-crystalline silicon has a much lower internal resistance, resulting in a higher conversion efficiency, typically reaching 20–25%. This is because electrons can move along specific low-resistance paths within the cell without being hindered by the unexpected presence of silicon nuclei or their positive repulsive charges.
- **Multi-crystalline wafers**
  Multi-crystalline silicon wafers are produced using a solidification process in which molten silicon is poured into a mould and allowed to solidify under specific thermal conditions. This manufacturing method is much cheaper compared to the process used for single-crystalline silicon.

  However, the efficiency of multi-crystalline solar cells is typically lower. This is because electrons moving through the junction must cross the boundaries between multiple individual crystals. Each boundary introduces concentrated energy losses, resulting in lower overall efficiency compared to single-crystalline devices.
- **Amorphous structures**
  Amorphous silicon solar cells are produced by depositing a thin film of silicon onto a glass substrate using a reactive gas, typically silane (SiH_4). This process is much faster and more cost-effective, but results in lower efficiency.

  However, the lack of a crystalline structure causes electrons to follow a more erratic path within the junction, leading to increased internal resistance. Consequently, the efficiency of amorphous silicon cells is lower.

  Despite this drawback, the process enables the production of low-cost solar cells that are well-suited for certain applications, such as flexible solar panels and devices where cost is a primary consideration.

II.2.2.3 Solar Cells Maximum Theoretical Efficiency

The radiative efficiency limit, also known as the Shockley–Queisser limit (SQ limit), represents the maximum theoretical efficiency of a solar cell with a single p–n junction. This concept was first introduced by William Shockley and Hans-Joachim Queisser in 1961.

Figure II.2.10 presents the results of the Shockley–Queisser calculations, focusing on: silicon solar cells with a bandgap $E_g = 1.12$ eV (left panel) and variations in efficiency for different bandgap values (right panel).

These calculations use actual measured solar spectra and estimate a maximum efficiency of 33.16% for single-junction solar cells made from materials with an optimal bandgap of 1.34 eV. This indicates that only about 33% of sunlight's energy (approximately 1000 W/m²) can theoretically be converted into electricity, equating to 330 W/m².

Silicon, the most widely used material for solar cells, has a bandgap of 1.12 eV, which limits its theoretical efficiency to around 33%. In practice, commercial monocrystalline silicon solar cells today achieve efficiencies of approximately 24%. Real-world losses stem from practical factors such as reflection, light blockage by surface wires, and some resistive losses, though internal resistance losses are minimal.

Referring to the left part of Figure II.2.10, the theoretical and unavailable losses fall into three main categories:

- Unabsorbed energy losses

  These losses occur because photons with energy less than the material's bandgap cannot excite electrons to jump from the valence band to the conduction band. For silicon ($E_g = 1.12$ eV), photons with wavelengths greater than 1.1 μm lack sufficient energy to generate photoelectrons. This results in the loss of about 19% of the total solar spectrum, represented by the pink-coloured area in Figure II.2.10.

  These losses are a function of E_g, obviously increasing with the band energy gap.

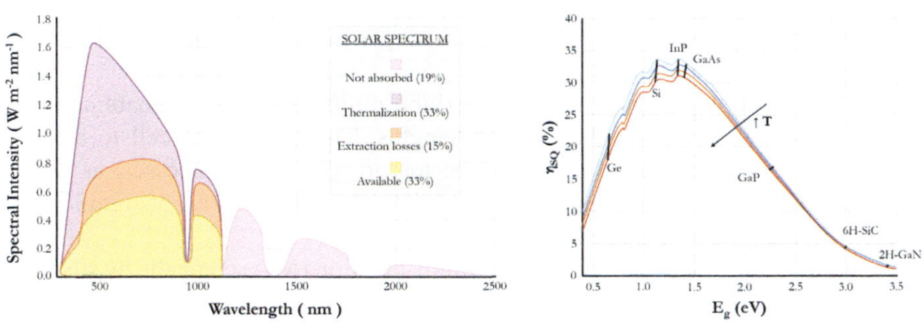

FIGURE II.2.10 Shockley–Queisser theoretical efficiency limits (η_{SG}) of single-junction PV cells: the left part reports the minimal theoretical losses calculated from silicon $E_g = 1.12$ eV. The right part shows η_{SG} variations with E_g and with the operative cell temperature.

- Thermalization losses
  These losses arise when photons have energy greater than the bandgap. The excess energy, beyond what is needed to excite electrons, is converted into heat rather than contributing to electricity generation.
  For silicon, these losses account for about 33% of the solar spectrum, as shown by the violet-coloured area in Figure II.2.10.
  Thermalization losses are higher for materials with smaller bandgap values because they absorb higher-energy photons that transfer more excess energy as heat. Together, unabsorbed energy losses and thermalization losses are referred to as spectrum losses, which account for about 52% of total sunlight energy, leaving a maximum theoretical efficiency of 48% for silicon.
- Extraction losses
  Extraction losses are more challenging to estimate. They arise due to inefficiencies in the internal and external circulation of photoelectrons, particularly in how the voltage–current (V–I) characteristics of the solar cell interact with the external circuit's impedance.
  Shockley and Queisser proposed an electrical model of the solar cell's behaviour to calculate these minimal extraction losses, which depend on the material's bandgap.
  For silicon, these losses amount to about 15% of the solar spectrum, represented by the orange-coloured area in Figure II.2.10. Consequently, the maximum theoretical efficiency of silicon-based PV cells is approximately 33%.

The Shockley–Queisser limit only applies to conventional solar cells with a single p–n junction. Solar cells with multiple layers and multiple junctions can exceed this limit. It can be demonstrated that in the case of a multi-junction solar cell with an infinite number of layers, the theoretical efficiency limit reaches 68.7% with normal sunlight, or up to 86.8% with concentrated sunlight.

In laboratory environments, multi-junction PV devices have already achieved efficiencies approaching 48%, although these technologies are still at low technological readiness levels (TRLs) and manufacturing readiness levels (MRLs). For now, most commercial installations use single-crystalline single-junction PV cells, which have a maximum efficiency of around 25%.

Cells with multiple material layers can exhibit multiple bandgaps, enabling them to respond to a wider range of light wavelengths. This allows the cell to capture and convert energy that would otherwise be lost to relaxation processes, as previously described. Interestingly, light with a specific wavelength does not significantly interact with materials that have larger bandgaps.

This property allows for the creation of multi-junction (MJ) solar cells by stacking materials with progressively smaller bandgaps. In such designs, materials with the largest bandgap are placed at the top, facing incoming light, with materials of smaller bandgaps layered underneath.

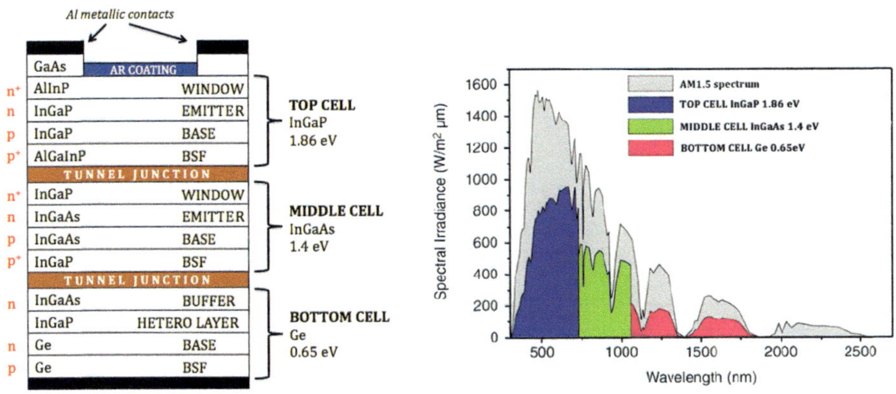

FIGURE II.2.11 Structure and behaviour of multi-junction cells (Fraunhofer Institute for Solar Energy Systems).

Figure II.2.11 illustrates the structure of a multi-junction solar cell. Photons penetrate through the layers until they reach a material with a suitable bandgap for absorption. Transparent conductors are essential in multi-junction cells to collect electrons generated at each layer while allowing photons to pass through to the lower layers.

This design maximizes energy conversion by reducing relaxation losses. However, it requires advanced materials and manufacturing techniques, which currently limit widespread adoption due to high production costs.

II.2.2.4 Solar Cells J-V Performances

The performance of solar cells can be fully described using a J-V Cartesian plane.

Figure II.2.12 reports the specific performances of a single cell, expressed for a unit surface exposed to solar light. It depicts in distinct colours the behaviour of the cell under different solar flux F values (orange – 1 Sun; green – 0.5 Sun; blue – 0.25 Sun; magenta – 0.1 Sun).

The x-axis defines the voltage ΔV measurable at PV cell poles, while the y-axis reports the current density (in mA/cm^2) that a single cell can feed when the circuit is close on an external utilizer.

Data show that open circuit voltage values ($\Delta V_{0,}$ $J = 0$, sometimes also referred to as ΔV_{OC}) are not particularly variable with F (ranging from about 0.57 to 0.65 for the represented cell).

Maximum "saturation" current density (J_{Sat}, ΔV=0, sometimes also referred to as "short-circuit", J_{SC}) increase linearly with F. This means that a PV cell basically functions as a real voltage generator.

The plotted curves remain nearly horizontal up to a specific voltage value, indicating that the cell is operating in a saturated state, feeding the maximum number of

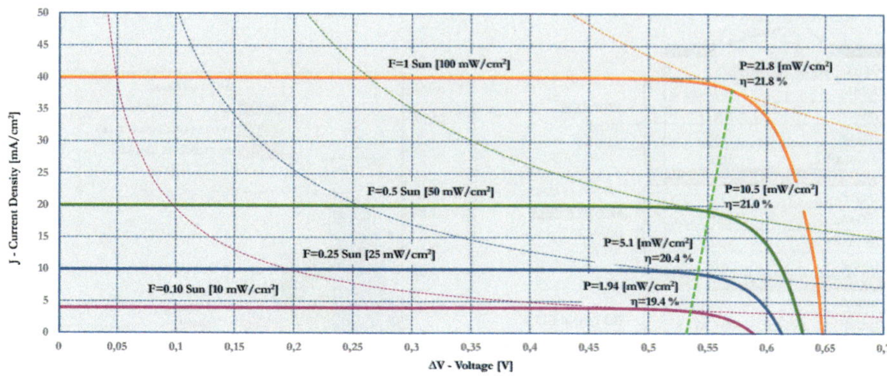

FIGURE II.2.12 Typical characteristics of single-junction solar cells.

photo-generated electrons per square centimetre that photons can excite at a given F value.

In the final portions of the curves, the current density J decreases linearly with voltage (ΔV), following the relationship:

$$\Delta V = \Delta V_o - r \cdot J$$

where r is the internal cell resistance for unit surface. In that part of the curve losses are proportional to J^2 according to the Joule law.

In the remaining part of the curves, the voltage losses are more than linear with J, due to the fact that the cell is dispensing a specific current which is approaching its saturation values.

The power output fed by the cell (for unit surface, in [mW/cm^2]) is given by the product $\Delta V{*}J$ of each possible operating point.

Among these, the maximum power output is reached somewhere within the knees of the single curves.

Maximum power points (normally referred to as MPPs) are also the highest efficiency points. Varying F, a maximum efficiency operating curve may be plotted (green dotted line in Figure II.2.12). This curve is usually almost linear, and can be expressed as:

$$\Delta V_{MPP} = \Delta V^*_{MPP} + r^* \cdot J$$

The maximum efficiency reaches 21.8 for the cell in this figure, and decreases only slightly to 19.4 for $F = 0.1$ Sun.

According to theoretical considerations, a single cell characteristic curve can also be normally approximated by an analytical formulation of the form:

$$J = J_{Sat} \cdot F - J_o \cdot \left(e^{\frac{\Delta V}{\Delta V_T}} - 1 \right)$$

where:

- J_{Sat} is the saturation current density of the whole cell for $F = 1$. For the example in Figure II-2.12, $J_{Sat} = 40\,\text{mA}/(\text{cm}^2 \cdot \text{Sun})$.
- J_0 is the "diode" saturation current, which depends on the type of cell, the doping density, as well as the quality of the semiconductor and the size of the p–n junction itself. For the example in this figure, $J_0 = 510^{-10}\,\text{mA}/\text{cm}^2$.
- $\Delta V_T = kT/q$ is the so-called "thermal voltage", where k is the Boltzmann constant (1.38065 10^{-23} J/k), T is cell operative temperature in K; and q is unitary electron charge (1.620218 10^{-19} C). For the example in this figure, $T = 300\,K$.

Solar cell performances are also sensitive to their operational temperature. In fact, both ΔV_{OC} (more) and J_{Sat} (less) decrease a bit as temperature increases: normal values for this decrement are, respectively, around 0.2–0.5%/K for ΔV_{OC} and 0.02–0.1%/K for J_{Sat}.

Due to the conversion losses within the cell, a significant amount of heat is generated. Under normal operating conditions, cell temperatures can reach 60–70°C, especially in summer. This contributes to additional performance losses.

To account for this, manufacturers often specify two sets of operative conditions to measure device performance:

- **STC, standard testing conditions**
  These values are prescribed by International Standards ISO 60904-3: Irradiance $F = 1$ Sun with a nominal Sun Spectrum (usually referred as "AM 1.5"); cell temperature 25 ± 2°C; no forced external air flow above the cell.
- **NOCT, normal operational cell temperature testing conditions**
  These values are defined autonomously by the manufacturer. Typical values may be: Irradiance $F = 0.8$ Sun; AM 1.5; air temperature: 20°C; wind speed 1 m/s; cell temperature 50 ± 2 °C.

II.2.2.5 Solar Generators Design and Expected Performances

As previously discussed, solar energy density is relatively low. To achieve significant levels of electric power generation (ranging from kilowatts to megawatts or more), extensive solar surfaces must be interconnected. But how is this accomplished? And how large can individual cells be?

Single mono-crystalline solar cells are typically standardized to a maximum circular surface area of 100 cm² but are often smaller. They are commonly cut into square or rectangular shapes for efficient placement side-by-side, allowing larger surfaces to be covered uniformly. For example, a single 64 cm² (8×8 cm) square solar cell has a maximum power output of about 1.3 W under $F = 1$ Sun, with an approximate voltage of 0.55 V and a current output of 2.4 A.

To increase the output, multiple cells are connected together. For instance, 200 cells could be connected to cover a total surface area of 1.28 m², producing an approximate output of 260 W under $F = 1$ Sun. These cells are usually configured to produce standardized voltage outputs that align with the requirements of typical DC devices, such as storage batteries.

Batteries are commonly standardized at 12 V or 24 V, and the solar PV system's output is typically designed to slightly exceed these values (e.g., 14 V or 28 V) to ensure charging for any possible solar radiation flux *F*. To achieve this:

- 25 cells connected in series produce approximately 14 V (for 12 V battery charging).
- 50 cells connected in series produce approximately 28 V (for 24 V battery charging).

These connected cells form modules, which are then assembled into larger configurations called PV panels. Panels generally have a rectangular shape, with an area of approximately 1.5 m^2 and a weight that allows for easy installation on roofs or the ground. These panels can be installed on fixed mounts or in sun-tracking devices, which enhance energy capture.

The output of a PV panel typically ranges from 150 W to 300 W, depending on the number, size, and efficiency of the individual cells.

For standardized voltage outputs around 14 V or 28 V, the nominal current output of a 12-V panel typically ranges from 12 to 24 A, while for 24-V applications, it ranges from 6 to 12 A. The total number of panels and their configuration depend on the size of the surface area available for installation. Panels that are mounted with the same orientation and receive a similar solar flux (*F*) are usually connected in parallel first, forming PV panel strings. This arrangement increases the generator's nominal current output proportionally to the number of panels connected in parallel.

To minimize Joule losses, the nominal current in a string is usually kept below 100 A. Consequently, strings typically consist of four to eight panels for 12-V configurations, resulting in total power outputs between 600 W and 2400 W. For 24-V configurations, strings often include 8–16 panels, providing total power outputs between 1.2 kW and 4.8 kW. In larger power generators, these strings are connected in series to further increase the overall voltage output.

For instance, a 44.5-kW PV generator can be designed by connecting 20 strings of eight panels each in series. Each panel in this setup is composed of four modules of 50 cells, with each cell measuring 64 cm^2, as previously described in Figure II.2.12.

The performance of the entire generator can be calculated step-by-step from the performance of its individual components. To this aim:

- <u>Single 64 cm^2 cell performance</u> is obtained simply by multiplying the J-axis by 64 and obtaining an I-axis expressing absolute current values in mA.
- The performance of the <u>50-cell module</u> is obtained by multiplying 50 × "Voltage" axis, with the cells being connected in series.
- The performance of the four-<u>module PV panel</u> is obtained by multiplying 4 × the "Current" axis, with the cells being connected in parallel.

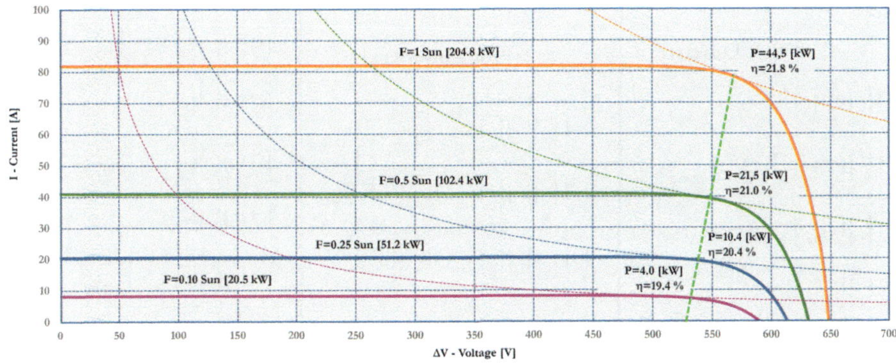

FIGURE II.2.13 Characteristic curves of a 42.4-kW nominal power output PV generator.

- The performance of a single eight-panel PV string is obtained by multiplying by 8 × the "Current" axis, with the cells being connected in parallel.
- The performance of the whole 10-string PV generator is obtained by multiplying 20 × the "Current" axis, with the cells being connected in series.

Summarizing the whole calculation, the y-axis must be multiplied by 64 × 4 × 8 = 2048, while the x-axis is multiplied by 50 × 20 = 1000. The overall number of single cells is 32,000, with an overall active surface of 204.8 m^2.

The overall characteristic curves of the resulting PV generator are reported in Figure II.2.13.

II.2.2.6 Other Typical Components of a Solar PV Power Plant

A solar PV power plant consists of more than just the PV generator. Several other essential components ensure proper operation, energy conversion, and grid compatibility.

- **DC/AC inverter**
  Since a PV generator typically produces DC electricity, a DC/AC inverter is required to convert the output into AC electricity. The AC power can then be used for local electric loads in stand-alone systems or fed into the grid through electricity metering devices.
  The inverter is designed to:
  - Convert direct current produced by the PV generator into alternate current, making its voltage and frequency compatible with local grid specifications.
  - Optimize PV generator operation, maximizing its energy yield, making it work as close as possible to its MPPs. This function is normally referred to as the maximum power points tracker, MPPT.
  - Monitor the entire system and ensure safety conditions.

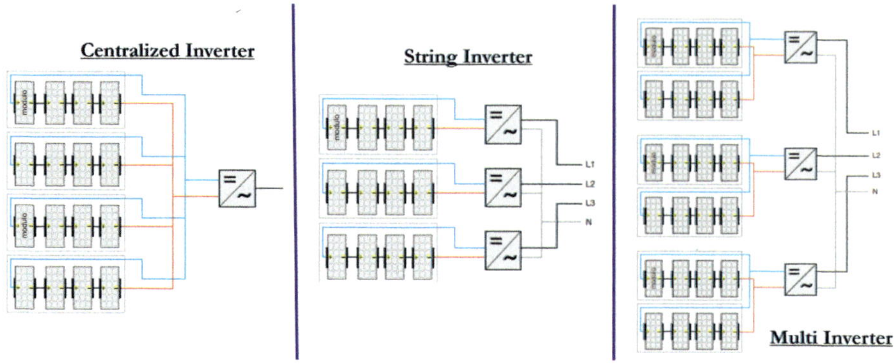

FIGURE II.2.14 Typical inverter configurations in system PV plants.

There are three main inverter configurations for connecting the strings that constitute a photovoltaic system:

- **Centralized inverter configurations**
  A single inverter manages the entire system (mono-inverter or centralized inverter system), as shown on the left-hand side of Figure II.2.14.
  This configuration is simple and cost-effective, making it suitable for small systems with modules of the same type and exposure. However, a failure in the inverter will halt energy production across the entire plant.
- **String-inverter configurations**
  In this setup, each string has its own inverter, as shown in the centre of Figure II.2.14. This approach is commonly used in medium-sized systems. It allows each string to operate independently at its own maximum power point, improving overall system performance. It also reduces issues arising from shading, differing exposures, or using modules with varying characteristics.
  This configuration enhances both performance and reliability.
- **Multi-inverter configurations**
  For large-scale systems (nominal power greater than 10 MW), multiple strings are connected to a single inverter (multi-inverter system), as shown on the right-hand side of Figure II.2.14. The photovoltaic field is divided into subfields, each served by its own inverter.
  This configuration reduces the total number of inverters required, lowering both investment and maintenance costs.

The inverter should ensure PV plant operation in all environmental conditions. To this aim, it must be verified that the maximum no-load voltage $\Delta V_{OC}\left(T_{min}\right)$ output from the PV strings at the minimum external temperature (–10°C) is lower than the maximum voltage that the inverter can withstand at its DC inlet at open circuit, $\Delta V_{OC_{max}}$:

$$\Delta V_{OC}\left(T_{min}\right) \leq \Delta V_{OC_{max}}$$

It is also necessary to verify that in all possible operating conditions (commonly –10°C to +70°C), the inverter operates within the voltage range declared by the manufacturer, corresponding to verify the following two conditions:

$$\Delta V_{MPP}\left(T_{max}\right) > \Delta V_{min}, \qquad \Delta V_{MPP}\left(T_{min}\right) < \Delta V_{max}$$

where:

$\Delta V_{MPP}\left(T_{max}\right)$ is the PV field output voltage at maximum temperature, and irradiance.
$\Delta V_{MPP}\left(T_{min}\right)$ is the PV field output voltage at minimum temperature and irradiance.
ΔV_{min} is the minimum operating voltage of the inverter MPPT.
ΔV_{max} is the maximum operating voltage of the inverter MPPT.

Figure II.2.15 illustrates the coupling between the PV field and the inverter, demonstrating how the conditions described above are met.

- **<u>Battery energy storage systems (BESS)</u>**
  Electricity generation from PV systems is non-programmable, meaning it doesn't necessarily align with grid demand. As renewable energy sources penetrate further into the electricity sector, this mismatch will become more significant. To address this, PV systems increasingly include battery energy storage systems (BESS) to store surplus electricity locally for later use.

  Electrical energy can be stored in large superconducting coils or converted into other forms of energy: kinetic energy stored in flywheels or compressed gases, gravitational energy in water reservoirs, chemical energy in synthetic fuels, and electrochemical energy in electrical accumulators (batteries). Among these, electrochemical batteries offer the highest level of technological maturity and are the most widely adopted in PV systems.

  Therefore, PV generators are increasingly required to be complemented by local BESSs.

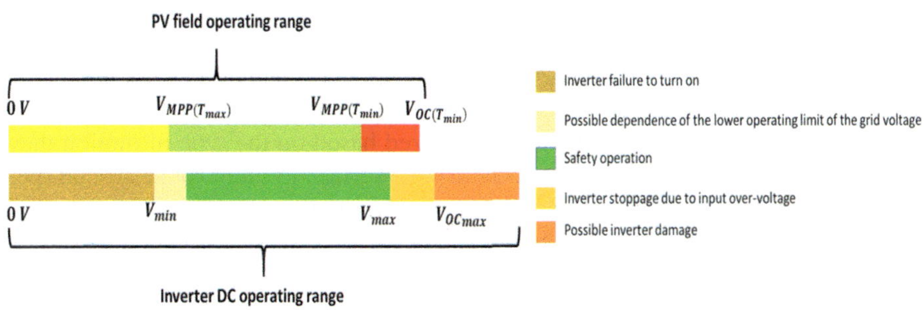

FIGURE II.2.15 Graphical representation of the coupling between the PV field and the inverter.

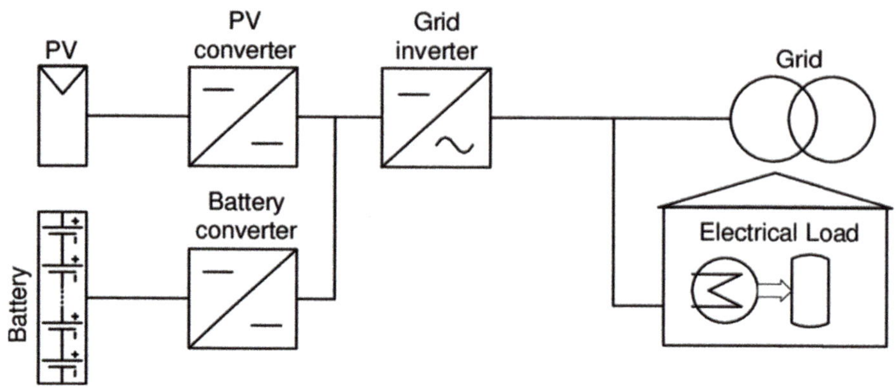

FIGURE II.2.16 Typical components of a PV plant and its grid connection scheme.

A sketch of the component normally included in a PV plant, together with their typical connection scheme, is reported for reference in Figure II.2.16.

Correct dimensioning must be done to efficiently couple the BESSs and the PV generator. This design phase, in some cases, can also lead to a direct parallel coupling of the PV generator and the BESSs, without the interposition of one or two AC/DC converters. In fact, a BESS has a charging characteristic curve which may be designed to fit very closely with the MPP curve of the PV generator.

II.2.2.7 Sun-Tracking Devices

The orientation of PV panels can be adjusted to track the Sun, keeping the surface as perpendicular as possible to the incoming solar beams. By maximizing solar irradiance, sun-tracking systems enable PV devices to produce more energy, thereby increasing their capacity factor.

While sun-tracking devices can enhance energy yields, they also increase the plant's CAPEX due to the cost of the tracking system itself.

The impact on OPEX varies depending on the type of tracking system installed; some systems may require significant maintenance, while others incur minimal additional costs. The overall effect on the plant's LCOE depends on both economic and site-specific considerations, as energy yield improvements may or may not justify the added investment.

Possible solar-tracking systems are categorized based on the type of motion they enable. These systems allow the photovoltaic (PV) surface to tilt or rotate around specific axes to achieve an optimal angle for capturing maximum sunlight.

When the PV surface rotates around a single axis, the system is called a "single-axis tracker". In contrast, when the movement occurs around two axes simultaneously, it is known as a "dual-axis tracker".

Single-axis trackers offer several benefits and are nowadays more often used than dual-axis trackers. Among these benefits are the following:

- These trackers are cost-effective, easy to set up, and have low operating costs.
- Single-axis systems are more reliable and durable.
- They generally have a longer lifespan than dual-axis systems.
- They are well-suited for regions with lower sunlight exposure.
- They may provide about 30% higher energy yields compared to fixed panels.

Most used single-axis Sun-trackers are:

- **Horizontal Single-Axis Solar Tracker (HSAT)**
  HSAT rotates east to west throughout the day on a fixed horizontal axis parallel to the ground. It is considered one of the most cost-effective configurations for PV applications. This system requires minimal construction material, as it is supported by multiple structures along the rotating axis. Its simplicity makes it a popular choice among tracking geometries.
- **Vertical Single-Axis Solar Tracker (VSAT)**
  These trackers rotate on a vertical axis and can be mounted in either a north–south or east–west orientation to better follow the Sun's vertical movement. They are commonly used in high-altitude or extreme-latitude locations.
- **Vertical-Tilted Single-Axis Solar Tracker (VTSAT)**
  VTSAT systems feature a tilt that aligns horizontally while rotating on a vertical axis. They are more efficient at generating energy compared to horizontal trackers. However, they are subject to higher wind loads, require a robust structure, and normally involve additional costs for concrete and steel foundations.

II.2.2.8 Low-Cost PV Devices to be Integrated in Building Materials

An exciting area of research and development in the photovoltaic (PV) sector focuses on integrating PV materials into conventional building construction materials and devices. This approach aims to seamlessly incorporate energy-generating capabilities into the built environment, enabling structures to contribute directly to renewable energy production.

Recent advancements have led to the launch of several innovative applications, including PV roof tiles, PV glass, and PV facades. These solutions allow buildings to generate electricity while maintaining their primary functions, such as providing shelter, insulation, or natural lighting.

The primary goal of these developments is to ensure that the cost of integrating PV materials into construction remains affordable. This means that the PV materials used in these applications must be inexpensive enough to keep the additional cost of the building components within acceptable limits.

In such applications, the efficiency of individual PV cells is not a critical factor. Instead, the focus is on ensuring that the cells are highly adaptable for installation and that they can be easily connected electrically to form larger, efficient systems. Flexibility in design and integration is a key requirement, as it allows the PV materials to meet the diverse demands of construction projects without compromising their energy-generating capabilities.

II.2.2.9 High-Efficiency Concentrated Multi-Junction PV Devices

An alternative development path in the PV sector focuses on the introduction of high-efficiency multi-junction cells. Unlike low-cost solutions, these cells are designed for maximum energy conversion efficiency. However, their high cost – often tens of times more than a silicon cell of the same size – limits their use to applications requiring smaller installation surfaces to control overall costs.

The only viable way to use multi-junction cells cost-effectively is to concentrate sunlight from a much larger area onto the smaller, high-performance cells. This is typically achieved using lens-based concentrating devices. (Glass-based concentrating devices, commonly used in solar thermodynamic applications, will be discussed later.)

Concentrating devices are designed to focus light accurately, but their effectiveness depends on the direction of incoming sunlight. Because of this, diffuse light, which constitutes a significant portion of sunlight on cloudy days or in humid climates, is often lost. To minimize these losses, concentrators are generally installed in dry climatic zones where direct sunlight predominates.

Additionally, solar concentration systems must be paired with tracking devices to maintain precise alignment with the Sun. As the concentration ratio increases, the tracking devices need to be increasingly accurate, adding to the system's complexity and cost.

Concentrating sunlight onto multi-junction cells can intensify solar input to hundreds of Suns. While this boosts electrical output, it also significantly increases thermal losses per unit surface area. To address this, multi-junction cells require active cooling systems, typically using water as a coolant. This design not only prevents overheating but also transforms the plant into a co-generative system, producing both electricity and thermal energy. The thermal energy in the coolant fluid, typically around 50°C, can be used for various local applications, such as heating or industrial processes.

Figure II.2.17 provides photos and a schematic representation of a concentrated multi-junction PV device. These systems can increase the capacity factors of the cells

FIGURE II.2.17 Photos and sketch of the basic behaviour of a concentrated multi-junction PV device.

by approximately 50%. The electric efficiency of the cells can reach around 35%, while an additional 40% co-generative thermal energy is made available through the coolant fluid. This combination significantly enhances the overall energy yield of the system.

II.2.2.10 Case Study: Description of a Real Plant in Operation

This subsection describes a real case study of a developing PV plant located in Lariano, Rome, Italy.

Data are kindly provided by E.S.C. (Energy System Consulting, an engineering company located in Avezzano, L'Aquila, Italy). The objective of this subsection is to briefly summarize the main engineering choices for the correct dimensioning of a PV plant.

The layout relating to the arrangement of the PV modules on the ground was developed following an accurate micro-siting study. Firstly, a clinometric analysis was conducted through a global clinometric survey to determine the loss of producibility of the PV system in relation to the horizon visible from the system itself. Clinometric shading is caused by the presence of hills, mountains, trees, and further obstacles located at a sufficiently large distance from the PV field, which is why, with good approximation, it can be assumed that their effects are uniform for each module and for each string.

Table II.2.1 reports the key features of the designed plant.

A local clinometric analysis was then conducted to evaluate the loss of producibility of the PV system due to mutual shading between the PV modules positioned on adjacent structures, as well as to the possible contribution of obstacles present within the system area.

TABLE II.2.1
Main Characteristics of Lariano PV Plant

Municipality	Lariano (Rome)
Total extension	11,909 m^2
Number of PV modules	1328
PV module peak power	710 W
Total active PV modules surface	3577 m^2
Total peak power	943 kW
Estimated annual production	1339 MWh
Inverters number	3
Inverter type	Multi-string
Module orientation	3-landscape
Tilt angle	30°
Distance N–S	7.5 m
Ground minimum height	0.8 m
Ground maximum height	2.8 m

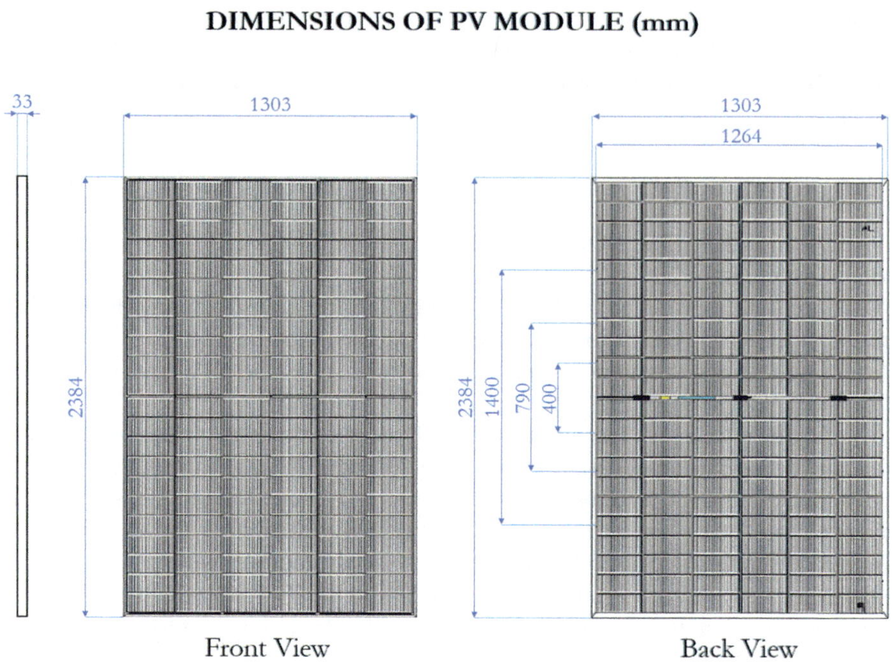

FIGURE II.2.18 Dimension of the single PV module (mm) of Lariano PV Powerplant.

TABLE II.2.2
Power Distribution among Modules and Strings

Inverter	Modules per string	Strings	Modules	Module power (W)	AC power (kW)
1	28	16	448	710	320
2	28	16	448	710	320
3	27	16	432	710	307

Following the above analyses, the modules were arranged in a three-landscape configuration, with a tilt angle of 30° from the horizontal. The modules are oriented towards the south, in order to optimize exposure to solar radiation.

The chosen PV modules are bi-facial type and single-crystalline silicon, with a peak power of 710 W each (see Figure II.2.18).

The PV generator consists of 1328 modules, electrically distributed into strings of 28 or 27 components in series, connected to three 320-kW inverters. Table II.2.2 summarizes the power distribution among modules and strings for proper inverter selection.

The correlations and calculations for inverter dimensioning were verified to ensure proper operation.

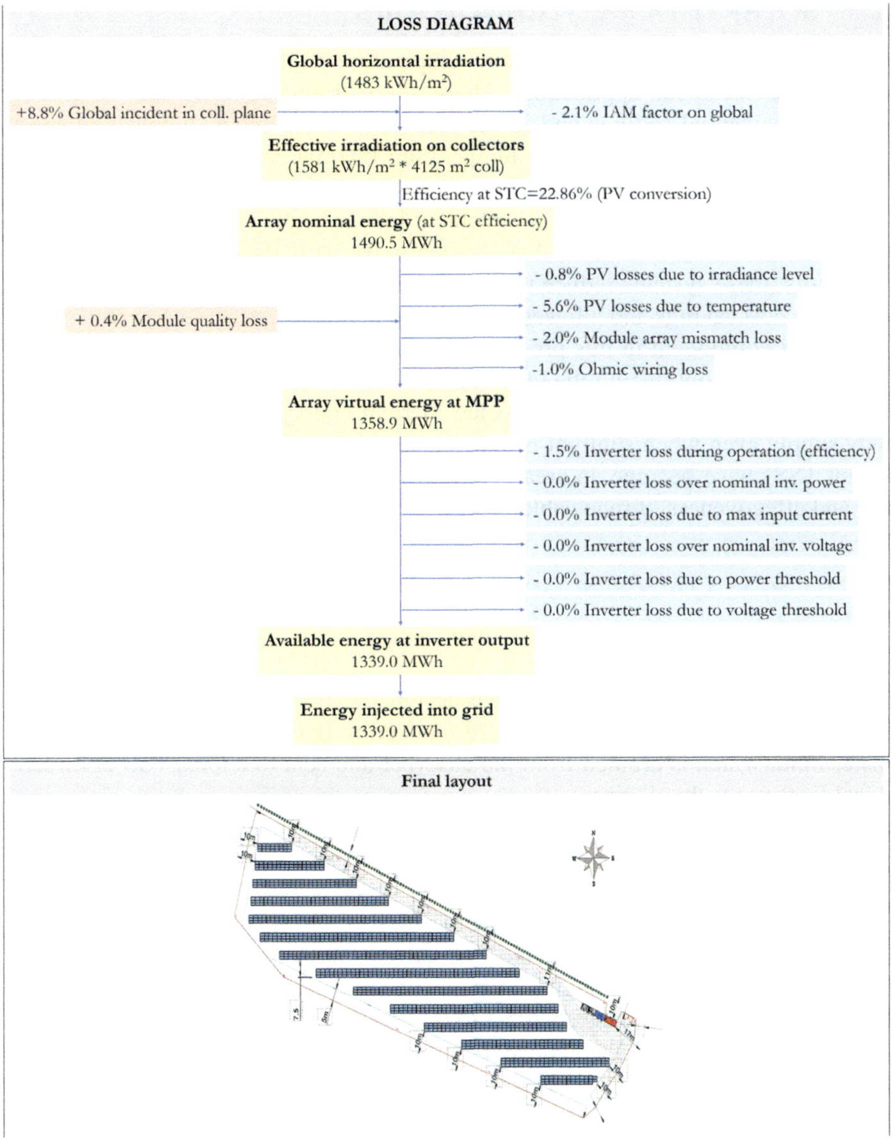

FIGURE II.2.19 PV plant loss diagram and final layout.

The photovoltaic generator is anchored to the ground by 14 fixed structures in a three-landscape configuration, with a north–south pitch of 7.5 m. These structures provide adequate stability and ensure optimal performance across varying environmental conditions.

The expected energy yield of the plant was simulated using PVsyst software, estimating an annual production of 1339 MWh. The energy flow data, along with the system's loss diagram and final layout, are illustrated in Figure II.2.19.

II.2.3 SOLAR THERMAL TECHNOLOGIES

The use of solar energy for heating low-temperature fluids represents a simple and cost-effective technical application. These thermal technologies are commonly utilized in residential settings and in facilities such as hospitals, hotels, and sports centres, where the demand for low-temperature thermal energy can be significant. Typical applications include domestic water heating and swimming pool heating.

Solar thermal energy can also complement indoor heating systems by serving as a preheater for the fluid used in these systems. By partially heating the fluid with solar energy, these systems reduce the consumption of electricity and/or fossil fuels, resulting in cost savings and a lower environmental impact.

A critical component in this process is the thermal storage system (TSS), which plays a key role in optimizing the integration of solar thermal energy with traditional heating systems. TSSs store the collected thermal energy for later use, ensuring a steady supply even when sunlight is unavailable. The design and implementation of efficient TSSs have become an area of increasing interest, as they enhance the viability and effectiveness of renewable energy integration.

II.2.3.1 Solar Collectors: Design and Types

A solar collector captures the Sun's radiant energy and transfers it to a heat transfer fluid, usually water.

To achieve this, an absorbing surface is designed to maximize solar energy absorption. Within the absorber, fluid paths are realized to permit its cooling by the heat transfer fluid, which is drained from the absorber and sent to a local TSS or directly to local final-use technologies.

The absorbing material is chosen depending on a combination of: cost constraints; ease of processing in realizing internal fluid paths; and high thermal conductivity towards the cooling fluid.

The choice of absorbing material depends on several factors, including cost, ease of manufacturing fluid pathways, and thermal conductivity. Energy losses in solar collectors occur primarily through convection and radiation between the absorber and the environment. Insulating the absorber to reduce these losses is challenging because excessive insulation could also reduce the incoming energy absorbed.

The entity of these losses (and therefore the efficiency of the absorber) therefore depends on the magnitude of convection and radiation losses. The convective losses linearly depend on the difference ($T_{abs} - T_{ext}$) between the temperature reached by the absorber and that of the external environment, while radiative losses depend on the fourth power of the absorber temperature.

Absorber temperature defines the maximum temperature which could be reached by the cooling fluid T_{out}, and therefore the "quality" of the thermal energy made available for final local uses.

This means that the efficiency of any given technology is expected to decrease as much as the cooling fluid temperature is increased. These energy losses (and their effect on device efficiency) obviously depend on the level of insulation realized between the absorber and the external environment. They are almost linear for low

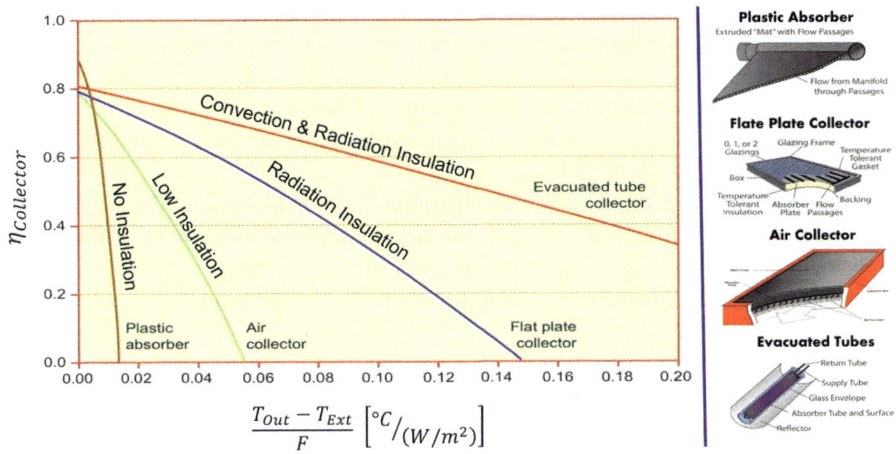

FIGURE II.2.20 Collectors' efficiency as a function of: insulation level; irradiance; and desired T_{out}.

values of $T_{out} - T_{ext}$ but tend to become more linear as the temperature of the cooling fluid is increased, since the radiative losses become more relevant.

Moreover, for a given device the coolant flow rate for unit surface is fixed. Therefore, according to simple energy balances of the devices, at a given efficiency, $T_{Out} - T_{Ext}$ would be linearly dependent on the sun flux F hitting the unit surface of the device.

Figure II.2.20 illustrates efficiency trends for different solar collector types, plotted against the ratio $(T_{Out} - T_{Ext})/F$.

Four curves are plotted in this figure which are relative to the four different collector types sketched on the right-hand part of the figure:

- Plastic absorber collectors, having practically no insulation, resulting in lower efficiency.
- Air collectors, low insulated collectors, with slightly better performance.
- Flat plate collectors, having a glass cover preventing re-irradiation from the collector. Those are the most commonly used collectors.
- Evacuated tube collectors, having a partially air-evacuated zone preventing convection towards the environment.

A value of 0.1 for the x-axis means a value of 100°C for $T_{Out} - T_{Ext}$ if F has its nominal value of 1 Sun (1 kW/m^2).

According to the data shown in the figure, for a nominal solar irradiance of 1 Sun, and minimally acceptable 70% efficiency conditions:

- Plastic absorber collectors can only reach a value of T_{Out} of 12–14°C higher than the external temperature.
- Air collectors can reach a T_{Out} of 20–22°C higher than the external temperature.

- Flat plate collectors can reach a T_{Out} 48–50°C higher than $T_{Ext.}$
- Evacuated tube collectors can reach a T_{Out} 90-–2°C higher than $T_{Ext.}$

To obtain these values we entered a given value of minimal efficiency (0.7) in the figure on the y-axis, determining the curves for different values for x (and therefore for T_{Out}, for given values of F and T_{Ext}).

In contrast, entering the graph with an x-axis value, for example 0.02 (which would mean a T_{out} of 45°C, if T_{Ext} = 25°, summer conditions, and F = 1 Sun), using a flat plate collector would determine an efficiency of about 0.72.

A simple energy balance can be stated for the heat transfer fluid:

$$F \cdot \eta_{collector} = Q_{H_2O} \cdot C_{p,H_2O} \cdot \left(T_{Out} - T_{Ext}\right)$$

where C_{p,H_2O} = 4.186 kJ/(l K), and Q_{H_2O} is the volumetric fluid flow rate.

It can be deduced that this would mean producing at most around 516 cl/min/m^2. At nominal operative conditions, approximately 1.93 m^2 of collectors would be needed to obtain 1 l/min of hot water at 45°C.

In summer, about 6 hours of equivalent peak irradiance are normally experienced, so that each m^2 could produce about 180 litres of hot water at 45°C (which is fine for hot domestic water [HDW] purposes), and coherent with the daily summer needs of a couple of domestic users.

If a greater temperature value is needed for the heat transfer fluid, the x-axis value increases. As an example, entering 0.04 would mean a T_{out} of 65°C, if T_{Ext} = 25° and F = 1 Sun.

The curves in this figure show that only three technologies could produce that x value, but with different efficiency values:

- Air collectors: in these conditions, would have an efficiency of about 0.3, so they would transfer 300 W/m^2 to the coolant. Approximately 9.3 m^2 of collectors would be needed to obtain 1 l/min of heat transfer fluid at 65°C.
- Flat plate collectors: they would have an efficiency of 0.65. Therefore, 4.3 m^2 of collectors would grant the same output performance (1 l/min of heat transfer fluid at 65°C).
- Evacuated tube collectors: their efficiency would be around 0.72. Therefore, 3.9 m^2 of installed devices would be enough.

These ratios are not linear. For example, if we need to produce a 100°C heat transfer fluid, for an external temperature of 25°C at F = 1 Sun, we should enter the graph with an x-axis value of 0.075. Only two technologies would possibly produce this output:

- Flat plate collectors would have an efficiency of 0.44; 11.9 m^2 of collectors would be needed to obtain 1 l/min of heat transfer fluid at 100°C.
- Evacuated tube collectors would have an efficiency of 0.64; 8.2 m^2 of collectors would be enough.

Lastly, if we want to produce a 150°C heat transfer fluid, the x-axis would be 0.125:

- Flat plate collectors' efficiency would only be 0.15; 58.1 m^2 would be needed to obtain 1 l/min of heat transfer fluid at 150°C.
- Evacuated tubes would instead maintain a "reasonable" efficiency of 0.52, so that 16.8 m^2 of collectors would be enough.

As we will see in the following paragraphs, much higher temperature values can be reached on the heat transfer fluid if the F values hitting the absorbing material are increased.

However, much higher insulation levels should be obtained so as not to lower efficiency.

With this aim in mind, various concentration and collecting technologies can be used, all using much more costly approaches and technologies, but with a completely different objective, since, in this case, the heat transfer fluid could be used to feed a high-efficiency thermo-electric power plant.

Let us now give some more details about the four different collector technologies already introduced:

- **Plastic Absorber Collectors**
  These only use plastic materials, limiting costs and eliminating corrosion risks. They are mostly used for pool-heating applications. Pool water circulates directly through the collectors without additional heat exchangers. They only produce low heating values (5–10°C more than the external temperature). Water is naturally circulating in the device (no pump is needed) due to water density changes with temperature.
- **Glazed Collectors**
  Both air collectors and flat plate collectors are normally glazed, but flat plate collectors are water cooled. They are slightly more costly but have much higher efficiency. As already said, they are the most commonly used amongst all solar thermal technologies. They consist of a modular container with a glass cover (typically about 1 m^2 in size). Collectors are similar in size to PV panels and are oriented and tilted at a fixed angle for maximum sunlight exposure. Water flows inside the absorber through U-shaped tubes via natural circulation. The absorber container is insulated laterally and at the back to minimize heat loss, while the glass cover mostly acts on reducing radiation losses through its "greenhouse effect" (being mostly transparent to incoming sunlight, but mostly opaque to the infra-red light re-emission of the absorber).
- **Evacuated Tube Collectors**
  This advanced technology involves placing the absorber material in a partially evacuated tube to reduce convective heat transfer mechanisms. Mirroring surfaces on the back plate redirect all incoming radiation onto the absorber material. This design significantly raises the process fluid's temperature, making it ideal for applications requiring temperature differences of more than 30–40°C above ambient, such as radiant heating or solar air-conditioning

systems. Their main application is in integrating building HVAC applications (indoor heating devices and/or conventional or absorption heat pumps for so-called "solar cooling" applications).

II.2.3.2 Solar Thermal Plants Configurations

Solar collectors are the main element of a solar thermal plant, but they must be complemented by a utilization circuit to transfer and manage the heat transfer fluid. Utilization circuits could also include pumps and/or circulating devices, TSSs and simple controllers depending on installation conditions and applications.

Common system configurations are normally divided into three categories (see Figure II.2.21):

- **Natural Circulation Systems**
  These only rely on density differences to move the heat transfer fluid. Normally, the utilization circuit is small in extension and mostly horizontal. A TSS is normally mounted directly on the top of the collector to be filled with the hotter lower density fluid.
- **Forced Circulation Systems**
  These use an external circulator to move the heat transfer fluid. All the pipes are slightly pressurized, so permitting the installation of the TSS many metres below the collector (i.e., collectors on a building roof and TSS in the basement). The circulator is controlled by quite simple on–off controllers sensing the temperature and level of the fluid inside the TSS. Solar collectors cannot be emptied during winter. Glycol/water mixtures are normally used to prevent the collector from freezing.
- **Drain-Back Systems**
  These use an external pump to move the heat transfer fluid. Pipes are not pressurized, so they permit emptying the collector in winter, preventing it

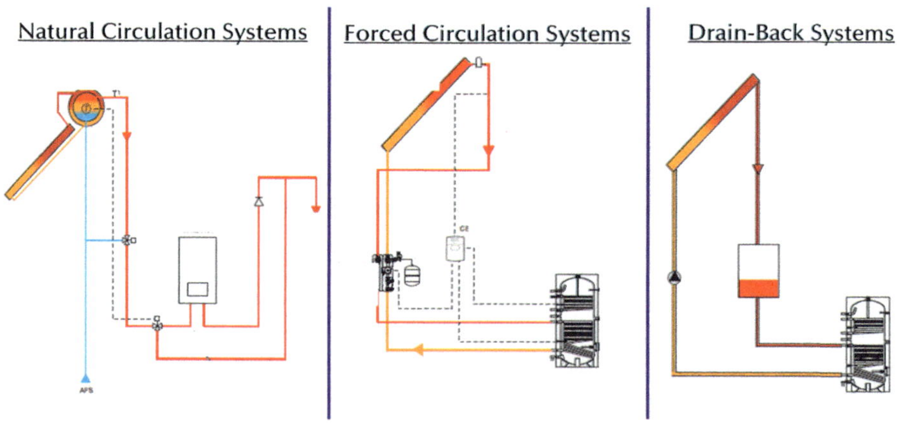

FIGURE II.2.21 Different plant configuration for solar thermal systems.

from freezing. In order not to use too much energy for pumping actions, the TSS is normally installed just under the roof of the building where the collectors are installed. The pump is controlled by simple on–off controllers sensing temperature and level of the fluid inside the TSS.

II.2.4 CONCENTRATED SOLAR THERMODYNAMIC TECHNOLOGIES

As previously mentioned, significantly higher temperatures can be achieved in the heat transfer fluid by increasing the solar flux (F) incident on the absorbing material. However, to maintain efficiency, these systems require much higher levels of insulation to minimize energy losses.

To this aim, various concentration and collecting technologies can be used, all using much more costly approaches and technologies, but with a completely different objective, since, in this case, the heat transfer fluid could be used to feed a high-efficiency thermo-electric power plant.

These plants would therefore be concentrated solar thermodynamic electric power plants, but are normally simply referred to as CPS plants.

CSPs are broadly categorized based on the way they collect and concentrate sunlight. Each category has unique characteristics, advantages, and limitations.

In the following, a brief overview of the main types is reported, referring to Figure II.2.22.

II.2.4.1 Parabolic Trough Systems

These systems use long, curved mirrors (parabolic troughs), about 4–5 metres wide and 20–25 metres long, which function as mono-axial solar concentrators to focus sunlight onto a receiver tube running along the focal line of the trough. A heat transfer fluid (commonly synthetic oil or molten salt) flows through the tube, absorbing the heat.

Their key characteristics are the following:

- Efficiency: Moderate to high thermal efficiency (up to 30% conversion efficiency) depending on the heat transfer fluid used. This is chosen to remain liquid in the whole solar field part of the plant. Using so-called diathermic synthetic oils normally means that its lower temperature can be environmental but the highest reached temperature values are not so high, so that plant efficiency is normally limited below 20%. Using molten salts, instead, increases the highest temperature (and therefore plant efficiency up to 30%) and more but makes it necessary to maintain the temperature of the salt always above around 300°C. Below that temperature the salt would crystallize, creating several issues for the plant.
- Scalability: These are well suited for medium-to-large-scale installations. A single string of collectors is normally built by five or six collectors in series. The final desired outlet temperature, T_{Out}, is reached by the heat transfer fluid for nominal F conditions. The heat transfer fluid flow rate can be then increased

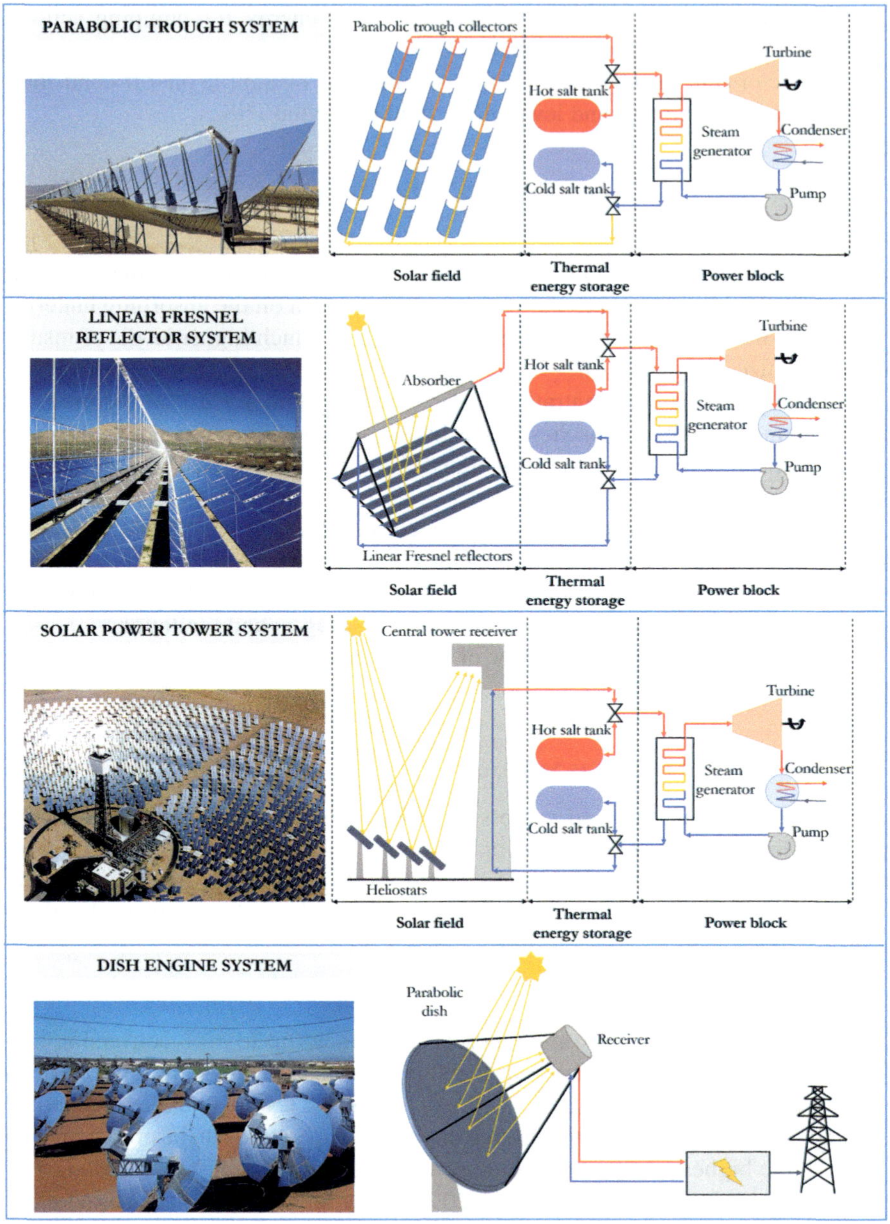

FIGURE II.2.22 Different plant configurations for concentrated solar plants.

as desired by connecting a few or several strings in parallel, depending on the size of the subaltern thermodynamic cycle.

- Energy Storage: They are often paired with molten salt storage for energy storage and delivery after sunset, increasing the capacity factor of the thermodynamic plant, and therefore also limiting its installed power and cost.
- Water Use: Depending on the installation site and plant size, water requirements for cooling in steam cycles may be significantly costly, though dry cooling options exist.
- Actual Installations: This is one of the most mature and widely deployed CSP technologies. SEGS plants in California, Shams 1 in the UAE, and ENEL Creen Power Plant in Priolo Gargallo in Sicily (Italy) are some examples of this technology.

II.2.4.2 Linear Fresnel Reflector Systems

These systems use flat or slightly curved mirrors (reflectors) arranged in a linear configuration all focusing sunlight onto a fixed receiver which is placed directly above the mirrors at a certain height. The receiver is an evacuated tube where the heat transfer fluid flows. Plant behaviour is similar, in fact, to parabolic trough systems.

The name of these reflectors comes from the French physicist and inventor Augustin-Jean Fresnel who, in the 19th century, proposed an effective technique to reduce lens size, making them practically flat.

Their key characteristics are the following:

- Cost-Effectiveness: Simpler and cheaper construction compared to parabolic troughs, but with lower efficiency.
- Compact Design: They have a more compact assembly, so they require less land area than other CSP systems.
- Water Use: Similar cooling needs than parabolic troughs, though dry cooling options exist.
- Suitability: Best f•r small-t•-medium-scale applicati•ns •r f•r industrial high-temperature heat pr•ducti•n.
- Actual Installations: Puerto Errado 2 in Spain.

II.2.4.3 Solar Power Towers (Central Receiver Systems)

These systems use a field of flat, sun-tracking mirrors (heliostats) to focus sunlight onto a central receiver located at the top of a tall tower. The concentrated heat is used to heat a fluid (e.g., molten salt) to generate steam.

Their key characteristics are the following:

- Efficiency: High temperatures (up to 800°C) enable greater thermal efficiency and improved power cycle performance.

- Energy Storage: They normally use molten salts as heat transfer fluid, so they have excellent storage capabilities due to the high heat capacity of molten salt. The difficulties in maintaining the salt liquid in the whole plants are much reduced with respect to parabolic trough plants.
- Land Use: Requires significant land for heliostat fields. Each heliostat must be mounted on a bi-axial tracker, increasing the costs and complexity of the technology, but they can be mounted also in non-flat sites.
- Flexibility: Capable of hybrid operation with fossil fuels or biomass.
- Actual Installations: Crescent Dunes (USA), PS10 and PS20 (Spain).

II.2.4.4 Dish/Engine Systems

These systems use small-size single parabolic dish-shaped mirrors (normally not more than 3–4 metres wide) to concentrate sunlight onto a receiver mounted at the dish's focal point. An air-fed thermodynamic Stirling engine is normally used to convert thermal energy into electricity. Each concentrator is a stand-alone electric generator, which can be connected in series and in parallel with all the others present on the same site.

Their key characteristics are the following:

- Efficiency: Around 30–35% at most.
- Modularity: Ideal for small-scale, decentralized power generation.
- Energy Storage: They may be coupled with TSSs but normally their energy storage capability is poor compared to other CSP systems.
- Land Use: Minimal land requirements, often suitable for off-grid remote locations.
- Actual Installations: Prototype projects such as the Big Dish in Australia.

II.2.4.5 CSP Technologies Comparison

Each CSP technology is suited to specific conditions and requirements. While power towers lead in efficiency and storage, parabolic troughs remain the most commercially established. Linear Fresnel systems and dish/engine designs, meanwhile, cater to cost-sensitive or niche applications.

Table II.2.3 reports a synthesis and brief comparison of the CSP technologies.

TABLE II.2.3
Comparison Table of the Various CSP which are Currently under Development

Category	Efficiency	Cost	Scale	Storage	Best applications
Parabolic	Mid-high	Mid	Medium	Good	Utility-scale plants
Linear Fresnel	Moderate	Low	Medium	Moderate	Utilities/heating
Solar tower	High	High	Large	Excellent	Utility-scale plants
Dish/engine	High	High	Small	Limited	Building/off-grid

In summary, solar energy technologies, including photovoltaics and concentrated solar power, are key to future energy solutions, especially considering the ample worldwide availability and distribution of the solar energy source. However, considering the very low solar energy density, increasing efficiency and reducing costs of the conversion devices is critical to expanding their role in global energy supply.

II.3 Wind Energy

II.3.1 SOURCE CHARACTERIZATION

The use of wind energy dates back centuries, with the earliest documented application being the Persian windmill around 650 BC. Over time, the utilization of wind energy has evolved significantly. Today, the focus has shifted to generating large-scale energy capable of meeting substantial portions of global energy demand and transporting it over long distances.

Modern wind energy systems aim to convert wind's kinetic energy into rotational mechanical energy using a shaft. While this mechanical energy can be used locally, its transport over long distances is inefficient. To address this, mechanical energy is now typically converted into electricity locally and fed into national electric grids via electric wind generators, making wind energy transportable and widely usable.

Wind energy originates from the movement of air masses, driven by differences in density, pressure, and temperature. These differences arise from meteorological and climatic phenomena, ultimately caused by uneven solar irradiation striking the Earth's surface.

It is estimated that approximately 0.2% of the Sun's energy reaching the Earth is converted into the mechanical energy of winds (and partially into ocean currents). This amounts to an enormous energy potential of about 350,400 TW, which is over 20 times the current global energy consumption.

In some areas, intense solar radiation causes local air to overheat, reducing its density and creating low-pressure zones (cyclonic areas). Conversely, cooler regions form high-pressure zones (anticyclonic areas). The pressure gradient between these zones generates wind. The variability of solar radiation across time and geography further affects wind patterns.

While some regions experience persistent air circulation, others are subject to sporadic winds influenced by microclimatic phenomena and unique geomorphological features. Large water bodies, such as lakes, rivers, and seas, also play a significant role as thermal accumulators, further shaping local wind behaviours.

Persistent winds, which are stable over time and space, are particularly valuable for energy generation. They ensure a consistent energy supply, enabling the design

DOI: 10.1201/9781003631125-18

of optimized wind energy devices tailored to specific flow conditions. These devices achieve higher capacity factors. In contrast, variable winds are more challenging to harness efficiently due to their inconsistent nature.

II.3.1.1 On Wind Power Density

The kinetic power (P_{IN}) made available by a given mass flow rate ($\dot{m}$) of air moving at speed u is expressed by the following relation:

$$P_{IN} = \frac{1}{2} \cdot \dot{m} \cdot u^2$$

The mass flow rate ($\dot{m}$) can be expressed as the product of the cross-sectional area (S), air density (ρ), and wind velocity (u). Substituting this into the equation allows the specific available power ($\tilde{P}_{IN}/S$, the power per unit area of the wind-converting device) to be written as:

$$\tilde{P}_{IN} = \frac{P_{IN}}{S} = \frac{1}{2} \cdot \rho \cdot u^3$$

This equation highlights that the available wind energy depends on the cube of the wind speed, the cross-sectional area, and the air density. The air density can be estimated by considering air as an ideal gas:

$$\rho = \frac{R \cdot T}{P} \approx 1.225 \frac{kg}{m^3}$$

in typical environmental conditions (15°C, 1bar).

Air density fluctuates due to changes in pressure, temperature, and humidity. Pressure variations depend on geographical location and climate. For instance, tropical regions typically experience annual pressure variations of 5–10 mbar, while mid-latitudes see variations of 20–50 mbar, and high latitudes can have up to 50 mbar. These fluctuations translate to a mean variation in air density of approximately ±1–2% around the average value.

Temperature also plays a significant role, with yearly oscillations reaching up to 40°C in some regions. Such changes can noticeably affect air density. Humidity variations, although less impactful than pressure and temperature, also contribute to density changes.

When all these factors combine, air density generally fluctuates within a range of ±5–10%. However, in areas with extreme weather conditions, such as deserts or polar regions, variations may rise to 10–15%. Despite these variations, the influence of air density on wind power density predictions is typically considered negligible in first-order analyses.

Wind power density is therefore cubically dependent on wind speed. It reaches for example:

- 100 W/m^2 if wind speed is around 5.6 m/s (20.3 km/h)
- 1 kW/m^2 if wind speed is around 12.1 m/s (43.7 km/h)

These values demonstrate that wind power density rarely exceeds solar power density (whose nominal maximum value is 1 Sun = 1 kW/m^2).

The critical question becomes how frequently and consistently wind speeds can reach the high values required for dense, high-capacity energy applications. This consistency, both in time and space, directly determines the feasibility and efficiency of wind energy exploitation.

II.3.1.2 On Wind Anemometry

The question of how frequently and consistently wind speeds reach the values required for energy applications can only be addressed statistically, using data from anemometric campaigns. These campaigns collect wind speed and direction data over extended observation periods at specific sites, organizing the information into joint frequency distribution functions (JFFs).

Figure II.3.1 illustrates a heatmap derived from a typical JFF, representing wind distribution at a specific site. JFFs are created by measuring the mean wind speed and direction over 10-minute intervals. Each observation is classified as an event belonging to a specific wind speed class and wind direction class.

Wind speed classes typically have an amplitude of 1 m/s and include about 20 categories (21 in Figure II.3.1). The last class groups all events with wind speeds exceeding the minimum threshold of that class.

	SPEED CLASSES																					
h/year	1	2	3	4	5	6	7	8	9	10	11	12	13	14	15	16	17	18	19	20	>20	TOT
E	37	33	50	50	53	51	41	36	30	24	15	12	8	4	4	2	3	2	2	2	0	5%
E-NE	17	42	28	35	29	21	12	11	7	6	4	1	3	1	2	1	1	1	1	2	2	3%
NE	36	57	39	31	19	9	4	2	1	1	1	3	2	1	1	3	3	2	0	0	1	2%
N-NE	15	34	21	30	26	20	20	15	10	6	5	5	3	3	1	3	1	2	2	2	2	3%
N	32	31	44	47	44	39	32	26	20	12	8	5	5	3	2	0	1	2	0	2	0	4%
N-NW	46	62	65	77	64	61	46	36	26	19	12	8	3	2	3	1	1	0	1	2	0	6%
NW	71	83	111	125	124	122	108	90	74	53	39	29	17	13	8	4	3	1	2	1	2	12%
W-NW	56	85	117	136	151	167	170	165	155	143	127	111	92	78	64	48	37	29	22	16	30	23%
W	28	65	70	83	97	98	100	98	96	88	77	66	56	48	39	31	24	19	13	10	19	14%
W-SW	49	47	56	72	81	79	75	65	55	48	39	28	21	15	11	9	5	5	3	2	1	9%
SW	15	27	38	48	38	41	37	29	25	17	14	9	6	5	3	2	3	1	1	1	1	4%
S-SW	30	41	42	48	44	38	31	25	16	12	8	6	4	2	1	2	1	2	1	0	1	4%
S	58	83	67	36	20	9	4	4	1	4	3	0	1	1	0	2	1	2	2	0	2	3%
S-SE	37	60	47	33	14	10	2	4	3	1	3	1	1	2	1	1	3	1	1	1	1	3%
SE	37	53	39	30	11	8	1	3	3	1	1	1	2	1	2	3	3	1	1	1	0	2%
E-SE	37	22	37	33	25	24	12	12	8	2	1	3	1	0	2	3	3	2	1	0	1	3%
TOT	7%	9%	10%	10%	10%	9%	8%	7%	6%	5%	4%	3%	3%	2%	2%	1%	1%	1%	1%	0%	1%	100%

FIGURE II.3.1 Typical anemometric data organization in a JFF wind frequency matrix.

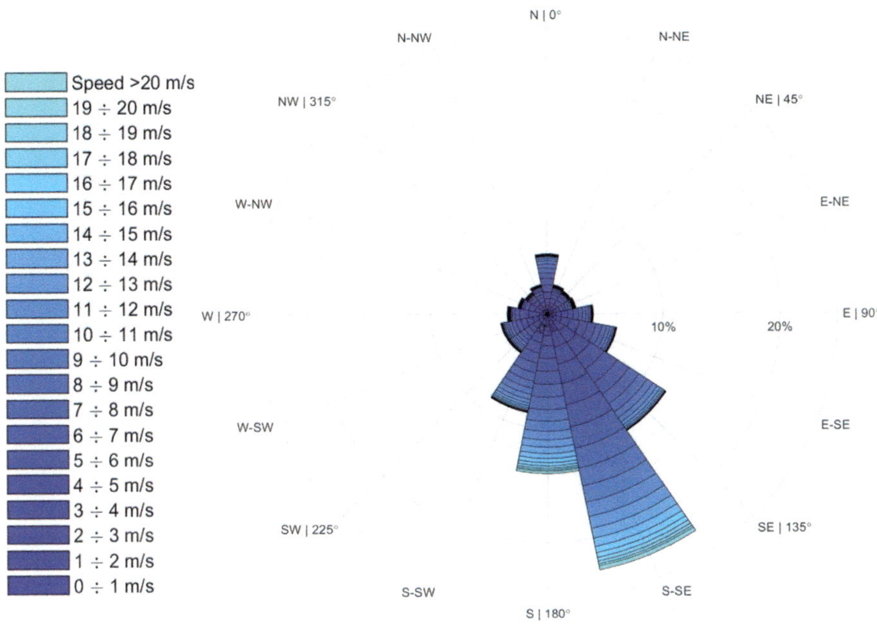

FIGURE II.3.2 Wind rose JFF graphical representations for the wind frequency JFF of Figure II.3.1.

Wind direction classes are usually defined by 16 sectors (the typical "Wind Rose"), each spanning 22.5°, centred on the cardinal directions.

Each six events in a specific joint speed and direction class represent a cumulative hour of observation. After one year, the total number of events in each joint class, divided by six, provides the hourly frequency (h/year) of occurrences in that class. The values displayed in the heatmap cells of Figure II.3.1 indicate these hourly frequencies, with red cells marking the most frequent joint classes.

On the right and bottom edges of the matrix, cumulative frequency data are provided, expressed as percentage occurrences for each direction class and speed class, respectively.

These distributions often reveal one or more prevailing wind directions. Wind speed distributions, on the other hand, tend to form a characteristic asymmetrical bell shape.

Anemometric data are also normally reported in graphical representations like those in Figures II.3.2 and II-3.3.

Figure II.3.2 condenses the JFF into a wind rose graph, where cumulative histograms show the progressive sum of frequencies for wind speed classes within each wind direction. This format clearly highlights the most probable wind directions.

Figure II.3.3 presents a cumulative histogram of wind speed classes. For each speed, cumulative bars represent the summed frequencies of all direction classes within that speed range. This visualization provides insight into the distribution of wind speeds across different categories.

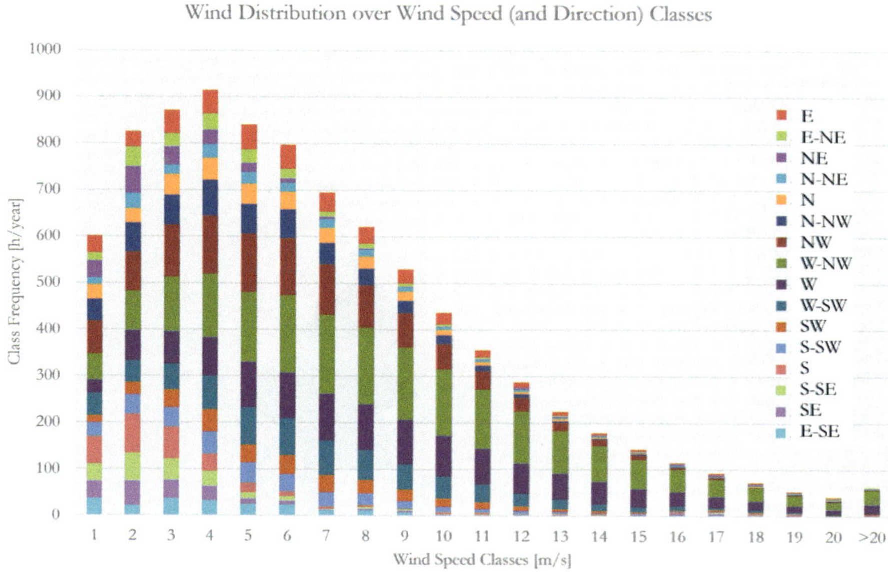

FIGURE II.3.3 Typical histogram representations for the wind frequency JFF of Figure II.3.1.

JFF data can be used to calculate the total energy passing through the unit area of the observation site. For a specific wind speed class (i) and direction class (j), with frequency n_{ij} expressed in hours per year, the total energy contribution is given by:

$$E_{ij} = n_{ij} \cdot \frac{\rho}{2} \cdot u_i^3$$

Where u_i is the characteristic speed value for the i-th wind speed class.

These energy values E_{ij} also determine a new JFF which describes the energy distribution across speed and direction classes.

As an example, the values of this new energy JFF matrix, directly implied by the previously described wind frequency JFF distribution, are reported in the colour heatmap in Figure II.3.4. The values in the cells are expressed in kW/(m^2 year).

The corresponding histogram of the energy JFF is reported in Figure II.3.5.

The energy distribution is notably skewed towards higher wind speeds (to the right part of Figure II.3.5), reflecting the disproportionately large energy contribution of faster winds.

The overall energy made available yearly at the measuring point is then evaluated by:

$$E_{ij} = \sum_{i=1}^{n}\sum_{j=1}^{m} E_{ij} = \frac{\rho}{2}\sum_{i=1}^{n} u_i^3 \sum_{j=1}^{m} n_{ij} = \frac{\rho}{2}\sum_{i=1}^{n} u_i^3 \tilde{n}_i$$

$$\tilde{n}_i = \sum_{j=1}^{m} n_{ij}$$

kWh/(m²*y)	1	2	3	4	5	6	7	8	9	10	11	12	13	14	15	16	17	18	19	20	>20	TOT
E	0	0	2	4	7	12	16	21	24	27	22	23	19	12	15	9	17	17	18	15	0	4%
E-NE	0	0	1	3	4	5	5	6	6	7	6	1	7	3	8	5	4	7	9	14	21	2%
NE	0	1	1	2	3	2	2	1	1	1	1	6	5	3	6	14	17	13	0	3	10	1%
N-NE	0	0	1	2	4	5	8	9	8	7	7	10	7	9	4	14	6	13	15	18	21	2%
N	0	0	1	3	6	9	12	15	16	13	12	10	12	9	8	0	6	13	0	18	0	2%
N-NW	0	1	2	6	9	15	18	21	21	21	18	15	7	6	11	5	6	0	8	18	0	3%
NW	0	1	3	9	17	30	41	52	60	60	58	56	42	40	30	18	17	7	15	9	21	8%
W-NW	0	1	4	10	21	40	65	95	127	160	189	215	226	240	242	220	204	189	169	143	311	38%
W	0	1	2	6	14	24	38	56	78	99	115	128	138	148	147	142	132	124	100	90	197	23%
W-SW	0	0	2	5	11	19	29	37	45	54	58	54	52	46	42	41	28	33	23	18	10	8%
SW	0	0	1	3	5	10	14	17	20	19	21	17	15	15	11	9	17	7	8	9	10	3%
S-SW	0	0	1	3	6	9	12	14	13	13	12	12	10	6	4	9	6	13	8	0	10	2%
S	0	1	2	3	3	2	2	2	1	4	4	0	2	3	0	9	6	13	15	0	21	1%
S-SE	0	1	1	2	2	2	1	2	2	1	4	2	2	6	4	5	17	7	8	9	10	1%
SE	0	0	1	2	2	2	0	2	2	1	1	2	5	3	8	14	17	7	8	9	0	1%
E-SE	0	0	1	2	4	6	5	7	7	2	1	6	2	0	8	14	17	13	8	0	10	1%
TOT	0%	0%	0%	1%	2%	3%	3%	5%	6%	6%	7%	7%	7%	7%	7%	7%	7%	6%	5%	5%	9%	100%

FIGURE II.3.4 Energy JFF matrix correspondent to wind frequency JFF of Figure II.3.1.

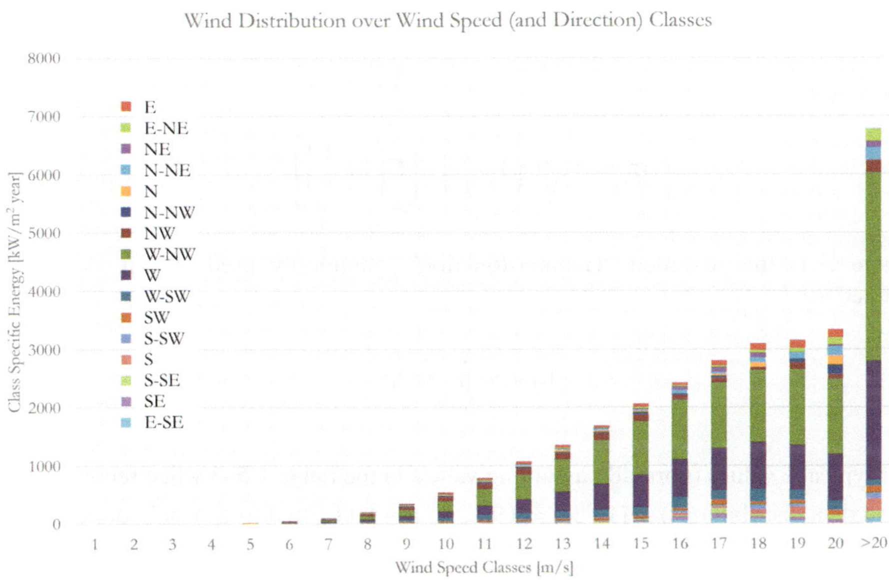

FIGURE II.3.5 Histogram graphical representations of the wind energy JFF of Figure II.3.4.

where $\tilde{n}_i$ are the cumulative frequencies of the wind speed classes obtained summing up the frequencies of all the wind direction classes with the same wind speed. Their distribution was shown in Figure II.3.3.

These latter single variable frequency distribution data ($\tilde{n}_i$, as a function of u_i) sometimes are also fitted by a continuous two-parameter statistical distribution, known as Weibull distribution, defined as a function of wind speed u by the following probability density function:

$$f(u) = \frac{k}{A} \cdot \left(\frac{u}{A}\right)^{k-1} \cdot e^{-\left(\frac{u}{A}\right)^k}, u \geq 0$$

where A is a scale factor (depending on the mean value of the measured data and expressed in m/s), while k is a shape-factor (depending also on the variance of the measured data). The cumulative probability density for a Weibull distribution is given by:

$$F(u) = 1 - e^{-\left(\frac{u}{A}\right)^k}$$

For each u value, F gives the cumulative frequency of the events with a wind speed lower than u.

The relations between A and k, and the correspondent mean $\underline{u}$ and standard deviation σ_u for measured data are the following:

$$\underline{u} = A \cdot \Gamma\left(1 + \frac{1}{k}\right)$$

$$\sigma_u = A \cdot \sqrt{\Gamma\left(1 + \frac{2}{k}\right) - \left[\Gamma\left(1 + \frac{1}{k}\right)\right]^2}$$

where Γ is the so-called "Gamma-function", commonly used in statistics, and defined as:

$$\Gamma(x) = \int_0^\infty t^{x-1} e^{-t} dt$$

For typical k values (normally assuming values in the range 1.5–3 when representing wind data distributions) $\Gamma(1 + 1/k) \approx 0.90$, $A / \underline{u} \approx 1.11$ (therefore A only depends on $\underline{u}$), while $\sigma_u / \underline{u} = 1.63 \div 3$ increasing almost linearly in the range 1.5–3, and being very similar to the k value.

For $k = 2$ the Weibull distribution is normally also called Rayleigh distribution. Figure II.3.6 depicts Weibull distributions and their variations with A and k values.

Anemometric data serve as the foundation for designing wind aerogenerators. The nominal operational wind speed (or speed range) of the turbine is determined based on the expected probabilities of wind speeds at the site. Once these nominal conditions are established, the generator's nominal power output, maximum power output, and its minimum (cut-in) and maximum (cut-off) operational wind speeds are also defined.

These design decisions directly influence the turbine's energy yield and capacity factor, as well as the balance between CAPEX and OPEX. Consequently, the LCOE

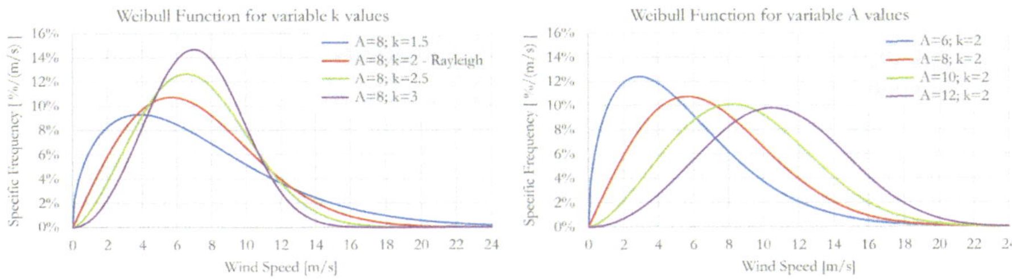

FIGURE II.3.6 – Weibull wind frequency distributions for various A and k values.

varies depending on the site's specific wind characteristics and the investor's tolerance for risk.

II.3.1.3 Some Critical Aspects of Standard Anemometric Campaigns

As mentioned earlier, traditional anemometric campaigns typically sample data every 10 minutes, classifying measurements into discrete wind speed and direction classes of predefined amplitudes. This approach has been in use for decades, originating in a time when data acquisition and analysis capabilities were limited. Its initial purpose was tailored to meteorological and aeronautical needs, which differ significantly from the requirements of designing energy-conversion devices. Therefore, it is essential to critically assess whether this method might overlook valuable information that could enhance the precision and effectiveness of energetic and economic design activities.

- About time sampling

  The standard 10-minute sampling period aligns with the need to monitor meteorological boundary conditions, which are the primary drivers of wind generation. However, this timeframe can miss important details when turbulent flows are present.

  In turbulent conditions, wind intensity and direction fluctuate locally, even under steady boundary conditions. These fluctuations have a mean value of zero, but their cubic mean (which determines the power content of the wind) does not cancel out.

  Consequently, turbulent flows can possess significantly more energy than laminar flows with the same mean wind speed.

  Turbulence levels depend not only on wind speed (and its associated Reynolds number) but also on geographical factors such as height above ground level, the mean height of ground obstacles, and meteorological conditions like air humidity and temperature gradients with altitude.

  Unfortunately, standard anemometry does not account for turbulence, and its effects must be estimated using semi-empirical correlations to correct energy yield predictions.
- About event cataloguing in classes

  The practice of grouping wind events into discrete speed and direction classes introduces a loss of precision that depends on the number of classes and their amplitude. While this method is acceptable for meteorological purposes, it can become problematic for energy evaluations, particularly when analysing speed classes.

For energy calculations, all events within a given speed class are treated as equivalent to a single representative speed value. However, as previously discussed, wind energy depends on the cube of the wind speed. Events near the upper boundary of a speed class contribute significantly more energy than those near the lower boundary.

To address this, representative speed values for each class should be based on cubic mean values rather than linear averages. Additionally, examining the gradients of the speed frequency distribution can reveal whether events within a class are evenly distributed or concentrated near the upper or lower boundary. These compensations become critical when the number of speed classes is limited, and their amplitude is large (e.g., 4–5 m/s).

- **About occurrence time of the events**

  Standard anemometric campaigns provide long-term data (typically spanning years), organized into annual wind frequency JFFs. However, these data do not include information about the time of day, week, month, or season when events occur. While this temporal information does not affect energy yield directly, it has significant implications for the plant's economic performance.

  The profitability of a wind energy plant depends on the value of electricity produced, which fluctuates throughout the year based on demand. For example, electricity prices vary predictably between seasons and during different times of the day.

  As shown in Figure II.3.7, the Italian electricity market in 2023 exhibited significant and predictable price oscillations both seasonally and daily.

  Knowing whether wind speeds are higher during daylight hours or in specific seasons can help estimate the average selling price for generated electricity. This price can vary by 20–30% above or below the yearly average electricity price, directly impacting the plant's revenue.

All these evaluations are not typically accounted for in a normal anemometric analysis but can be easily performed by properly customizing data treatment and analysis.

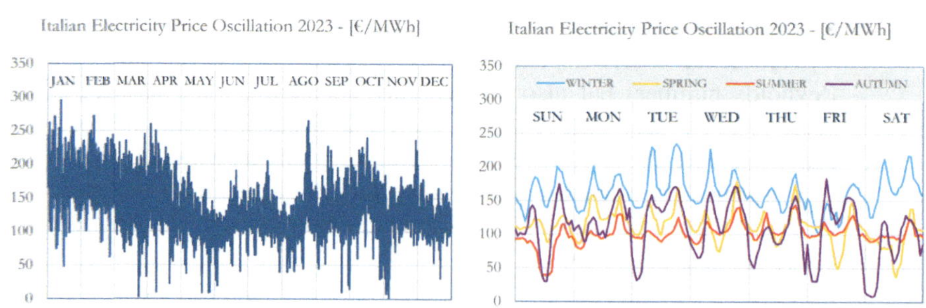

FIGURE II.3.7 Electricity buying price oscillations in Italy in 2023.

II.3.1.4 Expected Wind Non-Uniformities and CFD Simulations

Anemometric campaigns measure wind speeds at specific points within a much larger geographical area where wind energy conversion devices may be installed. This raises an important question: to what extent can measurements taken at single points be reliably extrapolated to represent the entire domain?

Two almost obvious considerations certainly arise:

- The presence of the ground influences air flow at a certain height.
  The ground significantly affects airflow at lower altitudes. According to fluid dynamics principles, the wind speed at ground level is effectively zero due to the boundary condition imposed by surface contact. The critical question is: how high above the ground does this influence persist? In other words, how thick is the boundary layer over the surface?

  The thickness of the boundary layer depends on the flow's Reynolds number and the roughness of the surface. For typical wind flows, the boundary layer often extends several tens of metres above the ground. Since most anemometers are installed at heights of 5, 10, or 20 metres above ground level (a.g.l.)—primarily due to cost constraints—they are often situated within the boundary layer. At these heights, the airflow is still influenced by surface roughness and may not be fully developed.

  In contrast, modern wind generators, especially large-scale devices, interact with airflows at much higher altitudes, often spanning 30–50 metres to 150–200 metres a.g.l. It is, therefore, critical to account for the variation of wind speed with height (z) when extrapolating data from measurement points.

  To estimate wind speeds at different heights based on measurements at a given height (z_m), semi-empirical correlations are commonly used. These correlations consider factors such as surface roughness, atmospheric stability, and turbulence. Two widely used formulations are:

  - **Power law formula**

    This is one of the most commonly used formulations, expressed as:

    $$u(z) = u_m \left(\frac{z}{z_m} \right)^n$$

    where n is the power law exponent, which depends on surface roughness and atmospheric stability. Typical values for n range between 0.1 (for smooth surfaces, like water) to 0.4 for rough surfaces (e.g., urban areas).

    This method assumes a neutral atmosphere, where the effects of thermal stratification are negligible. It is commonly used in wind energy studies to estimate wind speeds at different heights when detailed data are unavailable.

- **Logarithmic wind profile**

  The used formulation in this case is expressed as:

$$u(z) = \frac{u_s}{k}\, ln\left(\frac{z}{z_o}\right)^n, \; k \approx 0.4$$

where z_o expresses surface roughness by a characteristic dimension (expressed in metres) describing the mean density and height of the "obstacles" present at ground, and u_s is the friction velocity (m/s), which characterizes shear stress at the surface.

Friction velocity is a key concept in boundary layer meteorology that quantifies the rate of momentum transfer due to turbulent friction between the atmosphere and the Earth's surface.

Although it has units of velocity (m/s), u_s is not correlated with the actual wind speed. Instead, it reflects the intensity of the turbulent forces caused by wind interaction with the surface. It is defined by:

$$u_s = \sqrt{\tau / \rho}$$

where τ is the surface shear stress (N/m^2), which measures the force per unit area exerted by the wind on the surface, while ρ is air density (approximately 1.225 kg/m^3 at sea level under standard atmospheric conditions). Shear stress arises from the drag forces acting on air parcels due to surface roughness, obstacles, and molecular viscosity.

In calm atmospheric conditions, typical values for u_s are: 0.01–0.1 m/s for smooth surfaces (like water, ice, flat dunes); 0.1–0.3 m/s for moderately rough surfaces (e.g., grasslands, fields); 0.3–0.6 m/s for rough (e.g., forests, urban areas); >0.6 m/s for very rough terrains (e.g., dense urban environments).

However, u_s can also vary with atmospheric conditions. It can exceed 1 m/s in high winds (e.g., during storms), over rough surfaces.

It normally decreases in "stable atmosphere" (e.g., nighttime cooling) and increases in "unstable atmosphere" (e.g., daytime heating).

The logarithmic profile is particularly suitable for modelling wind speed variations in the lower boundary layer of the atmosphere, where surface effects dominate.

- **A complex orography site produces highly non-uniform speed distributions.** Sites with complex orography exhibit highly non-uniform wind speed distributions. While such sites generally have lower mean wind energy potential compared to flat regions, they may feature localized zones of high wind energy density. These sub-zones can offer significant productivity, making them attractive for wind energy development despite the challenges of overall non-uniformity.

  In complex terrains, wide-scale wind speed maps often prove inadequate for accurate energy yield predictions. They tend to overestimate productivity in some areas while failing to capture high-energy-density sub-zones.

Consequently, more anemometers must be deployed to capture local variations in wind speed and direction.

To address these challenges, data from multiple anemometers are used as boundary conditions in GIS-based CFD (computational fluid dynamics) applications.

These apps are specifically designed for wind generator and wind farm design and therefore import wind data through JFFs, solve a different CFD problem for each wind speed and direction class, and weight the resulting wind fields by their frequency. Starting from the raw measurements in single points, the wind frequency JFFs are therefore known in all the domains (x,y,z). The next step is placing the aerogenerators in the domain calculating their mutual interactions and their overall energy yield.

To meet this aim, SW imports their characteristic curves (in the form we will introduce and discuss in the next subsection) and newly solves a further CFD problem describing the wake effects produced by the single devices (i.e. the reduction in wind speed and increase in turbulence behind a turbine caused by energy extraction). For this they normally apply specific simplified fluid-dynamic models known as "park", or "eddy-viscosity" models.

II.3.1.5 On the Maximum Power Extraction from Wind: The Betz Limit

The so-called Betz theory determines the maximum power that can be extracted from wind. For this, it uses a simplified model of wind flow, assuming that a steady airstream with constant velocity moves through and around an ideal "actuator disk".

The actuator disk serves as a model for a real wind generator, extracting energy from the wind as it passes through a defined cross-sectional area S. As energy is extracted, the wind's linear momentum and kinetic energy decrease locally in a concentrated but non-dissipative manner.

While the model simplifies reality—ignoring angular momentum (despite turbine rotation and wake effects) and assuming no energy loss due to friction—it provides valuable insights into wind energy potential.

To briefly explain Betz theory and its results, we refer to the graphical sketch in Figure II.3.8, applying conservation equations.

Let us first apply the momentum conservation equation to the air flow (with an air flow rate $\dot{m}$) between section S_1 and S_2, sufficiently distant from, respectively, inbound and outbound section S.

Since the actuator disk subtracts energy from the flow, moving air applies to it a thrust force with a resultant F. By Newton's third law, the air applies an equal and opposite force on the actuator, which decelerates the flow. Consequently, the wind speed downstream (u_2) is lower than the upstream speed (u_1).

By Newton's second law, the force (F) exerted by the actuator on the air is proportional to the momentum change per unit time:

$$F = \dot{m}\left(u_1 - u_2\right)$$

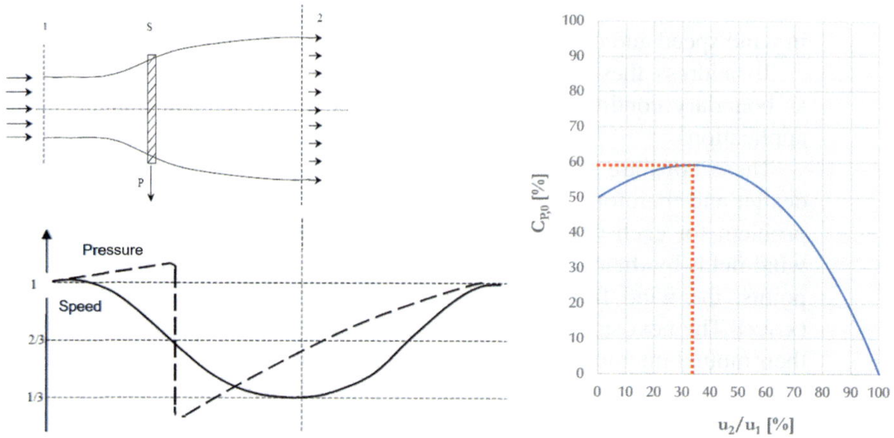

FIGURE II.3.8 Sketch reporting Betz theory on the maximum power extraction from a wind flow.

F is the only force acting on the fluid in the domain. It is equivalently applied to the flowing fluid in a concentrated way in section *S* where the fluid meanly flows with a horizontal speed *u*.

Therefore, extracted power *P* may be expressed as:

$$P = F\,u = \dot{m}\left(u_1 - u_2\right)u$$

Since the process is assumed to be horizontal, ideal, and adiabatic, the extracted power is equivalent to the loss of kinetic energy between sections S_1 and S_2:

$$P = \frac{\dot{m}}{2}\left(u_1^2 - u_2^2\right) = \frac{\dot{m}}{2}\left(u_1 - u_2\right)\left(u_1 + u_2\right)$$

Comparing the two equations reveals that the wind speed at the actuator section *S* is the average of the upstream and downstream speeds (u_1 and u_2):

$$u = \frac{\left(u_1 + u_2\right)}{2}$$

To further simplify, assume that the fluid (even though compressible) does not generate density modifications during the flow:

$$(\rho_1 = \rho = \rho_2).$$

This flow-type is called "incompressible flow" and formally occurs only if and when the fluid always remains sufficiently far from its local sonic conditions during the flow. This hypothesis, while not strictly valid, simplifies calculations without significantly altering results, but it may be removed, leading to a more complex theory.

Under this hypothesis, the wind kinetic power which would have entered the section in case of no-power extraction is expressed by:

$$P_{IN} = \frac{\dot{m}}{2}u_1^2 = \frac{\rho S}{2}u_1^3$$

Therefore, the power extracted from the wind flow may be re-expressed as:

$$P = \frac{\dot{m}}{2}\left(u_1^2 - u_2^2\right) = \frac{\rho S}{2}u\left(u_1^2 - u_2^2\right) = \frac{\rho S}{4}\left(u_1 + u_2\right)\left(u_1^2 - u_2^2\right)$$

Now, let us define the actuator power extraction ratio, or ideal power coefficient ($C_{P,0}$), as the ratio of extracted power to the available wind power:

$$C_{P,0} = \frac{P}{P_{IN}}$$

Then, $C_{P,0}$ could be expressed as:

$$C_{P,0} = \frac{P}{P_{IN}} = \frac{\frac{\rho S}{4}\left(u_1 + u_2\right)\left(u_1^2 - u_2^2\right)}{\frac{\rho S}{2}u_1^3} = \frac{1}{2}\left(1 + \frac{u_2}{u_1}\right)\left(1 - \frac{u_2^2}{u_1^2}\right)$$

The above equation shows that the actuator power coefficient, (in these ideal extraction conditions) solely depends on the overall braking ratio of the fluid between the inlet and outlet sections (u_2 / u_1) which can range from 1 (no braking) to 0 (complete fluid stop).

Within this range, C_P assumes its maximum value of 16/27 = 0.596 ≈ 0.6 when $u_2 / u_1 = 1/3$.

Betz theory demonstrates that no more than 59.6% of the kinetic energy in wind can be extracted by an ideal actuator. It also provides insight into the optimal design conditions for wind energy systems:

- **Optimal wind speed at the actuator**
  The wind speed at the actuator section (S) should ideally be two-thirds of the upstream speed ($u = 2/3u_1$).
- **Flow behaviour**
  In optimal conditions, the air flow reacts to the actuator's presence by converting part of its kinetic energy into pressure energy before entering the actuator. This behaviour resembles the function of a stator in a turbine.

Additionally, as the airflow decelerates and its pressure increases (in subsonic conditions), the flow area expands, as shown in Figure II.3.8. This expansion results in some air bypassing the actuator, leaving a portion of the wind energy unextractable.

Finally, the remaining energy in the flow exits the actuator section as kinetic energy in the downstream fluid, ensuring mass balance is maintained in the steady-state process.

II.3.2 WIND ENERGY CONVERSION TECHNOLOGIES: AEROGENERATORS

The simplest method to produce electricity from wind energy involves utilizing a portion of the thrust generated by the wind on one or more aerofoils (commonly referred to as "blades").

These aerofoils are strategically positioned within the airflow to generate momentum on a rotating axis. The resulting turbine rotor consists of several blades fixed in specific relative positions around the axis of rotation. As the rotor turns, it defines a cross-sectional area (S) through which wind interacts with the rotor, introducing inlet kinetic power: $P_{IN} = \dot{m} u_1^2 / 2$.

The rotor is engineered to convert a portion of this kinetic power into mechanical rotational power: $P_{Out} = I \omega^2 / 2$. This mechanical power is then transformed into electrical power via a connected rotating electric generator: P_{El}.

The complete system, which includes the turbine, generator, and all auxiliary devices, is collectively referred to as an **aerogenerator**.

Analogous to the Betz theory of power extraction, the performance of an aerogenerator can be evaluated using two key coefficients:

- **Rotor Power Coefficient (C_{PR})**
  This measures the efficiency of the rotor in converting the available wind power into mechanical rotational power:

$$C_{PR} = \frac{P_{Out}}{P_{IN}}$$

- **Overall Power Coefficient (C_P):**
  This represents the efficiency of the entire aerogenerator system, including the electrical conversion process:

$$C_P = \frac{P_{El}}{P_{IN}} = \frac{P_{El}}{P_{OUT}} \frac{P_{Out}}{P_{IN}} = \eta_{el} \cdot C_{PR}$$

  Here, η_{el} denotes the efficiency of the electric generator in converting mechanical power into electrical power.

The maximum theoretical value of both the rotor and overall power coefficients is determined by the Betz limit ($C_{P,0} \approx 0.6$).

This limit represents the maximum fraction of wind power that can be extracted under ideal conditions:

$$C_P \le C_{PR} \le C_{P,0} \approx 0.6$$

In practical applications, the efficiency of the electric generator (η_{el}) and other auxiliary components further reduce the achievable power coefficient, making it essential to optimize the aerogenerator's design to maximize power extraction.

II.3.2.1 Wind–Aerofoil Interactions, Rotor and Wind Turbine Types

When wind interacts with a stationary aerofoil, it alters the local kinetic and thermodynamic properties of the airflow. This interaction generates pressure differences across the aerofoil's surface, resulting in a net thrust force. This thrust has two components: one parallel to the wind direction, called drag (or resistance) force, and the other perpendicular to the wind direction, known as lift (or portance) force. The magnitude and direction of these forces depend on the aerofoil's shape and orientation relative to the wind.

When an aerofoil moves within the airflow, its behaviour changes predictably, as it is determined by the relative wind speed experienced at every point on the aerofoil. For a purely translating motion, the relative speed is uniform across the surface. However, when the aerofoil undergoes combined rotational and translational motion, the speed distribution becomes more complex but remains well-defined.

In wind turbines, aerofoils (or blades) are fixed to the rotor in a defined position during each rotation. The movement of the blades with respect to the incoming wind is primarily translational, influenced by the rotor's rotational speed and the distance of the blade section from the rotor hub. The relative translational speed also varies along the blade's length due to this distance.

All this considered, let us start recalling some typical characteristics of drag and lift forces produced on a still aerofoil:

- Drag force (F_D) acts parallel to the wind direction, opposing the aerofoil's motion. It arises from air friction and pressure differences between the front and rear surfaces of the aerofoil. Drag increases with: surface roughness, aerofoil thickness, and angle of attack (α), defined as the angle between the aerofoil's chord line and the incoming wind.
- Lift force (F_L) acts perpendicular to the wind direction, resulting from pressure differences above and below the aerofoil. According to Bernoulli's principle, the curved upper surface accelerates the airflow, reducing pressure, while the flatter lower surface experiences higher pressure. Lift depends on: aerofoil curvature (camber), angle of attack (α), and speed of the incoming wind.

To achieve optimal lift-to-drag ratios, engineers carefully design the aerofoil's geometry (specifically its camber, chord length, and thickness) and select appropriate angles of attack. If the angle of attack becomes too large, the aerofoil can experience stall, where airflow separation drastically reduces lift and increases drag.

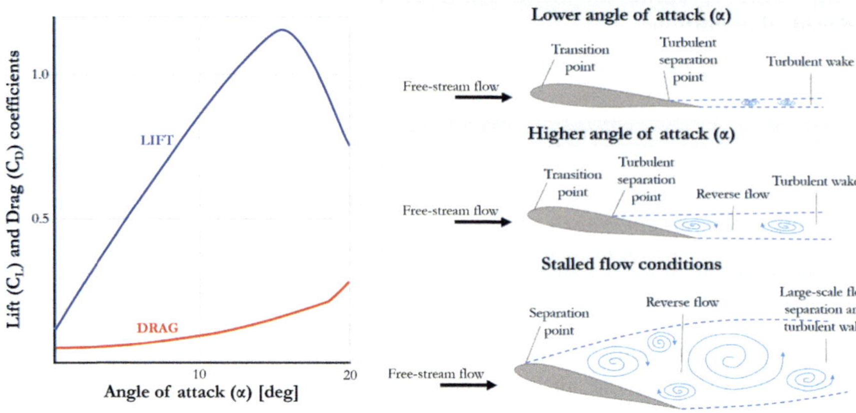

FIGURE II.3.9 Draft and lift behaviour of an aerofoil vs. its attack angle.

These aerodynamic principles guide the design of efficient aerofoils for wind turbine blades, as well as for applications like aircraft wings and sailboats.

Both the drag (F_D) and lift (F_L) forces are proportional to the kinetic energy of the wind interacting with the aerofoil, which depends on wind speed (u) and the aerofoil's cross-sectional area (S).

This approach permits the definition of two coefficients, respectively named drag (C_D) and lift (C_L) coefficients, expressed by:

$$F_D = C_D\left(\frac{1}{2}\rho S u^2\right), \quad F_L = C_L\left(\frac{1}{2}\rho S u^2\right)$$

The left part of Figure II.3.9 shows typical trends for C_D and C_L as functions of the angle of attack (α). The right part illustrates the flow separation phenomenon at increasing angles of attack, highlighting the conditions leading to aerofoil stall and its impact on lift.

Wind–aerofoil interactions are central to wind turbine efficiency, as their angle of attack, blade curvature, and material properties must be carefully designed to maximize energy extraction. These factors ensure smooth interactions between the wind and wing profile, enabling efficient conversion of wind's kinetic energy into mechanical rotation.

According to their different rotating axis and the way the blades generate a momentum over that axis, wind turbines are broadly classified into two types:

- Horizontal-Axis Wind Turbines (HAWTs)
  These turbines have blades that rotate around a horizontal axis. They are highly efficient at capturing wind energy and are the dominant type used in large-scale wind farms. HAWTs require alignment with the wind direction to operate in optimal conditions, achieved through yaw systems. These turbines rely exclusively on lift forces to generate torque on the rotor.

FIGURE II.3.10 Photos of HAWT- and VAWT-based aerogenerators.

- Vertical-Axis Wind Turbines (VAWTs)
  These turbines have blades that rotate around a vertical axis. They can capture wind from any direction, eliminating the need for alignment. While generally less efficient than HAWTs, VAWTs are advantageous in urban areas or regions with highly variable wind directions.

Figure II.3.10 presents some photographs of both HAWT- and VAWT-based aerogenerators.

Most utility-scale wind generators today are HAWTs due to their superior efficiency and widespread adoption in wind farms. Consequently, this section will primarily focus on HAWTs. However, many of the principles discussed also apply to VAWTs.

II.3.2.2 HAWT Blade Profiles

As evident in the photographs in Figure II.3.10, the blades of HAWTs extend fully in the radial direction. This design implies that, for a given rotational speed (ω) of the rotor, the linear velocity at any point along the blade increases linearly with the distance (R) from the axis of rotation, given by the relation: $v = \omega R$.

When the blade rotates, an observer situated on the blade would experience a relative wind speed (w) whose direction and magnitude depend on the radial position (R).

According to these considerations, the left part of Figure II.3.11 describes how the aerofoil is commonly designed, depicting different cross-sections of the blade at various distances R from its rotation axis, and showing that the angle difference between w and the rotation plane (γ) also depends on R.

To maximize the **blade lift coefficient** (C_L) and its corresponding contribution to the torque around the rotor axis – and thus optimize energy conversion efficiency – it is essential to maintain the **optimal angle of attack** (α_{Opt}) for the aerofoil. The angle α_{Opt} is determined by the aerofoil's aerodynamic properties. The difference between γ (the angle of *w* relative to the rotation plane) and α is known as the **pitch angle** (β).

The pitch angle is defined by the twisting of the aerofoil along the blade, i.e., the angular distance between the rotation plane and the chord line connecting the inlet and outlet of the aerofoil.

A series of non-obvious consequences of these aerodynamical considerations arise, regarding blade design, manufacturing, and operational control of wind turbines:

- Blades have a very complex 3D geometry.
  For given wind (u) and rotational (ω) speeds, the angle γ decreases with increasing R. To maintain $\alpha = \alpha_{Opt}$ across all values of R, the **pitch angle** (β) **must decrease with** R**.**

  As a result, blades are intentionally **pitch-twisted**, meaning their geometry is designed with a twist along their length to ensure the optimal angle of attack relative to the wind.

  Additionally, blades are often tapered along their length. This tapering serves two purposes:
  - It reduces material usage and weight, mitigating the impact of centrifugal forces on large utility-scale aerogenerators.
  - It accounts for the increasing energy available in the wind as *R* grows, due to progressively larger annular sections of the rotor swept area interacting with the wind.

  The combined effects of pitch twisting and tapering result in a highly complex 3D geometry, making blade design and manufacturing intricate and costly. Refer to the right side of Figure II.3.11 for an illustration.
- Turbines' speed should be variable with wind speed.
  Pitch-twisted blades have a pitch angle varying with R to maintain $\alpha = \alpha_{Opt}$ at all R values. But the blade will be in its optimal aerodynamical conditions only for one specific γ value. Therefore, the ratio between u and ω, should be constant as well.

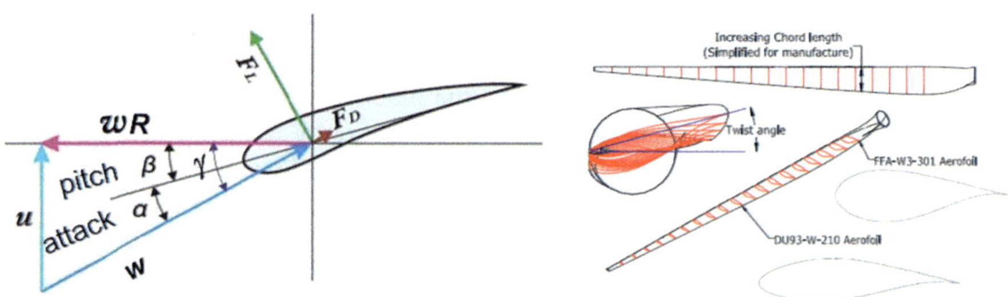

FIGURE II.3.11 An aerofoil with lift and draft forces, and pitch and attach angles.

To ensure optimal performance across varying wind speeds, the turbine's rotational speed should ideally vary linearly with wind speed. However, as will be discussed in subsequent sections, this is challenging to implement due to constraints imposed by the turbine's connection to the electrical grid.

- Blades could vary their pitch mounting angle to optimize efficiency.
If turbine speed variation is not feasible, another option is to allow the blade pitch angle to vary dynamically. By mounting each blade on a rotating axis, the entire blade can rotate slightly to adjust the pitch angle (β) uniformly along its length as wind speed changes.
While this technique allows for some optimization, it typically results in suboptimal performance compared to variable-speed operation.

II.3.2.3 HAWTs Performances and Power Coefficient Curves

The power coefficient (C_p) is a critical parameter that defines the performance of wind turbines, reflecting their ability to extract energy from the wind. Understanding the factors that influence C_p is essential for optimizing turbine design and operation.

According to all the previous considerations, the ratio between ω and u is the most relevant turbine optimization parameter, as it determines the critical fluid-dynamic angles at the blade's inlet sections.

A turbine optimization parameter is therefore introduced, commonly referred to as the tip-speed ratio (or also "fundamental kinetic ratio"), and defined as:

$$\lambda = \frac{\varpi R_{Max}}{u}$$

where R_{Max} is the distance from the axis at the blade tip.

According to previous considerations, there will be only one value of λ to optimize the power performance of a wind converter (λ_{Opt}).

Figure II.3.12 depicts typical expected C_p performances for a non-controlled and optimized HAWT. The data refer to a real 2-MW electric power output wind converter, having an $R_{Max} = 45\,\text{m}$.

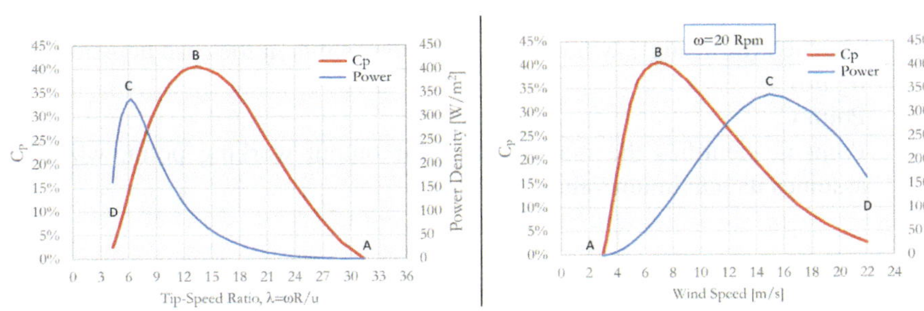

FIGURE II.3.12 Typical C_{PR} values for a non-controlled and optimized HAWT.

The left part of the figure reports C_P values (red curve, main left axis) and power output density (blue curve, secondary right axis, in W/m^2) as a function of λ.

If we assume that the generator rotates at a fixed speed (e.g. 20 rpm, in the figure), with $\lambda = \varpi R_{Max} / u$, the curves in the left part of the figure may be reverted in the corresponding curves (right part of the figure) depicting C_P values (red curve, main left axis) and power output density (blue curve, secondary right axis, in W/m^2) as a function of u in m/s.

Some points may be identified on these curves which are important for plant behaviour:

- Point A – (Cut-In Wind Speed)
  Point A identified the minimum wind speed (and maximum λ value) which may justify keeping the generator in rotation having a net output ≥ 0. For winds under this speed, the generator would extract from wind an energy output that would not be sufficient to win the frictional power losses due to its rotation at its nominal speed. Wind speed at point A is then also referred to as the minimal "cut-in" wind speed.
- Point D – (Cut-Off Wind Speed)
  Point D identifies the opposite end point of C_P and power output density curves with respect to A. Therefore, it also identifies the maximum wind speed (and minimum λ value) which may justify keeping the generator in rotation. For winds over this speed (also referred to as the wind cut-off speed) excessive loads would be exerted on the rotor, increasing the risk of structural damage or mechanical failure, so that the HAWT must be shut down.

  In this case, as will be discussed in more detail later, the turbine's control system should engage safety measures such as:
  - Feathering the blades (disengaging them and permitting free rotation on their own axis). In this way, the blades are no longer capable of transmitting any motor momentum to the hub.
  - Actively braking the rotor with a safety brake directly acting on the hub axis.
- Point B – (Optimal Power Coefficient)
  Point B identifies the optimal turbine working point, where C_P has its maximum value. This happens only when $\lambda = \lambda_{Opt}$ so that all blade cross-sections work at their optimal design attack angle α_{Opt}. If the turbine must rotate at a fixed speed (20 rpm in the case depicted in the figure), $\lambda = \lambda_{Opt}$ can only be obtained for a single wind speed (u_B, about 7 m/s in the case depicted in the figure).
- Point C
  Point C identifies the maximum power output working point, where P_{Out} assumes its maximum value.

With:

$$P_{Out} = C_P \cdot P_{IN} = C_P(u) \cdot \frac{\rho}{2} \cdot S \cdot u^3$$

its maximum can only happen for wind speed values (u_C) greater than u_B, where C_P is maximum, and therefore $\dot{C}_P = 0$.

In fact, for $u = u_C$, where P_{Out} is maximum, its derivative must be null:

$$\dot{P}_{Out} = \frac{\rho S}{2} \cdot u^3 \left[\dot{C}_P u + 3C_P \right] = 0$$

which can only happen if $\dot{C}_P < 0$, for the u_C value satisfying:

$$u_C = -\frac{3C_P(u_C)}{\dot{C}_P(u_C)} > u_B > 0$$

Let us now look at the effect on C_P and P_{Out} of the possible control actions implemented on HAWTs.

II.3.2.4 Turbine Speed Variation

Allowing the turbine operating speed (ω) to vary within a certain range ($\varpi_{min} - \varpi_{Max}$) provides significant advantages. This adjustment enables the turbine to maintain an optimal tip-speed ratio ($\lambda = \varpi R_{Max} / u$, e.g. its optimal λ_{Opt} value), over a range of wind speeds (u).

This possibility and its consequences on the resulting C_P and P_{Out} curves are graphically depicted in Figure II.3.13.

The left part of the figure reports the C_P trend as a function of λ. If ϖ can be varied (e.g. from 10 to 30 rpm), this curve can be transformed in different curves in a C_P vs. λ plane. For the case depicted in the figure, all these possible curves would be in between the green and the violet curves in the central part of the figure. As shown, varying turbine speed also permits to implement a lower cut-in speed but does not increase the cut-off speed which is fixed for safety reasons (22 m/s in this figure).

Figure II.3.13 shows that continuously increasing ϖ linearly from 10 to 30 rpm as wind speed increases from 3.3 to 10 m/s, the HAWT can maintain its maximum energy conversion efficiency.

Consequently, the energy yield would increase, enhancing both the maximum power output of the device (as shown by the curves in the right part of the figure), and also the capacity factor of the aerogenerator.

II.3.2.5 Pitch Angle Variation

As previously discussed, if the turbine's rotational speed (ω) cannot be varied, another optimization approach involves adjusting the pitch mounting angle (γ) of the blades according to wind speed intensity.

Figure II.3.14 illustrates the impact of pitch angle variation on the C_P and P_{Out} curves. While this approach achieves results similar to those obtained through speed variation, its effectiveness is slightly lower.

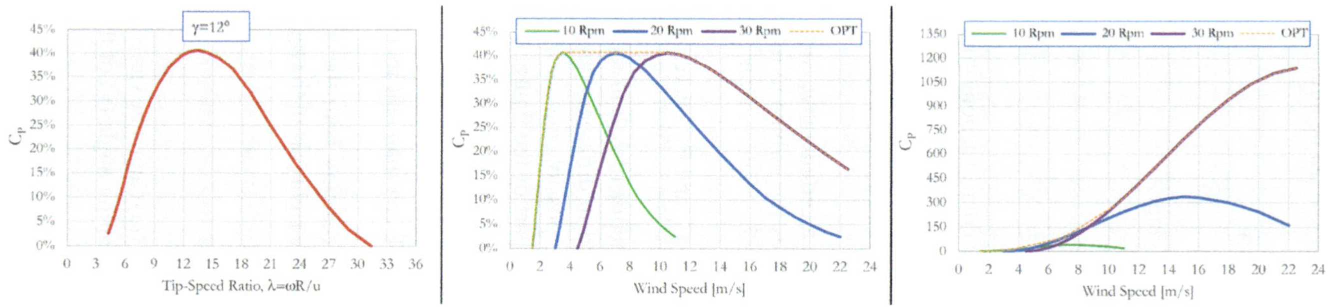

FIGURE II.3.13 Effect of turbine speed variations on C_{PR} curves.

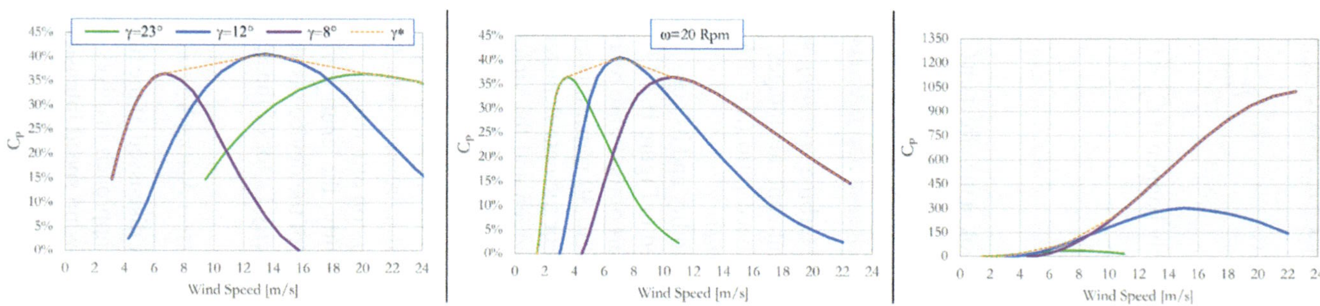

FIGURE II.3.14 Effect of pitch angle variations on C_{PR} curves.

In fact, varying γ would mean having the possibility to move λ_{Opt} values within a certain range, expanding or compressing the whole C_p curve in the C_p vs. the λ plane. However, this action would lower slightly the maximum obtainable C_p values, because even though the inlet section of the blade may be maintained at its optimal attack angle, the rotation of the whole aerofoil moves other parts of the profile away from their optimal design point.

All this considered, properly varying γ, making reference to the left part of the figure, all the curves between the green and the violet ones can be obtained.

These can be reverted (while maintaining ϖ value fixed at 20 rpm) in correspondingly variable curves on the C_p vs. ϖ plane, "emulating" the effect of the ϖ variation from 10 to 30 rpm as depicted in the previous case.

By this control action, the C_p of the HAWT can be relevantly increased in the range from 3.3 to 10 m/s maintaining its energy conversion efficiency above about 90% of its maximum value.

Consequently, also in this case, the energy yield increases, enhancing both the maximum power output of the device (as shown by the curves in the right part of the figure), and also the capacity factor of the aerogenerator.

II.3.2.6 Site-Dependency of HAWTs Sizing, Capacity Factor Curves

We have explored the optimization capabilities of HAWT aerogenerators and later discuss the extent to which the two main control techniques can be implemented.

Now, let us focus on a critical aspect of HAWT sizing and optimization: the relationship between the turbine's design and site-specific wind conditions.

Previous discussions highlighted that it is possible to optimize turbines over a wide range of wind speeds ($u_{Min} - u_{Max}$). Having the possibility to optimize the turbine for a wide range of wind speed is certainly interesting, but certain trade-offs must be considered:

- Effects of Increasing u_{Max}:
  Due to power dependency with the cube of wind speed, increasing the optimization window in terms of its u_{Max} would result in a significant over-sizing of the HAWT.

  The turbine capacity factor is commonly referred to as its maximum output power (P_{Max}). In this assumption, increasing u_{Max} would mean increasing P_{Max}, and therefore lowering the capacity factor of the installed generator.

  Also, consider that an increase in u_{Max} would lead to an over-dimensioning of all machine components needed to cope with increased P_{Max} (mechanical transmission gearings, electric generator, electric power converters).

  As a consequence, the cost of all these components will be substantially increased while their mean load factor will be lowered, also determining probably a certain decrement in their efficiency at lower load working points.

  Summarizing, asking for a higher widening of the optimization windows in terms of its u_{Max} substantially increases investment CAPEX cost, reduces the machine capacity factor, but increases energy yield (in some cases also wind cut-off speed may be substantially increased).

Therefore, the overall effect on LCOE is not clearly foreseeable and should be evaluated on a case-by-case basis, looking at the site-dependent wind frequency and energy JFFs. Mean values and standard deviations of these distributions would be the crucial parameters for the correct sizing and optimization of the HAWT.

- Effects of Decreasing u_{Min}:
  Similar but less critical considerations may be done about requests for a widening of the optimization windows in terms of u_{Min}. This request produces an increase in control equipment costs (and therefore in the CAPEX of the investment), but would not result in an over-sizing of the HAWT: its nominal power would remain the same, so their capacity factor would be increased as a consequence of the increased energy yield (in some cases also wind cut-in speed may be substantially decreased).

Previous considerations show that HAWT design and customization possibilities can be very relevant and crucially site dependent.

Therefore, similarly sized machines (with the same P_{Max}) could have quite different costs and performance curves and consequently lead to very different capacity factors, depending on the installation site, and the wind JFFs.

These considerations also modify substantially the whole design path for a wind farm. Normally, once a designer has completed his anemometry analyses, they contact the HAWT manufacturers, sharing data about local JFFs, so that they can perform an on-demand pre-customization of the machine, generating the expected customized HAWT's C_p vs. u performance curve.

Wind farm designers can then use these site-specific performance curves in finite-element CFD apps to design the number and positioning of the generators of the wind farm, calculating its energy yield, finalizing the project, and consequently ordering the machine to be manufactured.

As a further consequence of what we discussed, for a given size of the HAWTs, their cost increases for those sites having a higher standard deviation (σ) of wind speed and energy distributions. Their capacity factor is commonly in the range 1700–2500 h/year.

Sites characterized by a lower σ are advantaged, leading to higher capacity factors, reaching over 3000–3500 h/year for off-shore installations, and in the range 2000–3000 h/year in more common on-shore wind farms.

Summarizing, when comparing sites with the same mean values ($\underline{u}$) but different standard deviation (σ) of their wind speed frequency distributions, it must be considered that they will have different mean values in terms of their energy distributions (because these values increase with σ).

Therefore, the possibility of widening the optimization window in terms of the engine speed range ($u_{Min} - u_{Max}$) is particularly interesting in high σ sites.

II.3.2.7 Types and Typical Components of HAWTs

As already explained, HAWTs are the most common type of wind turbines, especially for large utility-scale purposes and applications. Here, we discuss their possible configurations in more detail. While most considerations in this section apply primarily to HAWTs, they can also extend to VAWTs.

HAWTs operate with blades rotating around a horizontal axis aligned with the wind direction. Their efficiency and scalability make them ideal for applications ranging from small-scale residential systems to large wind farms.

They can be classified in various ways. The first is depending on rotor mounting position with respect to the tower built to "sustain" its rotating axis at a certain height a.g.l.:

- Upwind HAWTs
  The rotor is positioned upwind of the tower, meaning the blades interact with the wind before it reaches the tower. This configuration prevents the tower from obstructing airflow and reduces wake turbulence on the blades, enhancing efficiency. However, aligning the rotor with the wind direction requires a motorized yaw system, adding complexity to the design. Upwind designs are the most efficient and widely used for large-scale turbines.
- Downwind HAWTs
  The rotor is located downwind of the tower, allowing the wind to pass through the tower before hitting the blades. This arrangement simplifies alignment with wind direction, often reducing the need for a yaw system. It is particularly suitable for high-wind environments, where blade bending can dissipate loads. However, the tower creates turbulence, leading to reduced efficiency and increased noise and vibration.

A second typical categorization of HAWTs is their size, which can be very variable. As a general consideration, it can be stated that in the last decades, HAWTs in the utility sector have become much bigger.

Single HAWTs are now reaching up to 10 MW for a single HAWT and consequently enormous, "giant", dimensions (see Figure II.3.15). Depending on their size, HAWTs could be categorized as:

- Small-size HAWTs
  These turbines are designed for residential or small commercial use, typically with a rotor diameter under 10 metres. They are easy to install and maintain and ideal for low-power applications or off-grid systems.

  They may have a substantially lower power coefficient with respect to larger machines. Moreover, they only make use of a small portion (in elevation) of the flowing wind kinetic energy, so largely increasing soil use for a needed energy output.

  To solve these issues, multiple HAWTs may be mounted on the same tower, increasing tower height, but civil construction costs normally do not encourage these applications with respect to single larger size rotors.

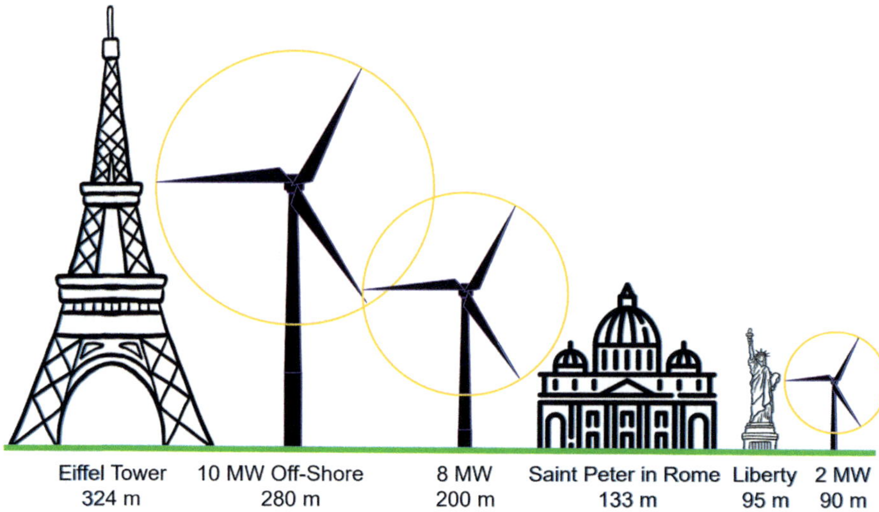

FIGURE II.3.15 Large HAWT heights compared for those of some of the highest world structures.

- Large-size HAWTs
  As already said, these turbines are the most common in wind farm applications. These turbines may have rotor diameters exceeding 100 metres and generate up to 10 MW of electric power. They have high efficiency and energy output and are specifically designed for grid integration and utility-scale power generation.

An HAWT basically consists of several essential components that work together to convert wind energy into electricity.

The more common components in a HAWT are as follows (see Figure II.3.16 as a reference):

- **Rotor Blades**
  The blades are the most critical part of a wind turbine, designed to capture the kinetic energy of the wind. They use the lift component of the wind-generated thrust to generate motor momentum over a rotation axis.

  As already explained, they are typically tapered and twisted to maintain an optimal angle of attack along the blade length. They are normally made of lightweight yet durable materials such as fibreglass-reinforced plastic or carbon fibre. Aluminium alloys can also be used.

  The number of blades on an HAWT rotor significantly impacts its aerodynamic efficiency, structural design, and application suitability. The number of blades affects the tip-speed ratio (λ), and the efficiency of energy conversion. Fewer blades result in a higher optimal λ, favouring faster rotation, while more blades lower the optimal λ, prioritizing torque over speed.

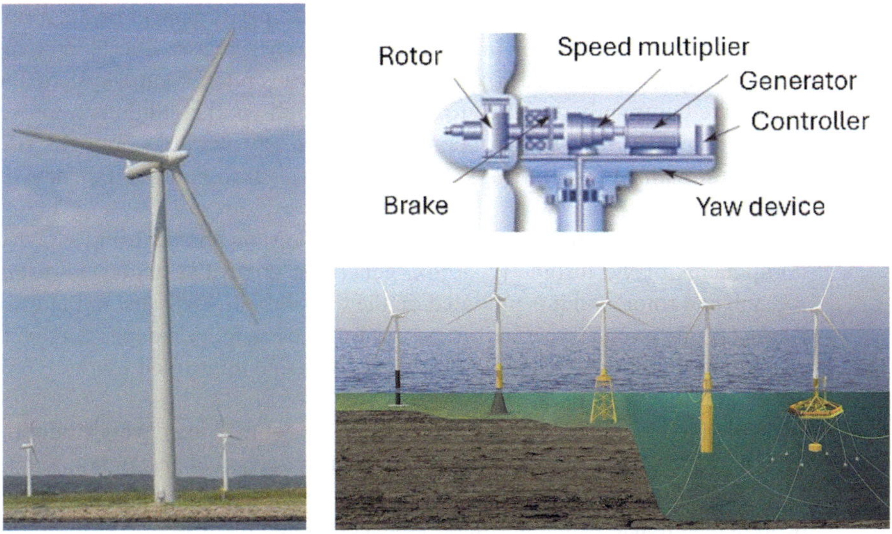

FIGURE II.3.16 Typical HAWT components.

All this being considered, for utility-scale electricity generation, three blades are the optimal choice, balancing efficiency, durability, and cost. The most common HAWT configurations include:

- <u>Two-Bladed Rotors</u>
  These are lightweight, implying less material cost, and normally permit faster rotation (which, as we will see later, can reduce gearbox size). However, they normally have a lower aerodynamic efficiency with respect to some other rotors and may increase vibration due to imbalances caused by uneven wind loading. They are often used in mid-sized turbines.

- <u>Three-Bladed Rotors</u>
  These are the most common choice in large and very large HAWTs since they reach a balance between efficiency, stability, and mechanical reliability. The three-blade configuration ensures smooth rotation, reduces vibrations, and optimizes energy capture. They have slightly higher material and manufacturing costs compared to two-bladed designs.

- <u>Multi-Bladed Rotors</u>
  These more easily generate high torque, making them ideal for low-wind-speed environments and mechanical applications like water pumping. Their increased drag reduces overall efficiency, and the design is not suitable for high-speed, large-scale energy generation.

- **<u>Hub</u>**
  The hub connects the rotor blades to the turbine's drivetrain. It transmits the rotational motion of the blades to the low-speed shaft. The hub hosts pitch-control devices, when present. As already explained, these devices may be used

to continuously control blade angles to optimize performance, or, in simpler cases, to feather the blades in emergency shutdown phases due to very high winds. Simpler small-sized HAWTs do not have any pitch-control system: their blades are commonly designed to stall under high-wind conditions to prevent turbine damage. This design approach is normally possible only for small machines, with a low centrifugal load, and also leads to poorer efficiency.

- **Nacelle**
  The nacelle is the enclosure that houses the key mechanical and electrical components of the turbine (gearbox, generator, control systems). It sustains the hub and rotor, and it is mounted on the top of the HAWT sustaining tower. It must rotate on the tower axis to permit turbine orientation depending on the inlet wind direction. This movement is defined and controlled by the HAWT control and yaw systems.

  In large-size HAWTs, the nacelle with its content may weigh hundreds of tons and has an overall size which permits physical inspection for turbine mounting and maintenance purposes. It is accessible through a properly designed staircase which normally is placed inside the tower.
- **Electric Generator**
  This converts rotational energy into electrical energy. The most frequent choice is to use an AC generator and to connect it directly to the grid. In some other cases, as better explained below, a DC generator may be used, or the AC generator may not be directly connected to the grid, but through properly designed power electronic converters. In both these cases, some power electronic converters should be added to the HAWT, but the electric generator rotor can be assembled directly on the wind rotor, with no need for any transmission device. These machines are referred to as "direct-drive" HAWTs and permit the variation of rotor speed on a very wide range so as to optimize rotor efficiency. They are particularly of interest in sites with a widely distributed wind (high σ JFFs).
- **Gearbox – Speed Multiplier**
  When present (in most large-scale devices), this is designed to increase the rotational speed of the shaft to match the electric generator's operating speed. It is needed if no power electronics (a DC/AC inverter, or two coupled AC/DC-DC/AC converters) are used to adjust the electricity produced by the generator to make it compatible with the electric grid. Most electric grids worldwide, in fact, are AC grids with a fixed and stable frequency within the range $f = 50 - 60\,Hz$.

  If an AC electric generator is used to produce electricity and it is directly connected to the grid, its electromagnetic rotor must rotate at a speed which can only be an integer sub-multiple of 3000 rpm (corresponding to a grid frequency $f = 50\,\text{Hz}$), or 3600 rpm (if $f = 60\,\text{Hz}$). This integer division factor would be the number of rotating dipoles in the electric machine, which is normally lower than 5–6.

  This means that the electromagnetic rotor must rotate steadily at a fixed speed of at least several hundreds of rpm. On the contrary, as already discussed, for structural reasons, the wind rotor of a large-size HAWT must rotate lower than about 50–60 rpm (or even less, depending on machine size).

Moreover, this speed should be possibly varied in a certain range ($\varpi_{min} - \varpi_{Max}$) to optimize rotor efficiency.

Therefore, in these machines the need for an efficient speed multiplier is straightforward. This gearbox has a high power (up to 10 MW) and normally must realize an increasing higher transmission ratio as the machine size increases (due to the decrease in maximum turbine rotor speed).

This is normally obtained through cascade gearings, also using epicyclic gearing in some cases (having higher efficiency for high transmission ratios).

The final transmission gearing may be conic, so defining a variable transmission ratio (at least to some extent) makes it possible to implement rotor speed machine optimization techniques in these machine types also.

On the contrary, if the gearbox has a constant (or discretely variable) transmission ratio, further optimization possibilities are normally implemented on the HAWT through pitch-control devices. Some direct-drive turbines eliminate the gearbox for simplicity.

- **Electric Power Converters**
  As already explained, some electronic power controllers may be needed in HAWTs to make electricity compatible with the electric grid. Normally, these power controllers, when needed, are installed at ground level in proper housing. They may also be shared between several HAWTs depending on the application size and grid connection conditions.
  For large-size generators, and most national contexts, electricity is transmitted to the grid at a mid-voltage level, in AC at a fixed and stable frequency within the range $f = 50 - 60\,\text{Hz}$. Normally, for direct-drive HAWTs, electric grid compatibility is obtained through:
    - DC/AC inverter, if a DC generator was coupled with the turbine rotor.
    - Two coupled AC/DC DC/AC cascade converters, if an AC generator was coupled with the turbine rotor.
- **Control System**
  This monitors wind conditions (intensity and direction) and interacts with the electric generator and the yaw system, eventually modifying their preset internal setpoints according to efficiency and safety factors.
- **Yaw System**
  This is the control and actuation system which ensures that the turbine is aligned with the wind direction for maximum energy capture. In downwind HAWTs, alignment is normally achieved through a passive system that uses wind thrust on tail-aligning surfaces, while IN upwind HAWTs a yaw motor is typically connected to the hub axis via a clutch and a transmission system.
- **Brake System**
  This is a critical passive safety feature that slows or stops the rotor in high winds or during maintenance. In normal conditions, in fact, the rotor is only braked electrically through the electric generator. The main braking action is given by disk or drum brakes, but also blade pitch control may assist rotor braking.

- **Tower**
  The tower supports the nacelle and rotor blades at a height where wind speeds are greater and more consistent. It is commonly made in tubular steel for utility-scale turbines, while lattice or hybrid towers may be used for smaller turbines. Its height varies depending on the turbine size and maximum power output, typically ranging from 30 metres for small turbines to over 120 metres for large turbines.
- **Foundation**
  The tower foundation anchors the HAWT turbine to the ground, providing stability to withstand wind loads and vibrations (see lower right part of Figure II.3.13). Commonly, steel-reinforced concrete foundations are used for onshore land-based turbines, while pile foundations are more common for offshore turbines.

  In some cases, offshore HAWTs may be buoyant. In fact, these are designed to float on the water's surface rather than being fixed to the seabed. Unlike fixed-bottom turbines, buoyant foundations can be installed in waters deeper than 60 metres. They can be placed far offshore, where winds are stronger and more consistent.

  A floating platform provides the structural base for the turbine and keeps it buoyant. Platforms are typically made of steel or reinforced concrete to withstand the marine environment, and may be of distinct types:

  - Spar Buoys: Long, cylindrical structures that extend deep below the water's surface, providing stability through ballast.
  - Semi-Submersibles: Wide, multi-column platforms that achieve stability through a combination of buoyancy and mooring systems.
  - Tension Leg Platforms (TLPs): Platforms tethered to the seabed with mooring lines, minimizing vertical movement.

  A Mooring System anchors the floating platform to the seabed, ensuring it stays in position despite waves, currents, and wind. It is composed of mooring lines (chains, steel cables, or synthetic ropes connect the platform to the seabed) and anchors (drag anchors, suction anchors, or gravity anchors, depending on the seabed type: rocky, sandy, …).

  A Ballast System is used to lower the centre of gravity and improve the platform's stability: heavy materials, such as concrete or water, are added to the lower parts of the platform (e.g., in a spar buoy or semi-submersible). Their weight counterbalances the forces acting on the turbine, reducing tilting or excessive motion.

II.3.2.8 On the Sizing of High-Voltage Cables Inside Cable Ducts

HAWTs are normally installed in rural areas, often far from the local electric grid. Therefore, the dimensioning of electric cables connecting the power plant to the high-voltage distribution cabin may become a crucial part of the windfarm project.

When installing large-size HAWTs, the electrical connections between the different wind towers, and between them and the electrical transformation station are all made at high voltage (e.g. 36 kV), in order to minimize power losses. The sizing of cables is performed in compliance with the reference regulations.

For example, in the EU, the reference technical regulations are: IEC 60840 (about power cables with extruded insulation and their accessories for rated voltages above 30 kV up to 150 kV); EN 60909 (short-circuit currents in three-phase AC systems); and IEC 60287 (electric cables and calculation of the current rating).

The cable section is selected to meet the following conditions:

$$I_b \leq I'_z; \quad \Delta V \leq 4\%; \quad \Delta P \leq 5\%,$$

where:

- I_b represents the load current, representing the maximum current flowing through the cable line. It is calculated as a function of maximum power for each track (P_n), nominal voltage of the plant, V_n, and power factor, $cos\,\varphi$:

$$I_b = \frac{P_n}{cos\,\varphi V_n \sqrt{3}}$$

- I'_z represents the actual current-carrying capacity of the cable. It depends on the nominal carrying capacity of the cable and on the relative laying conditions along the entire route (usually provided by the cable manufacturer).
- ΔV represents the maximum voltage drop along the cable line. It is a function of I_b, $cos\,\varphi$, cable phase resistance R_f, and longitudinal phase reactance of the cable, X_f, according to:

$$\Delta V = I_b \left(R_f\, cos\,\varphi + X_f\, sin\,\varphi \right) \sqrt{3}$$

- ΔP represents the power loss by Joule effect for each of the subfields, taking into account the length of the considered line section, L:

$$\Delta P = 3 R_f L \cdot I_b^2$$

The process of dimensioning HV cables begins with selecting the type and diameter of the cable. Once all three conditions ($I_b \leq I'_z$; $\Delta V \leq 4\%$; $\Delta P \leq 5\%$) are satisfied, the iterative sizing process is considered complete.

II.3.2.9 Main Environmental Impacts of Large-Scale HAWTs

Large-scale HAWTs are generally considered environmentally sustainable energy solutions, but their immense size and operational requirements can lead to notable disturbances in the ecosystems in which they are installed. These impacts, while not inherently severe, must be carefully assessed and mitigated to ensure sustainable integration into their surroundings. Environmental effects arise during the construction, operation, and eventual decommissioning phases, potentially affecting ecosystems, wildlife, and local communities.

The main impacts include the following:

- Impact on Wildlife
  The rotor blades can pose significant risks to flying species, especially migratory birds and bats. These animals may alter their behaviours, avoid certain areas, or suffer fatalities due to blade strikes. Effective site selection and advanced detection systems, such as radar-based monitoring or acoustic deterrents, can help reduce these risks. Furthermore, the construction of wind farms in sensitive ecosystems can disrupt habitats, potentially leading to the displacement or decline of certain species.
- Significant Land Use and Visual Impact
  While turbines occupy relatively little ground space themselves, wind farms require substantial land for turbine placement, access roads, and supporting infrastructure. This can result in changes to land use patterns, potentially interfering with agriculture, forestry, or other natural landscapes. Additionally, the visual impact of tall turbines with rotating blades can alter scenic views, which may negatively affect tourism or lead to reduced local acceptance of the wind farm.
- Noise Pollution
  Wind turbines produce mechanical and aerodynamic noise, which can be disruptive to nearby communities. Mechanical noise arises from turbine components such as the gearbox and generator, while aerodynamic noise is caused by the movement of the blades through the air. Modern turbine designs incorporate noise-reduction features, but careful site planning and minimum distance regulations remain critical to minimizing these disturbances.
- Marine Ecosystem Effects
  For offshore HAWTs, the installation of foundations and mooring systems can disrupt seabed habitats, potentially affecting marine life. However, in some cases, turbine foundations can act as artificial reefs, providing habitats for new marine ecosystems. Proper planning and ecological assessments can minimize negative effects and, in certain instances, even promote biodiversity.
- Electromagnetic Interference
  The operation of turbines may interfere with radar systems or telecommunications due to the physical presence of turbine structures and the rotation of blades. This issue is more significant in locations near airports, military installations, or telecommunications infrastructure. Advanced technologies and strategic placement of turbines are typically used to minimize these effects.

II.3.2.10 High-Altitude Wind Turbines

High-altitude wind turbines, also known as airborne or high-altitude wind energy systems, are an innovative approach to wind energy generation. These systems harness the strong and consistent winds found at higher altitudes, typically ranging

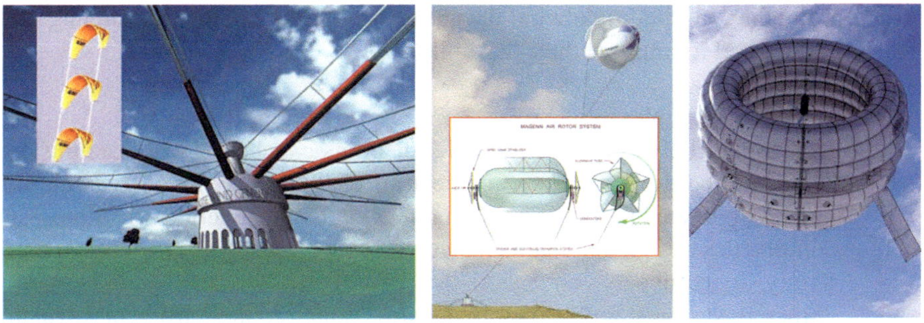

FIGURE II.3.17 Different design approaches for high-altitude wind turbines.

from hundreds to thousands of metres above the surface. By operating at these elevations, they tap into faster and more stable wind resources, presenting a transformative potential for renewable energy generation, especially in areas with limited surface wind resources.

Unlike conventional turbines, which are fixed to the ground or sea floor, high-altitude systems float or fly in the air, utilizing advanced technologies to remain aloft. At heights of 300–1000 metres or more, wind speeds are often two to three times faster than those near the ground. This exponential increase in wind power density with altitude stems from the reduced influence of surface friction.

High-altitude wind energy systems are still in their developmental stages and are characterized by low technological and manufacturing readiness levels (TRLs and MRLs). Despite their potential, several challenges must be addressed before they can become commercially viable.

Several innovative design approaches have been proposed for high-altitude wind energy systems (see Figure II.3.17):

- Kite-Based Systems
  These systems employ tethered kites or wings to generate lift and extract energy from high-altitude winds. The kite moves in a controlled figure-of-eight or circular pattern, pulling on a tether that is connected to a ground-based generator. Some designs feature onboard generators, where power is transmitted to the ground through conductive tethers.
- Flying Rotor Systems
  These systems consist of airborne turbines, resembling traditional rotor blades, suspended by tethers. The rotors generate electricity directly in the air, transmitting it to the ground via specialized cables. Stability at high altitudes is often achieved using helicopter-like designs or blimps, which employ part of the wind-generated lift to remain aloft through the Magnus effect (a lift effect generated by the rotation of an aerofoil in airflow).
- Helium-Filled Platforms
  Helium-filled balloons or airships are used to elevate lightweight turbines to high altitudes. The turbines, mounted on the floating platforms, capture wind energy and transmit electricity to the ground via tethers.

While promising, high-altitude wind turbines face several significant challenges that must be addressed to improve their feasibility:

- Technical Complexity:
  Developing lightweight yet durable materials capable of withstanding high-altitude conditions is a key engineering challenge. Tethered systems must handle high tension forces, wear and tear, and atmospheric variability, which complicates design and maintenance.
- Safety Concerns:
  Tethered systems pose risks to aviation and require stringent regulatory oversight to ensure safe deployment. A failure in the tethering mechanism could lead to catastrophic damage to the turbine, nearby structures, or the environment.
- Energy Transmission:
  Transmitting electricity from high altitudes to the ground involves specialized cables or wireless systems, adding complexity and cost.
- Regulatory and Legal Hurdles:
  Airspace usage and interference with aviation must be addressed through strict regulations. Deployment may face resistance due to concerns over air traffic, noise, or aesthetics.
- Weather Dependency:
  High-altitude turbines must withstand extreme weather conditions, including strong gusts, lightning, and icing.

In summary, wind power provides a scalable and sustainable energy solution. Advances in turbine technology and grid integration strategies are essential for maximizing wind energy's potential. In particular, off-shore applications could play a relevant role in many contexts.

11.4 Hydro Energy

The utilization of hydro energy dates back thousands of years. Its application for electricity generation is also far from recent, as it was one of the pivotal drivers of the Second Industrial Revolution in the late 19th century.

For these reasons, hydro energy must be regarded as a fully established and conventional energy source. Among all such sources, it stands out as the most sustainable and environmentally compatible, although its conversion processes can sometimes lead to significant environmental impacts.

Hydro energy typically refers to the energy contained in liquid water as it flows freely across the Earth's surface. This energy is harnessed in both its potential (gravitational) and kinetic forms. From an energetic perspective, water is particularly advantageous due to its high energy density.

Water density, under the same environmental conditions, is approximately 1000 times greater than air density. This difference has a substantial impact on the design of turbines: water turbines can achieve the same energy output as air turbines with significantly smaller sizes, with the size ratio often reaching 1:1000.

Water on Earth participates in a natural cycle, alternating between liquid and vapour forms. In this cycle, it absorbs solar energy and drives energy transformation processes that we can intercept for human use.

Among these processes, the most energetically significant is the transformation of potential energy into kinetic energy in river basins. As water flows from mountains and hills towards the sea, it naturally dissipates energy, creating opportunities for energy capture.

Beyond river basins, saltwater bodies such as oceans also contain significant energy potential. This potential arises from processes driven by temperature and density gradients, as well as gravitational interactions with celestial bodies.

Ocean currents, sea waves, and tidal waves represent promising sustainable energy sources. Although these have not yet been widely exploited, they are likely to play an increasingly significant role in the future, particularly in specific regions of the world.

This chapter will first examine the technologies typically employed in conventional hydroelectric plants situated within river basins. Following this, we will explore how similar methods and technologies might be adapted to harness energy from other hydro energy sources, such as sea waves.

DOI: 10.1201/9781003631125-19

II.4.1 RIVER BASINS ENERGY

II.4.1.1 Source Characteristics

Water flowing in a river undergoes highly dissipative flows. Since it remains in constant contact with the surrounding atmosphere, which is incompressible, it cannot increase its pressure. As a result, water continuously transforms its potential gravitational energy into kinetic energy as it flows downstream, with much of this energy dissipated during the process.

By applying the energy conservation law between two sections of a river, the power available for energy conversion processes (P_{IN}) is fully defined in terms of the water's volumetric flow rate (Q, also referred to as the river's "discharge") and the geodetic altitude difference (ΔH, often called "head"). The relationship is expressed as:

$$P_{IN} = \rho \cdot g \cdot Q \cdot \Delta H = 9.81 \cdot Q \cdot \Delta H \left[\text{kW}\right]$$

where Q is expressed in m^3/s and ΔH in m.

River basins exhibit a wide range of discharges. Average discharges typically vary between 1 and 100 m^3/s, although larger rivers can far exceed these values (e.g., over 10^5 m^3/s for the Amazon River and over 10^4 m^3/s for the 25 largest rivers worldwide).

For each metre of head harnessed by energy conversion technologies, the potential power inlet (P_{IN}) could reach: approximately 1 GW for the Amazon River; 100 MW for the largest 25 rivers globally; 10 kW to 1 MW for most small to medium-sized river basins.

The number of metres of head available for energy harvesting depends heavily on local river orography. In certain regions, rivers rapidly lose altitude as they flow from mountains or hills, while in others, they run almost flat.

These variations underscore the vast local and global potential of river basin energy.

However, extracting all the water flow from a basin for energy purposes is not permissible. Regulatory frameworks mandate a minimum vital outflow (MVO or MVF), also referred to as the ecological or environmental flow. This represents the amount of water that must remain in the river to sustain its ecological health and functionality.

The MVF ensures the survival of aquatic ecosystems, fish habitats, and riparian vegetation, even when water is diverted for energy production or other uses. Determining this flow requires balancing ecological preservation with economic interests, accounting for factors such as seasonal variations, local biodiversity, and community needs.

Failure to maintain the MVF can result in reduced water quality, biodiversity loss, and long-term environmental degradation.

Establishing the MVF and understanding river behaviour require detailed hydrographic knowledge, typically obtained through prolonged flow rate measurements at various river sections.

These measurements are often represented graphically in a river hydrograph, which shows how a river's discharge (flow rate) varies over time at a specific location.

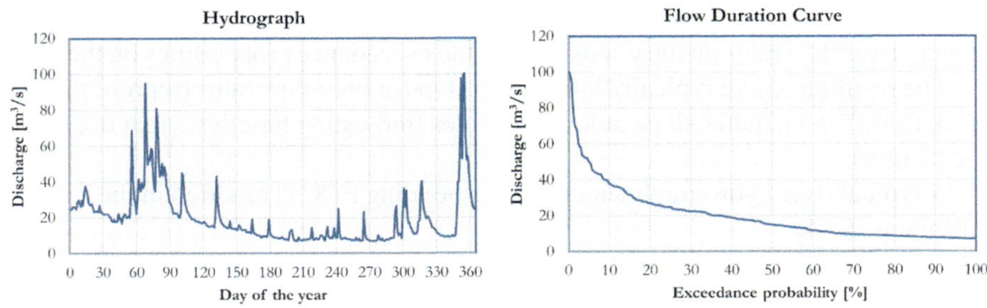

FIGURE II.4.1 A typical river hydrograph and its corresponding flow duration curve.

It provides information on how a river responds to different events, such as precipitation, snowmelt, or upstream water release.

The left part of Figure II.4.1 illustrates a typical hydrograph of a river section.

If the river is not fully seasonal (i.e., not dry at certain times of the year), the hydrograph identifies the baseflow, which is the portion of the discharge sustained by groundwater seepage during dry periods.

The hydrograph also shows typical river discharge seasonal and/or impulsive variations. For example, it permits the identification of the so-called rising (where the discharge increases due to an input event, such as rainfall or snowmelt) and recession limbs (the drainage of water from the catchment and return to baseflow levels), as well as river peak flow (the maximum discharge recorded, for example during a rainfall event). It is also use to determine the river's typical lag time (for example between the peak of the rainfall event and the peak river discharge), which reflects the catchment's response time of the river with respect to rainfall events in the territory and is therefore influenced by factors like soil type, land use, and slope.

Hydrographs can represent single events (event hydrographs) or summarize annual or multi-annual observations.

The latter are particularly valuable for energy-harvesting designs. Data from annual or multi-annual hydrographs are used to construct flow duration curves (FDCs).

An FDC graphically represents the relationship between the magnitude and frequency of streamflow in a river or watercourse over a specified period. It provides a summary of the variability in flow and is widely used in hydrology, water resource management, and environmental studies. The FDC is crucial for understanding a river's behaviour, planning water usage, and managing both ecological and human water needs.

The FDC represents the percentage of time (duration) that a specific flow rate is equalled or exceeded in a river. It is constructed by sorting the flow data by the river hydrograph in descending order, from the highest to the lowest values.

When flow rate events are so reordered, they easily permit the evaluation of the percentage of time in which each given flow rate was exceeded, which is the so-called exceedance probability. To achieve this aim, a simple re-scaling of the x-axis is needed.

Therefore, the FDC is a graphical statistical representation of river flow rate events having river flow rates on the y-axis against the exceedance probabilities on the x-axis.

The resulting curve typically has a steep slope at high flow rates (representing rare peak flows) and a flatter slope at low flow rates (indicating baseflows that occur more frequently).

A typical river hydrograph with its corresponding FDC is reported in the right side of Figure II.4.1.

It graphically permits the identification of:

- High flows level
  The upper curve portion represents rare, large flows critical for flood risk assessments.
- Median flows level
  The mid-section of the curve indicates typical flow conditions, often used for assessing water availability for irrigation, domestic, or industrial use.
- Low flows level
  The lower curve portion shows the baseflows, essential for understanding drought conditions and maintaining ecological flows.

Hydropower engineers use FDCs to estimate potential energy generation by analysing flow variability.

The curve highlights flow ranges that occur most frequently, enabling the design of turbines optimized for these conditions. It also informs plant sizing, balancing investment costs with capacity factors.

FDCs are critical for: setting thresholds for MVO to balance power production with environmental sustainability; assessing water availability for competing uses (e.g., agriculture, drinking water, and industry); planning storage requirements (e.g., reservoir sizing); and optimizing control algorithms to manage seasonal and annual flow variability.

II.4.1.2 Hydropower Plant Types and Components

Hydropower plants are categorized based on how they harness the energy of flowing water.

The primary distinction is between two types of plants (see Figure II.4.2), with the choice depending on environmental, economic, and geographic considerations:

- **Impoundment Hydropower Plants**
  Impoundment plants (illustrated on the left side of Figure II.4.3) utilize a dam to store water in a reservoir, creating potential energy. Water is released from the reservoir through turbines to generate electricity. These systems are highly versatile, offering consistent energy output and the ability to regulate power generation by controlling water discharge. Additionally, they can support other applications such as flood control, irrigation, and recreational activities. However, impoundment systems can significantly alter river ecosystems

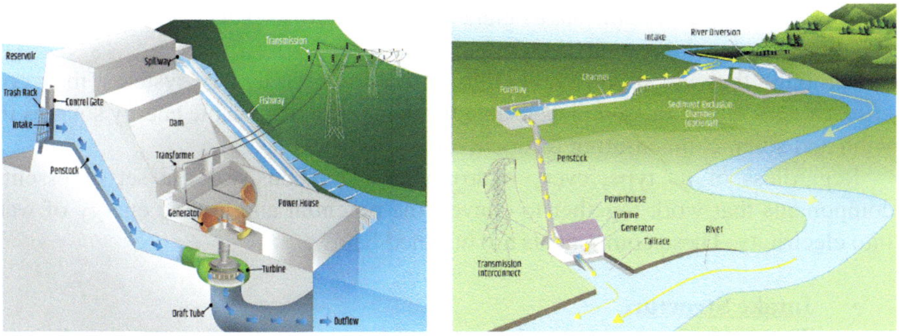

FIGURE II.4.2 Types of hydropower plants.

FIGURE II.4.3 Typical turbine rotors design for various turbine specific speed N_s values.

and may displace local communities, making their environmental and social impacts substantial.

- **Diversion (Run-of-River) Hydropower Plants**
  Diversion plants (illustrated on the right side of Figure II.4.3) channel part of a river's flow through a canal or penstock to drive turbines, often without requiring a large reservoir. These systems rely on the river's natural flow, making them less intrusive to ecosystems compared to impoundment plants. However, their reliance on seasonal and flow variations can result in less consistent power output.

Hydropower plants can also be installed between two water reservoirs, whether or not they are part of a river basin. These systems are typically used to store electricity temporarily, such as during periods of low demand (often overnight). This is the case of **pumped storage hydropower plants**.

Pumped storage hydropower plants function as large-scale energy storage systems. Excess electricity from the grid is used to pump water from a lower reservoir to an upper reservoir during off-peak times when electricity is cheaper. When demand rises and electricity prices increase, water is released from the upper reservoir back to the lower reservoir, passing through turbines to generate power.

Pumped storage systems act as giant batteries, storing energy and enhancing grid stability by balancing supply and demand.

Although they are slightly less efficient than modern battery systems, their cost-effectiveness makes them indispensable for integrating renewable energy sources like wind and solar, compensating for their variability.

Regardless of its type, every hydropower plant consists of a set of essential components that work together to convert the potential and kinetic energy of water into electricity. These components are outlined below:

- **Intake Structure**
  The intake controls the flow of water entering the system and directs it towards the turbines. Trash racks are installed to prevent debris from damaging the turbines, and adjustable gates regulate the volume of water entering the system. The intake is designed to minimize turbulence and energy losses while maintaining a steady water flow.
- **Penstock**
  The intake feeds the penstock, a large, reinforced pipe or tunnel that channels water to the turbines. Penstocks are typically made of steel or concrete to withstand high pressures. Their diameter and length are carefully selected to minimize frictional losses while maximizing flow efficiency. To manage pressure fluctuations and prevent damage during sudden flow changes, penstocks are often equipped with surge tanks.
- **Turbines**
  At the turbines, water from the penstock retains its original flow rate but has an increased pressure. Turbines are engineered to convert this energy into mechanical rotational power before returning the water to the riverbed. Turbine design is tailored to the specific combination of head and flow rate specifications for the site. In the next section, we discuss the types of turbines in more detail.
- **Electric Generator**
  A synchronous AC generator is typically used to produce electricity compatible with the electric grid. In most cases, this generator is directly connected to the turbine hub, avoiding the need for gearings or transmission systems and thereby reducing energy losses.

  As previously discussed in relation to wind turbines, global electric grids predominantly operate on alternating current (AC) at stable frequencies (e.g., $f = 50\,\text{Hz}$ or $f = 60$ Hz depending on the country).

  For compatibility, the electromagnetic rotor of the generator must rotate at a fixed speed that is an integer sub-multiple of 3000 rpm (for $f = 50\,\text{Hz}$ grids) or 3600 rpm (for $f = 60\,\text{Hz}$ grids).

  The specific speed (N) is determined by:

$$N = \frac{f}{\#_{dip}} \cdot \frac{2\pi}{60} \cdot [\text{rpm}]$$

- **Draft Tube**
  The draft tube is located at the turbine's outlet and directs the water back into the river or tailrace. Its design slows the velocity of water exiting the turbine, reducing energy losses and improving turbine efficiency. At the end of the draft tube, a tailrace is typically constructed. This channel or pipe carries water away from the draft tube to the downstream river or water body, minimizing backpressure on the turbine.

II.4.1.3 Hydropower Turbines

The design of hydropower turbines is primarily influenced by the pressure head (ΔH) and the discharge (Q) they need to process. These parameters define the turbine's inlet power and, consequently, its nominal power output (P_{OUT}), considering a certain efficiency (η) of the device. The nominal power output can be expressed as:

$$P_{OUT} = 9.81 \cdot \eta \cdot Q \cdot \Delta H \left[kW\right]$$

The rotational speed of the turbine (N) also plays a significant role, but as previously noted, it is usually determined by the electric grid frequency (f) and the number of magnetic dipoles ($\#_{dip}$) in the generator's rotor.

Using Buckingham's Π Theorem, these parameters are shown to be dynamically interrelated, which simplifies turbine design.

A specific parameter known as the **turbine specific speed** (N_s) is commonly employed to characterize design conditions. It is defined as:

$$N_s = N \cdot P_{Out}^{\frac{1}{2}} \cdot \Delta H^{\frac{5}{4}}$$

Turbines with the same N_s are dynamically analogous, meaning they share similar geometric proportions, dynamic behaviour, and efficiency characteristics.

Therefore, N_s is a key factor in selecting appropriate turbine designs for specific applications.

It is essential to note that N_s is determined based on the parameters ($N, P_{Out}, \Delta H$) of each individual turbine, not the entire hydropower plant. While ΔH, and N are fixed for a certain plant and river configuration (at most varying depending on $\#_{dip}$), P_{Out} for a single turbine depends on the total number of turbines chosen (and correspondent gates and penstocks).

This provides flexibility for designers to vary P_{Out} and, consequently, N_s, to select the most suitable turbine design.

Depending on N_s ranges, the main consolidated approaches for hydro turbine design are the following (see Figure II.4.3 for reference):

- **Low Specific Speed – Pelton Turbines**
  According with the N_s definition, low N_s values ($N_s < 50$, if N is expressed in rpm, P_{Out} in kW, ΔH in metres) are obtained for high heads and low flow rates.

These conditions more commonly happen in diversion than in impoundment hydropower plants.

So-called "Pelton" turbines are a consolidated design for these N_s values (see Figure II.4.4). In particular, these turbines (first designed by Lester Allan Pelton in the 1870s) are pure impulse turbines (converting the whole inlet head into kinetic energy in the stator section) which are ideal for steep waterfalls or high-elevation reservoirs.

The Pelton rotor is isobaric: water flow is ejected through one or more nozzles on distinct buckets mounted on a wheel, which are designed to produce a high thrust properly deviating the fluid.

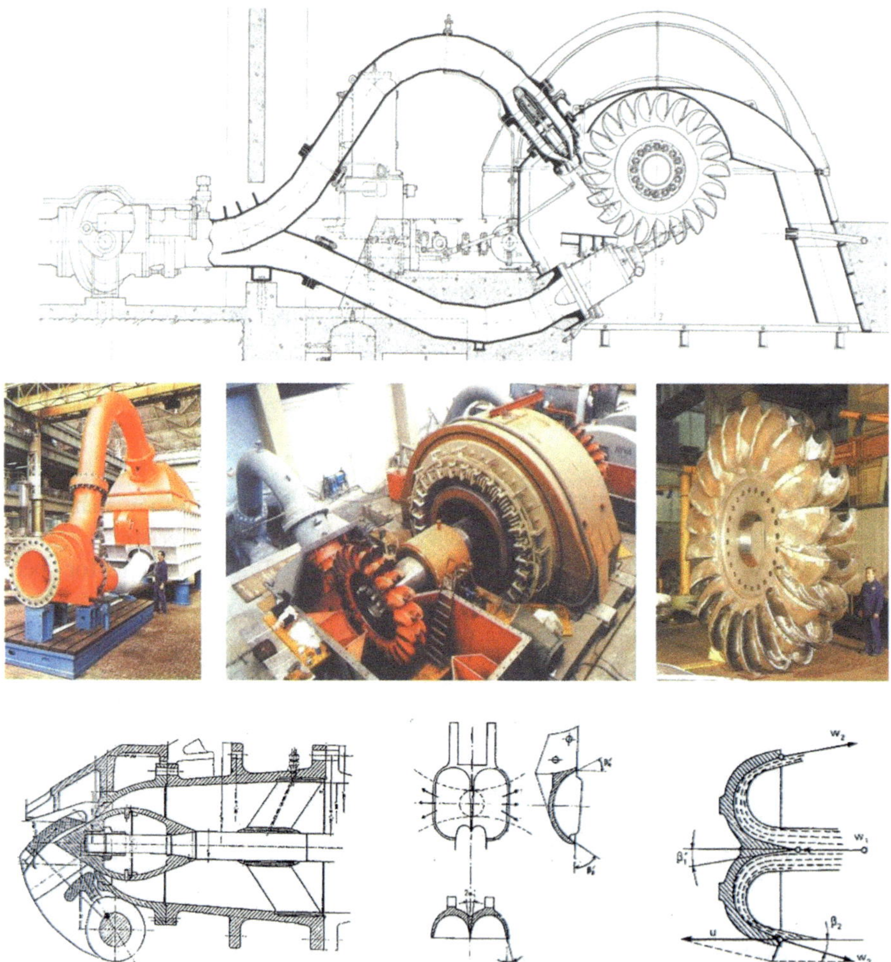

FIGURE II.4.4 Photos and schematic diagrams of a Pelton turbine and its parts.

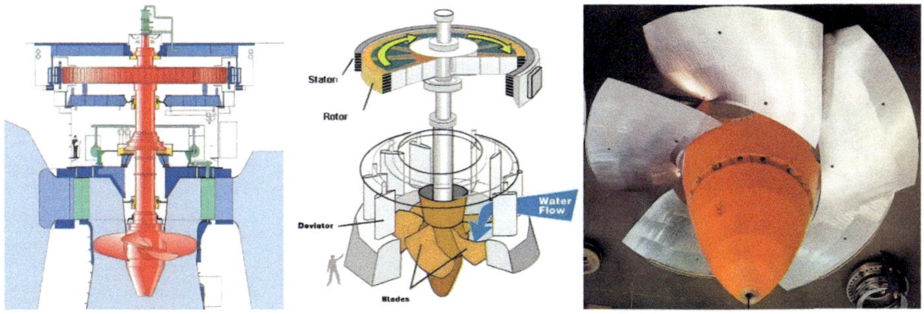

FIGURE II.4.5 Photos and schematic diagrams of a Kaplan turbine and its parts.

Pelton turbines are regulated through statoric devices to adjust head and flow rate variations:

- "Double" Spines: Movable components that adjust the nozzle's convergence ratio to adapt to variations in the head conservatively.
- Divergence Plates: Dissipative obstacles that divert part of the water flow, preventing it from impacting the bucket wheel during sudden flow changes.

Pelton turbines are versatile in size and output, ranging from 10 kW to over 10 MW, while maintaining a similar design approach.

- **High Specific Speed – Kaplan Turbines**

  So-called "Kaplan" turbines (see Figure II.4.5) are the more consolidated turbine design for high N_s values of $N_s > 200$. An alternative is the so-called "propeller" turbines. Both are used for low-head, high-flo w situations, such as in run-off-river hydropower plants. They have a high degree of reaction, using mainly the pressure energy of water to produce thrust on rotor blades.

  They are axial turbines and feature variable-pitch blades, which can be adjusted to optimize performance based on the inlet flow rate.

  This adaptability makes Kaplan turbines highly efficient in conditions with fluctuating water flow.

- **Medium Specific Speed – Francis Turbines**

  So called "Francis" turbines (see Figure II.4.6) are the more consolidated turbine design for medium N_s values of $50 < N_s < 200$, making them suitable for medium-head and medium-flow conditions, such as those found in many run-of-river plants or dams with moderate elevation differences.

  Francis turbines use both the pressure and kinetic energy of water to produce thrust on rotor blades.

II.4.2 WAVE ENERGY

II.4.2.1 Source Characteristics

The energy content of a water wave is a critical concept in oceanography and renewable energy engineering, particularly for wave energy conversion systems. Waves carry significant energy due to the interactions of wind, water, and gravity. This

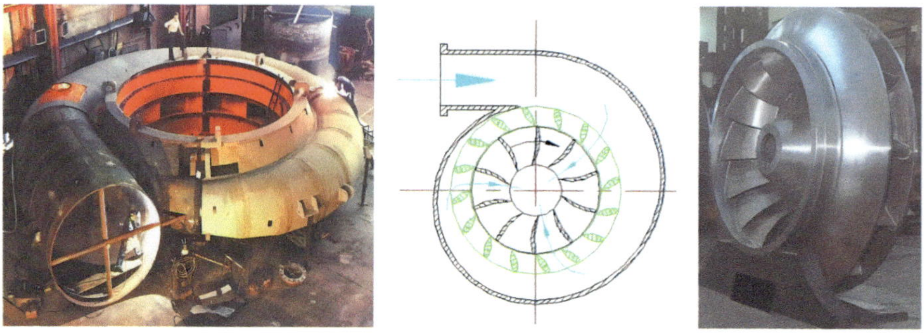

FIGURE II.4.6 Photos and schematic diagrams of a Francis turbine and its parts.

energy is distributed between kinetic and potential components and can be harnessed for power generation.

This section employs the linear wave theory, a widely used mathematical model for describing small-amplitude waves propagating on a fluid surface.

This theory provides reasonably accurate predictions under many practical conditions and is computationally straightforward, making it particularly useful for engineering, coastal studies, and wave energy applications.

Linear wave theory relies on several simplifying assumptions that limit its applicability but make the problem mathematically tractable:

- **Small Waves in Deep Water**
  The wave height (H) is assumed to be small compared to both the wavelength (λ) and the water depth (h). This ensures that wave motion can be treated as a small perturbation, enabling the linearization of the Navier–Stokes equation.
- **Incompressible and Ideal Water**
  The water is treated as incompressible and non-viscous, meaning no energy is lost due to friction.
- **Irrotational Flow**
  The flow of water particles is assumed to be irrotational, meaning there is no vorticity or rotational motion in the fluid.

Under these assumptions, according to Figure II.4.7, it can be demonstrated that the power content of the wave per unit width can be evaluated as:

$$P = \left[\frac{\rho \cdot g^2}{32 \cdot \pi} \cdot tanh\left(\frac{2\pi}{\lambda} h\right)\right] \cdot H^2 \cdot T \approx 0.955 \cdot H^2 \cdot T, \text{ in}\left[\frac{\text{kW}}{\text{m}}\right]$$

where: ρ is the water density; g is the gravitational acceleration; H is the wave height; T is the wave period; λ is the wavelength.

Therefore, so-called wave energy flux (P) is proportional to the product of the wave period and the square of the wave height, so that long-period waves (greater T)

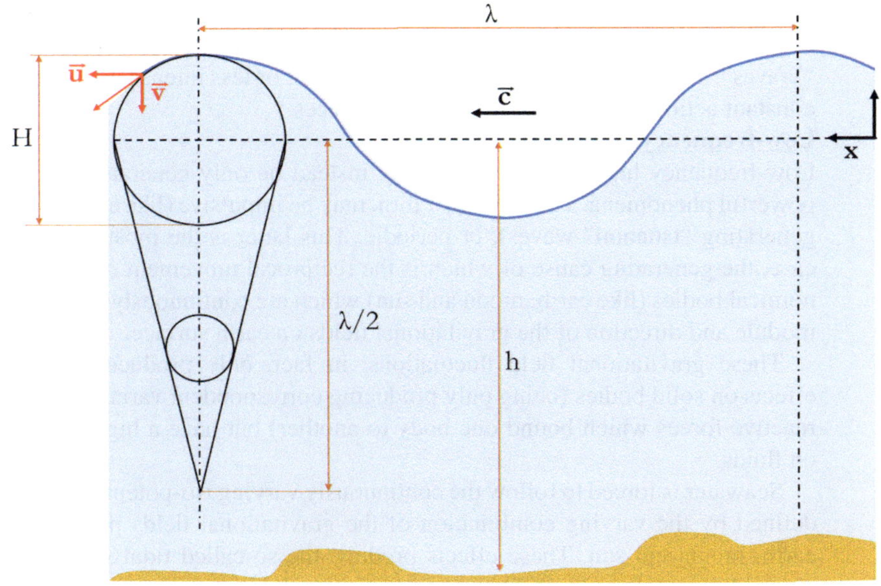

FIGURE II.4.7 Graphical schematic representation of a linear ideal wave.

typically carry more energy than short-period waves because they represent longer oscillations and greater energy transfer over time.

However, even short-period waves (on the scale of tenths of seconds) contain significant energy densities (on the order of tenths of kilowatts per metre of wave front and height, which, as an example, is several times higher than the peak energy density of solar radiation). This makes wave energy a highly promising renewable resource, especially in areas where waves are frequent and predictable.

In an ideal linear wave, each water particle exhibits an orbital motion that is spatially and temporally out of phase with neighbouring particles. This phase difference facilitates the steady movement of the wave.

For deep-water waves, the water particles move in circular orbits, with the radius of these orbits decreasing linearly with depth.

The wave energy density is the total energy stored per unit surface area by its water particles. Therefore, it consists of both kinetic energy (associated with the motion of water particles within the wave) and potential energy (due to the elevation of the water surface above its mean level).

Summarizing, in a wave there are no water transport phenomena, but an energy flow is steadily transferred outside the examined dominium. How is this possible? Because the presence of a wave is only justified by the corresponding presence of generating causes which somehow feed the wave with an equivalent steady energy content, making its steady particles' movement possible as well.

In this sense, the amplitude and period of the generated waves are a direct consequence of their generating energies:

- **High-frequency, low-energy waves**
  High-frequency low-energy waves are commonly simply referred to as "Waves". They are normally generated by a more or less intense, uniform and constant action of the wind on wide sea surfaces.
- **Low-frequency, high-energy waves**
  Low-frequency high-energy waves may instead be only generated by more powerful phenomena. These, in their turn, may be impulsive (like earthquakes, generating "tsunami" waves), or periodic. This latter is the most interesting case, the generating cause of which is the reciprocal movement of big astronomical bodies (like earth, moon and sun) which are continuously varying the module and direction of the gravitational fields on earth surface.

  These gravitational field fluctuations, in fact, only produce negligible effects on solid bodies (being only producing correspondent variations on the reactive forces which bound one body to another) but have a high influence on fluids.

  Seawater is forced to follow the continuously varying iso-potential surfaces defined by the varying combination of the gravitational fields produced by earth, moon and sun. These effects produce the so-called tidal waves (also referred to simply as "tides").

  Tidal waves are far from being described as ideal water waves due to the constraint to possible water movements caused by the emerging lands and to the large non-uniformity of water depth in the various seas and oceans.

  As a general rule, tidal waves are much less evident and powerful in closed seas (e.g. the Mediterranean Sea) while they are more visible and powerful in the open ocean.

  Local effects, however, can excite the tidal movements and intensify their energy contents making them more interesting from an energy-harvesting perspective.

  Linear wave theory cannot be used to estimate quantitatively the power content of a tidal wave, which is, instead, more correctly described by semi-empirical correlations like:

$$P = \frac{\rho}{2} \cdot h \cdot U^3 \approx 5 \cdot h \cdot U^3, \quad \text{in} \left[\frac{\text{kW}}{\text{m}} \right]$$

where h is local sea-water depth in metres, and U is mean tidal speed in m/s. Tidal waves, in fact, are very different from small waves because they are locally associated with large water masses moving with tidal currents.

The speed of tidal currents varies based on location and time, typically ranging from 1 to 3 m/s, but in some extreme locations (like narrow channels) it can reach 5 m/s or more.

Correspondingly, tidal power density in coastal areas (e.g. $U = 2$ m/s, $h = 20$ m) may reach up to 80 kW per unit width.

However, harnessing tidal energy requires more than high power densities. Tidal waves are slow, with periods of about 12 hours, making impoundment hydropower plants the most practical solution. These plants store water during tidal movement and release it for power generation during the return phase.

The feasibility of such systems depends on the construction of large basins (onshore or offshore) at reasonable environmental and economic costs.

In these cases, the harvestable power is directly linked to the volume of water impounded and released during each tidal cycle, which in turn depends on local tidal height and site-specific environmental conditions.

II.4.2.2 Ocean Wave Energy Conversion Technologies

Let us now gain a brief insight into the main technologies which are currently under development to harvest energy from high-frequency sea waves: these technologies are usually cumulatively referred to as wave energy converters (WECs).

WECs are still not conventional devices, being characterized by low technological and manufacturing readiness levels (TRLs and MRLs). However, some of them have reached higher TRLs and have already demonstrated their feasibility in full-scale demonstration plants.

WECs vary significantly in design and operational principles. They are typically classified based on their working mechanisms and the type of water particle motion they exploit. Below are the most common types of WECs, as illustrated in Figure II.4.8.

- **Point Absorbers**
  Point absorbers are floating or submerged devices that capture energy from the vertical motion of waves at a single point. They typically consist of a floating buoy connected to a fixed structure or anchored to the seabed. The relative motion between the buoy and the stationary structure drives a mechanical system (e.g., a piston or hydraulic pump) to generate electricity.
- **Oscillating Water Columns**
  Oscillating water columns are partially submerged chambers that capture wave energy through the oscillatory motion of water columns: an underwater opening allows waves to enter and displace the air above the water column. As waves enter and exit the chamber, the rising and falling water column compresses and decompresses the air, which drives a turbine connected to a generator.
- **Attenuators**
  Attenuators are long, segmented floating devices that lie parallel to the direction of wave propagation and capture energy through their bending motion. These devices are typically composed of multiple hinged sections connected by joints. As waves pass, the relative motion between the segments drives hydraulic or mechanical systems to produce electricity.
- **Oscillating Wave Surge Converters**
  These devices harvest energy from the horizontal motion of waves, particularly in shallow waters. A pivoting flap is anchored to the seabed and moves

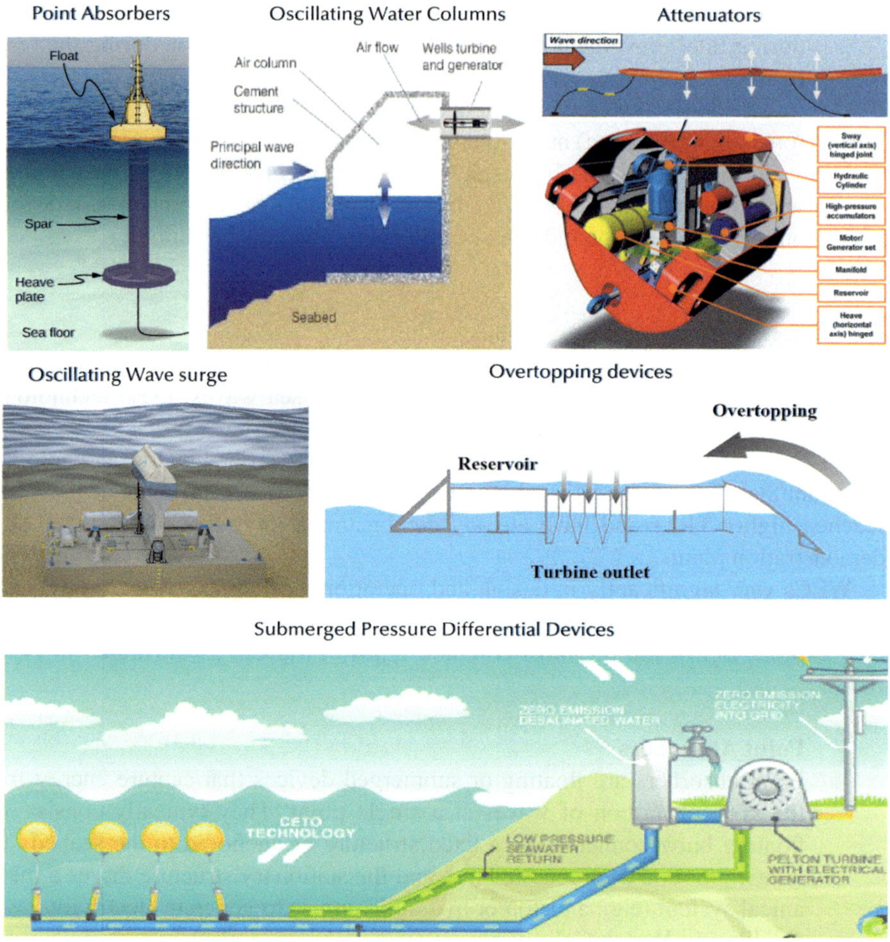

FIGURE II.4.8 Graphical summary of some WECs.

back and forth with wave surges. The motion of the flap drives hydraulic pistons or other mechanisms to generate electricity.

- **Overtopping Devices**
  Overtopping devices capture wave energy by channelling water into a reservoir and then using the stored water to drive turbines. These devices have a ramp or structure that allows waves to spill over into a raised reservoir. The gravitational potential energy of the trapped water is converted into electricity as it flows back to the sea through turbines.
- **Submerged Pressure Differential Devices**
  These systems capture energy by exploiting pressure variations beneath the waves. Submerged chambers or membranes are installed just below the water surface. As waves pass, the pressure differences cause a flexible membrane to move, driving a pump or generator.

II.4.2.3 Tidal Power Plants

Let us now look at the hydropower plants designed to harvest energy from tidal waves. Tidal power is a relatively mature technology compared to wave energy converters (WECs). The first tidal hydropower plant, built in 1965 near Saint-Malo, France (illustrated in Figure II.4.9), exemplifies this technology's potential.

Tidal power plants basically are impoundment hydropower plants, using a dam and a large water reservoir.

Local tide heights are normally perfectly known or can be derived according to measurements and astronomic considerations. The tidal time variation typically exhibits three main oscillation periods, respectively about: 12.5 hours; 14 days; 182.5 days. These water oscillations, in fact, are respectively due to: Earth's rotation; Moon's revolution; Earth's revolution.

The effects of the first two periods on tidal height are depicted in Figure II.4.10.

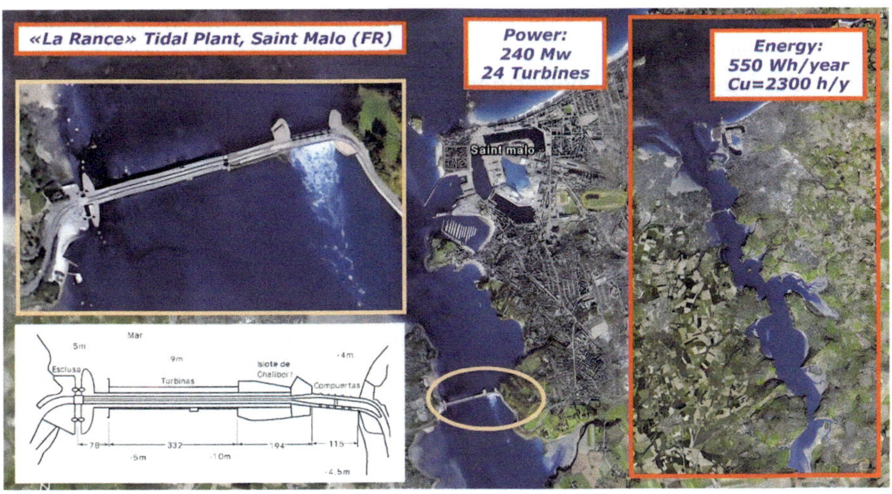

FIGURE II.4.9 La Rance Tidal Hydropower Plant.

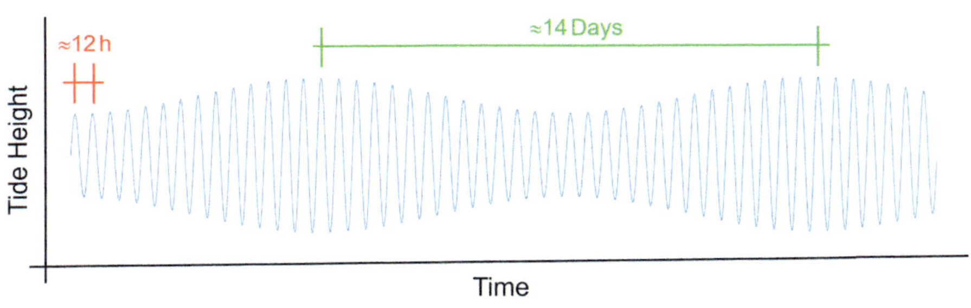

FIGURE II.4.10 Typical tidal oscillation periods.

Tidal waves may be used to harvest energy filling and/or emptying the reservoir with a calculatable inlet/discharge flow rate depending both on the tidal time law, and on the local orography of the reservoir.

Therefore, a theoretical water flow rate passing back and forth to the fully opened dam can be calculated and used as a reference for the design of the turbines to be used for energy harvesting.

To realize energy-harvesting conditions, in fact, the dam should be closed at least partially during specific tidal phases to create a height difference (and therefore a pressure head) between the tide and the reservoir to be used to generate thrust on a turbine generating mechanical energy.

This can be done using the tidal rising phase, its falling phase, or both of them, respectively, in:

- **Single-effect flood-type tidal powerplants**
  Water during generation phases flows in the turbines from the sea to the basin.
- **Single-effect ebb-type tidal powerplants**
  Water during generation phases flows in the turbines from the basin to the sea.
- **Double-effect tidal powerplants**
  Water may flow in the turbines in both directions depending on the generation phases, so the turbines should be "reversible". This means they should produce energy independently from the direction of the flow inside them.

  This flexibility of use of the turbines is obtained, for example, by turbines rotating on a perpendicular axis with respect to the water flow direction ("watermill"-type turbines).

  This gain in flexibility of the plant may be certainly advantageous but is normally "paid" for by a reduction in turbine efficiency with respect to those designed for single-effect plants. According to this expected efficiency reduction, double-effect plants are expected to produce a little less energy than single-effect ones.

  Double-effect plants can double the generation time. Consequently, the power of the turbines installed could be halved by limiting investment CAPEX costs. Therefore, the choice of the most convenient plant is not straightforward and mainly depends on economic considerations, depending on the cost of technologies as well as on the selling price of the electricity produced.

  From this point of view, another important aspect should be considered, which would certainly favour one of the single-effect plants with respect to the double-effect one. In fact, the electricity selling price would depend on the daytime hours during which electricity will be generated. These are repeatable and foreseeable such as the tidal wave, so that limiting the producing hours to those of maximum selling price, choosing one of the single-effect configurations, could more than balance the additional cost for the over-dimensioned turbines.

II.4.2.4 Design Approaches for Tidal Hydropower Plants

Let us look now at the typical design approaches for tidal hydropower plants.

Figure II.4.11 depicts the design approach to be used to choose and size turbines, respectively, in single-effect flood-type tidal power plants (upper part), single-effect ebb-type tidal power plants (middle part), and double-effect tidal power plants.

Two lines are reported in each of the sketches drawn, which respectively describe the sinusoidal trend of the tide vs. time (identified by the blue dotted curve), and the way the basin level should correspondingly oscillate vs. time as a consequence of the correctly designed size of the turbines and gates operation (identified by the red dotted curve).

Turbine operation phases and gates opening phases are respectively identified by the blue and red ranges at the bottom of each sketch.

All the sketches start from a point A, where the basin and tide level are equalized and describe the way to operate during a whole tide cycle (lasting approximately 12 h) to have the whole plant in the same conditions at the start of the following cycle.

To be more precise, cycle-to-cycle variations are expected according to Figure II.4.11 and should be considered for a more correct design and energy yield prediction of the power plant.

Referring to single-effect flood-type tidal power plants (upper part of Figure II.4.12), the cycle starts during a tide rising phase (point A). Gates are left open to flood the basin as much as possible, up to the level in point B (reached a bit delayed with respect to maximum tide time). Gates are then closed, but the generation phase is not promptly started.

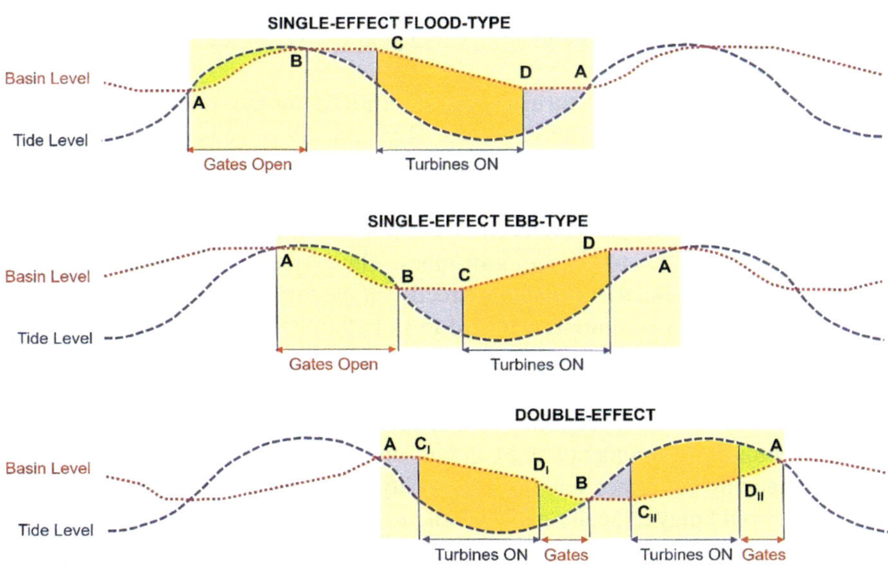

FIGURE II.4.11 Typical control schemes of tidal plants.

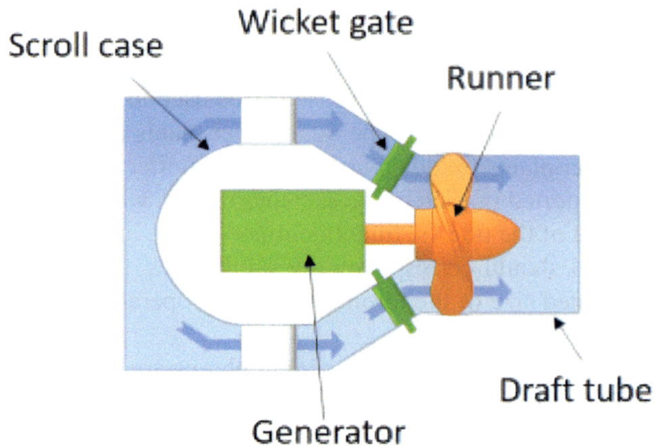

FIGURE II.4.12 Schematic representation of a bulb turbine. Credit: US DOE – Dept. of Energy.

The purpose is to bring the plant back to point A by discharging the necessary basin volume. Under this constraint, in order to stabilize the turbines' working point while optimizing their efficiency, it is convenient to keep gates closed before and after the energy-harvesting phase, maintaining the pressure head on the turbines as stable as possible and always above a minimal design value. A period of turbine operation is thus derived (segment C–D in Figure II.4.11).

Similar approaches apply to single-effect ebb-type plants (middle part of Figure II.4.11) and double-effect plants (lower part of Figure II.4.11), with specific adjustments for their operational modes. In this latter case, two distinct generation phases are identified (segments C_I–D_I and C_{II}–D_{II} in the lower part of Figure II.4.11).

The number and size of turbines are determined by the volume of water to be discharged during the generation period. High specific speeds are typical for tidal turbines, making Kaplan turbines a preferred choice for single-effect systems.

Kaplan turbines used in tidal power plants are often very large, with capacities ranging from 5 MW to 20 MW each. Their rotors can have radii of several metres, necessitating large penstocks and draft tubes. These spaces often include accessible rooms to house the electric generator directly on the turbine axis.
These turbine types, commonly referred to as bulb turbines (see Figure II.4.12), are designed for high efficiency and integration within the penstock structure, simplifying installation and maintenance.

In summary, hydropower remains a reliable and widely used renewable energy source. Future advancements must focus on minimizing environmental impacts while enhancing efficiency and grid flexibility. Tidal applications and wave energy converters could play a locally relevant role in some cases.

11.5 Geothermal Energy

11.5.1 SOURCE CHARACTERIZATION

Geothermal energy, derived from the Earth's internal heat, has been used for centuries for low-temperature applications. However, its significance in modern energy systems lies in its potential for thermoelectric power generation, which requires locally high energy density and thermal value to justify the installation of dedicated power plants.

Hence, let us now characterize the source to see how frequent these conditions are on our planet and what role this source could play, both locally and globally, to enhance energy sustainability.

11.5.1.1 Earth's Internal Heat

Earth's internal heat originates primarily from three sources, established during and after the planet's formation:

- **Primordial Heat**
  Primordial heat was generated during Earth's formation ~4.5 billion years ago, arising from the kinetic energy of colliding planetesimals and celestial bodies. The resulting energy, trapped within the Earth, constitutes the main contributor to its internal thermal energy.
- **Radiogenic Heat**
  Radiogenic heat comes from the decay of radioactive isotopes, such as uranium-238, thorium-232, and potassium-40, within the Earth's mantle and crust. This process continues today, providing a steady supply of heat that contributes to Earth's geothermal gradient. The decay of these isotopes releases particles and energy, which are absorbed by surrounding materials, generating heat.
- **Gravitational Differentiation (Core Formation)**
  This heat source is due to the way various elements were gravitationally separated. As Earth formed, heavier elements like iron and nickel sank toward the centre, forming the core, while lighter elements rose to the surface. This process, known as gravitational differentiation, released gravitational potential

DOI: 10.1201/9781003631125-20

energy, which was converted into heat. This process contributed to Earth's molten state during its early stages, some of which remains in the core today.

Additional smaller contributing factors to Earth's internal heat are: tidal heating due to tidal forces (caused by the gravitational interactions between Earth, the Moon, and the Sun) which can generate some movements and consequent heat generation within Earth's interior; latent heat generated by Earth's outer core's slow crystallization at the boundary with the solid inner core.

The heat from all these sources is distributed unevenly within the Earth. The core retains much of the primordial heat and generates additional heat through the slow solidification of the inner core. The mantle and crust contain heat from both primordial and radiogenic sources, with radiogenic heat being most prominent in the crust.

II.5.1.2 Earth's Thermal Gradient

The thermal gradient of Earth varies significantly, influenced by heat sources and geological features:

- **Crust**
  - Thickness: 30–60 km (continental), 5–10 km (oceanic).
  - Average gradient: ~25–30°C/km but varies based on geological and tectonic settings.
  - Active geothermal regions may exhibit gradients exceeding 50°C/km due to magma intrusion or hydrothermal activity, while stable continental areas may have gradients of ~15–20°C/km.
- **Mantle**
  - Lower gradient compared to the crust due to convective heat transfer.
  - Temperatures: ~500–900°C at the base of the crust to ~4000°C near the core–mantle boundary (~3000 km depth).
- **Core**
  - Steep gradient, with temperatures rising from ~4000°C in the outer core to ~6000°C in the inner core.
  - Heat transfer is dominated by conduction and convection.

II.5.1.3 Earth's Surface Heat Flux

Surface heat flux refers to the amount of heat escaping from the Earth's interior to its surface and is normally measured in milliwatts per square metre (mW/m^2).

The global average heat flux is approximately 87 mW/m^2, but this value varies widely based on location and geological conditions (see Figure II.5.1).

Oceanic heat flux is higher, mainly due to the reduced thickness of the Earth's crust, reaching up to 200 mW/m^2 in regions of mid-ocean ridges, where magma rises and new crust forms. Heat flux decreases as oceanic plates cool and move away from spreading centres.

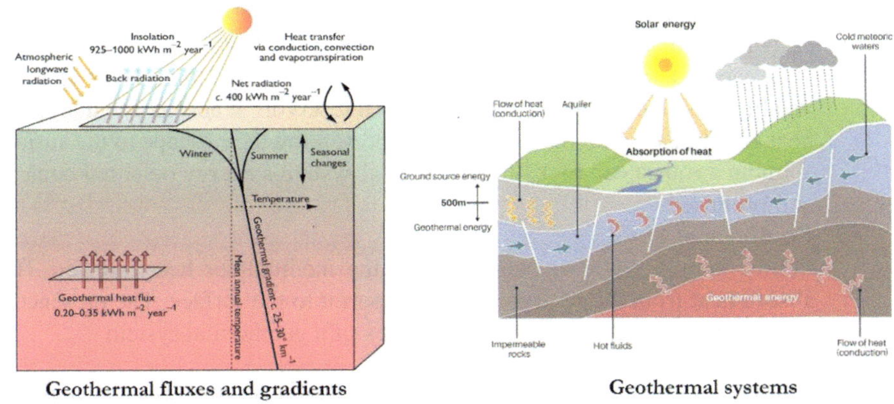

FIGURE II.5.1 Sketch illustrating geothermal processes: fluxes and gradients in the left part; geothermal systems behaviour in the right part.

Continental regions, instead, generally show lower heat flux values (40–80 mW/m²), except in areas with geothermal activity or thin crust. Radiogenic heat production in the continental crust can also contribute significantly to the surface heat flux.

Localized geothermal anomalies, such as volcanic regions or geothermal fields, can have heat flux values exceeding 1 W/m². These anomalies indicate the presence of shallow heat sources, such as magma chambers or hydrothermal reservoirs.

Local surface fluxes, as well as thermal gradients are also influenced by local thermal properties of Earth materials.

High-conductivity materials, such as crystalline rocks, efficiently conduct heat, while sedimentary rocks and fluids have lower thermal conductivities.

On the other hand, rocks with high heat capacity, such as basalt, can retain significant thermal energy.

II.5.1.4 Geothermal Systems

A geothermal system harnesses Earth's internal heat for various anthropic applications, such as electricity generation, direct heating, and industrial processes.

It consists of three key components that work together to make heat extraction and utilization possible (see right side of Figure II.5.1):

- **Heat Source**
  The heat source is the driving force of a geothermal system. Typically, this consists of a magma chamber, or a hot rock formation located deep within the Earth's crust. The heat source provides the thermal energy necessary to raise the temperature of surrounding rock and fluids.
- **Reservoir**
  The geothermal reservoir is a subsurface zone where heat is stored. It is typically composed of porous and permeable rocks filled with fluids such as water

or steam. The reservoir acts as a natural storage system, retaining heat and allowing it to be tapped for energy use.

A cap rock may partially close the reservoir upwards. It is a layer of impermeable rock that sits above the geothermal reservoir. It acts as a seal, trapping heat and fluids within the reservoir and preventing their escape to the surface. Without a robust cap rock, the geothermal system's energy potential could be compromised.

- **Geothermal Fluid**
  Fluids, mainly water or steam, act as the medium for heat transfer. They absorb heat from the reservoir and transport it to the surface. The presence of sufficient fluid is critical for the efficiency of a geothermal system.

Geothermal systems can be classified based on their temperature, physical state, and geological setting.

These characteristics determine their potential applications and technical requirements.

- **Temperature Classification**
  - Low-Temperature Systems (<90°C)
    Suitable for direct heating applications, such as residential heating, greenhouse cultivation, and aquaculture.
  - Medium-Temperature Systems (90°C–150°C)
    Used for both direct heating and binary cycle electric power generation.
  - High-Temperature Systems (>150°C)
    Primarily used for electricity generation in flash steam or dry steam power plants.
- **Classification by the Physical State of the Geothermal Fluid**
  - Liquid-Dominated Systems
    Contain hot water as the primary medium. These are the most common types of geothermal systems and are used for both heating and power generation.
  - Vapour-Dominated Systems
    Contain steam as the dominant phase and are highly efficient for electricity generation.
- **Classification by the Dominant Heat Transfer Mechanism**
  - Convection-Dominated Systems
    Occur in regions with active volcanism or tectonic activity, where heat and fluids are transported to the surface by convection.
  - Conduction-Dominated Systems
    Found in areas without significant tectonic activity, where heat is transferred through conduction in rocks.

Incomplete natural geothermal systems can be artificially enhanced (e.g., by injecting water into hot, dry rocks).

These systems, referred to as enhanced geothermal systems (EGSs), improve heat extraction potential but increase costs.

Geothermal energy is normally considered a fully renewable source, because the renewability of its heat source is granted (not being expected to decrease globally and locally in an entity within a human generation).

The sustainability and renewability of geothermal systems utilization is not straightforward. In fact, a geothermal system over time, due to its utilization, can lose its characteristics, depending on its utilization rhythm. Especially, the circulating flow rate of the geothermal fluid in the area could significantly lower or even become null, if its extraction and utilization are too intense.

Therefore, the size of geothermal utilization plants should be carefully designed as a result of both geological and geoclimatic considerations referring to an ample area around the existing geothermal system.

In fact, the area involved in the collection (through precipitation) and subsurface convective movement of the geothermal fluid can be very large (extending over thousands of square kilometres) and must be well understood in order to determine whether and where to re-inject the geothermal fluid after use, with the aim of stabilizing and/or enhancing its natural flow rate.

It must be considered, in fact, that the distance from the utilization point and fluid reinjection could also reach tens of km, strongly impacting on the overall costs of prospective utilization devices.

II.5.2 GEOTHERMAL ELECTRIC POWER PLANTS

As already said, geothermal energy sources are mainly interesting only if the energy density and thermal value of the source are locally sufficiently high to permit the installation of properly designed thermoelectric power plants.

This may occur both for medium (90°C–150°C) and high-temperature systems (>150°C)

Geothermal electric power plants are classified into three main types (see Figure II.5.2) based on the method of extracting and converting geothermal energy into electricity.

II.5.2.1 Dry Steam Power Plants

Dry steam power plants are the simplest and oldest type, with the first such plant built in 1909 in Larderello, Italy (see Figure II.5.3).

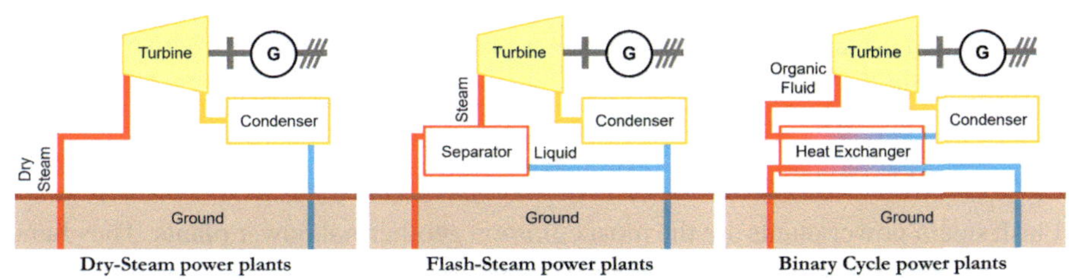

FIGURE II.5.2 Geothermal power plant types.

FIGURE II.5.3 The first geothermal power plant for electricity production (Larderello – Italy) – 1909.

They are only applicable in geothermal reservoirs that can produce steam directly. The steam is piped into a turbine, which drives a generator to produce electricity.

They are used for high-temperature vapour-dominant geothermal systems, such as The Geysers in California, and Larderello in Italy.

The key thermodynamic processes are:

- **Direct Steam Extraction**
  Superheated or saturated steam from a geothermal reservoir is extracted and directed to a turbine. The steam must already exist in vapour form within the reservoir, typically at temperatures exceeding 150°C.
- **Expansion Process (Turbine)**
  The steam expands through a turbine, performing work as it drives the turbine blades, which are connected to a generator to produce electricity. The expansion process follows a near-isentropic path in ideal conditions, maximizing energy conversion efficiency.
- **Condensation and Recycling**
  After passing through the turbine, the steam is cooled and condensed into liquid water in a condenser. This step is essential to extend expansion up to water saturation pressure at nearly environmental conditions (about 0.05 bar at 30°C), so increasing energy production for unit geothermal fluid.
- **Reinjection**
  The condensed water is extracted from the condenser and reinjected into the geothermal reservoir in the correct place to re-feed the convective fluid cycle within the terrain.

II.5.2.2 Flash Steam Power Plants

Flash steam power plants are the most common geothermal power plants. They handle high-pressure hot water from geothermal reservoirs. When the water is brought to the surface, it "flashes" into steam due to the sudden drop in pressure. The steam is then directed to a turbine to generate electricity.

Flash steam power plants are designed for high-temperature geothermal systems which are not vapour-dominant, where the geothermal fluid is a mix of water and steam at high pressure and temperature. The thermodynamic processes include:

- **Extraction of High-Pressure Fluid**
  Geothermal fluid is brought to the surface through production wells. At this stage, it is typically a mixture of liquid water and steam under high pressure.
- **Flashing Process**
  The high-pressure fluid enters a flash tank, where the pressure is reduced to a lower level. The sudden pressure drop causes part of the water to vaporize (or "flash") into steam. This process is isenthalpic, meaning the enthalpy remains constant during the flash.
- **Steam Expansion in Turbine**
  The flashed steam is directed to a turbine, where it expands and drives a generator to produce electricity. Similar to dry steam plants, the turbine expansion ideally follows an isentropic process for maximum efficiency.
- **Condensation and Recycling**
  After exiting the turbine, the steam is condensed in a condenser. The remaining geothermal fluid, along with the condensed steam, is reinjected into the reservoir to maintain pressure.

II.5.2.3 Binary Cycle Power Plants

Binary cycle power plants are a versatile and efficient technology designed to generate electricity from moderate-temperature geothermal resources, typically ranging from 85°C to 150°C (185°F to 302°F).

Unlike flash or dry steam power plants, which rely on high-temperature steam directly from the geothermal reservoir, binary cycle systems employ a closed-loop process that utilizes a secondary working fluid with a low boiling point. This enables the extraction of energy from geothermal sources that would otherwise be unsuitable for conventional methods.

Hot water or steam from a geothermal reservoir is extracted via production wells and directed into a heat exchanger. In this process the geothermal fluid does not directly contact the turbine or the generator components.

After transferring its heat to the secondary working fluid, the geothermal fluid is reinjected into the reservoir through reinjection wells. This reinjection process ensures reservoir sustainability by maintaining pressure and enabling the long-term use of the resource.

The secondary working fluid, often a hydrocarbon or organic compound with a low boiling point, is circulated through a closed-loop system. Key steps include:

- **Heat Absorption**
  The working fluid absorbs heat from the geothermal fluid via a heat exchanger, causing it to vaporize.

- **Turbine Expansion**
  The vaporized working fluid expands through a turbine, converting thermal energy into mechanical energy to drive an electric generator.
- **Cooling and Condensation**
  After passing through the turbine, the working fluid is cooled and condensed back into a liquid state using an air-cooled or water-cooled condenser. The liquid working fluid is then recirculated to the heat exchanger, ensuring continuous operation.

The choice of working fluid is critical to the efficiency and performance of binary cycle power plants. Suitable fluids include:

- **Isobutane**
  A commonly used hydrocarbon with excellent thermal properties for moderate-temperature geothermal resources.
- **Pentane (n-pentane or isopentane)**
  Often selected for its low boiling point and high vapour pressure, making it ideal for lower-temperature applications.
- **Ammonia–Water Mixtures**
  Occasionally used in advanced systems to optimize heat transfer and energy conversion.

Key properties considered in the selection process include:

- **Boiling Point**: The fluid must vaporize efficiently at low temperatures.
- **Thermal Stability**: To ensure reliability under operational conditions.
- **Non-Corrosive nor Toxic Nature**: Minimizing environmental and equipment risks.

Cooling is a crucial step in all geothermal plants to condense the working fluid back into its liquid state.

The choice of cooling method depends on the plant's location and coolant resource availability and temperature:

- Air-cooled systems are the only possible choice in arid regions, where water resources are scarce. Fans are used to dissipate heat into the atmosphere.
- Water-cooled systems are more efficient than air cooling but, requiring an abundant and sustainable water supply, they are only used in regions with access to lakes, rivers, or other water sources.

Increasing diffusion of binary cycle power plants (even for medium-low power outputs) is largely broadening geothermal energy utilization, since they can operate with lower-temperature geothermal resources, expanding the geographical range of potential installations.

Therefore, binary cycle power plants represent a transformative approach to geothermal energy generation, enabling the efficient use of moderate-temperature resources that were previously untapped.

With their low environmental impact, broad applicability, and sustainable operation, they play a critical role in advancing the adoption of geothermal energy worldwide. Continued innovation in working fluids, heat exchange technologies, and cooling systems will further enhance the viability and efficiency of binary cycle power plants, contributing to a cleaner and more sustainable energy future.

Advancements in technology and exploration are further expanding the potential of geothermal energy. Enhanced geothermal systems (EGSs) are being developed to create artificial reservoirs in hot dry rock formations, significantly increasing the geographical range of geothermal energy. Additionally, hybrid systems combining geothermal with solar or wind energy offer promising solutions for enhancing energy efficiency and reliability.

II.5.2.4 Potential Environmental Impacts of Geothermal Power Plants

Every natural geothermal system has a certain impact on its territory, mainly due to the exit of hot water from deep underground which may have been contaminated by heavy metals and/or dissolved gases.

These impacts may all be amplified when the area is further drilled for energy applications. Some geothermal reservoirs also release hydrogen sulphide (H_2S), which has a characteristic rotten egg smell and can be hazardous at high concentrations. Advanced gas scrubbers and mitigation technologies are often employed to manage these emissions.

Discharge of hot water or steam into nearby water bodies can increase temperatures, potentially disrupting local ecosystems.

As already underlined, the extraction flow rate of geothermal fluid from underground is a very critical design parameter when designing a geothermal power plant. In the same way, re-injected flow rates and re-injection locations should be carefully chosen.

Design errors from these points of view may in fact lead also to land subsidence, causing structural damage to nearby buildings and infrastructure.

Geothermal fluid extraction can also occasionally enhance the frequency of small earthquakes, a phenomenon known as induced seismicity. This is most common in enhanced geothermal systems (EGSs), which involve hydraulic fracturing to increase permeability in geothermal reservoirs.

The magnitude of these events is typically low, because drilling activities do not mechanically interact at all with the main seismogenic structures which are much deeper in Earth's crust with respect to drilling operations, but seismic activity has to be carefully monitored to minimize risks.

11.6 Biomasses, Biofuels, and e-Fuels

As the world strives to decarbonize energy production and the transportation sector, synthetic fuels have emerged as a promising component of the global energy transition. In fact, these energy carriers offer several potential advantages, including reduced greenhouse gas emissions, improved energy density, enhanced energy security, and decreased dependence on fossil fuel-exporting nations.

With global energy demand rising and climate change reshaping the energy landscape, refining the definition of "synthetic fuels" has become increasingly important to evaluate all the possibilities we have to implement a full transition toward a sustainable energy system.

A comprehensive definition should encompass both current and future technologies, distinguishing synthetic fuels as a unique category separate from petroleum-based fuels.

History and Definitions

Historically, synthetic fuel production was driven by the need to diversify energy supplies, particularly during times of crisis, such as World War II. Early efforts focused on converting fossil-based feedstocks (like coal or oil shale) or biomass into gas or liquid fuels to mitigate geopolitical and supply risks.

By the 20th century, the term "synthetic fuels" broadly referred to non-petroleum-derived fuels produced by converting carbonaceous materials. Initially, economic priorities focused on maximizing energy consumption and profitability. As energy supply constraints emerged, the focus shifted towards efficiency and reduced energy use, though environmental impacts and carbon emissions were largely overlooked. Today, growing climate concerns have redefined synthetic fuels, with an emphasis on low-carbon technologies and sustainability.

So, recently, the category of synthetic fuels has expanded to include so-called e-fuels, produced by reacting a carrier molecule (often CO_2) with hydrogen (H_2). In this context, synthetic fuels can be redefined as fuels produced chemically or biologically: this may happen either by linking together molecules like carbon dioxide or nitrogen to create energy carriers, or by facilitating the breakdown of large molecular structures (both of fossil and of biological origin), as in pyrolysis.

This broad definition encompasses fuels created through processes such as hydrogen addition or carbon rearrangement.

DOI: 10.1201/9781003631125-21

Classification of Synthetic Fuels

Synthetic fuels can be classified into three main categories based on their feedstocks:

- **Fossil-derived synthetic fuels**
  These are produced from hydrocarbons, carbon, and hydrogen via chemical processes. While they do not involve crude oil distillation, they still contribute to carbon emissions and are not a long-term solution for environmental sustainability.
- **Biofuels**
  These are derived from biological sources (usually cumulatively referred to as biomasses) such as crops, algae, or organic waste. Biofuels are produced through thermochemical or biochemical processes.
  Theoretically, their CO_2 cycle is net zero emitting, on a life cycle basis: the amount of CO_2 which is absorbed by the atmosphere during biological sources growth is then remitted through the final use of the biofuel.
  In real processes, this could not be the case, due both to additional CO_2 emissions due to the energy consumption of the whole production cycle, but eventually also to potential CO_2 sinks due to the application of CCS (carbon capture and sequestration) technologies to the final users. The CO_2 balance of these fuels has therefore to be calculated carefully case by case.
- **e-Fuels**
  e-Fuels are produced combining hydrogen with CO_2, CO, or N_2 to create gaseous or liquid fuels. Theoretically, energy used in all processes can come either from RES or fossil sources, but only using RES-based e-fuels is a really feasible option for sustainable energy systems.

Production processes often rely on synthesis reactions powered by renewable or other energy sources, making their environmental impact vary widely, depending on the raw materials, production methods, and energy inputs involved.

The overarching goal of synthetic fuels is to provide a sustainable alternative to fossil fuels, reducing greenhouse gas emissions and reliance on finite and localized resources. Beyond producing electricity or heat, synthetic fuels offer distinct advantages for transportation, particularly where batteries face limitations (in the hard-to-abate sectors).

Moreover, while electric vehicles are clean and efficient, they depend on grid infrastructure, which is vulnerable to disruptions and requires costly upgrades. In contrast, synthetic fuels can be stored and transported like traditional fossil fuels, using existing infrastructure, offering a reliable and scalable energy source.

Despite their potential, most of these new and sustainable synthetic fuels remain in the early stages of commercial deployment. Scaling up production is hindered by high costs and the limited number of operational pilot facilities. However, declining renewable energy costs and advances in technology could enable wider adoption. The future of sustainable synthetic fuels will depend on technological advancements and supportive policies, but they are expected to play an increasingly significant role in reducing emissions and improving energy security.

In the following sections, we will delve deeper into biofuels and e-fuels. These could represent sustainable technologies that align with the global transition towards cleaner energy solutions.

II.6.1 ON BIOMASS AND BIOFUELS

The term **biomass** refers to a wide range of materials of organic origin, excluding plastics and fossil-based substances. Scientifically, it includes any material derived from biological processes, inherently linked to carbon chemistry, and originating directly or indirectly from photosynthesis.

This fundamental process is activated by sunlight (part of the electromagnetic waves constituting solar energy) and allows plants and algae to convert water, and carbon dioxide into carbohydrates, $C_n\left(H_2O\right)_m$, and oxygen through the reaction:

$$CO_2 + H_2O + \text{Solar energy} \rightarrow C_n\left(H_2O\right)_m + O_2$$

The energy stored in the chemical bonds of carbohydrates is significantly higher than in their precursor molecules, CO_2 and H_2O, which are stable and less reactive.

Therefore, photosynthesis can be seen as an energy conversion process, directly converting solar energy into chemical potential energy.

Summarizing, chlorophyll-based photosynthesis is characterized as:

- Photoactivated: Occurs only by means of solar radiation at certain wavelengths.
- Endoenergetic: Transforms solar energy into stored chemical energy.

Autotrophic organisms, primarily plants, are the only living entities capable of performing this synthesis, producing the carbohydrates necessary for their growth and functioning. This process marks the starting point of the carbon cycle, wherein plants transform atmospheric CO_2 into a form usable by other organisms.

The carbon cycle continues beyond the death of plants and animals. Over time, the carbohydrates contained within these organisms undergo a transformation.

When buried under sediment layers, the combined effects of pressure, temperature, and oxygen deprivation can convert these organic materials into fossil hydrocarbons (a process that unfolds over millions of years).

Plants, as the first complex life forms on Earth, were instrumental in shaping the primordial atmosphere. Originally rich in CO_2, the atmosphere was gradually transformed as plants absorbed carbon dioxide and released oxygen, enabling the evolution of animal life. However, human activities have disrupted this balance, contributing to rising CO_2 levels and global warming.

By reducing reliance on fossil hydrocarbons and promoting the sustainable use of biomass, it is possible to mitigate this trend and move towards a more balanced and sustainable carbon cycle.

As already said, chlorophyll-based photosynthesis utilizes solar energy to transform inorganic carbon (CO_2) into organic carbon ($C_n\left(H_2O\right)_m$), releasing oxygen as a by-product. This reaction is inherently endothermic, with sunlight providing the required radiant energy.

A fraction of the solar energy is converted into chemical energy stored in the bonds of carbohydrates. Unfortunately, the energy efficiency of photosynthesis is relatively low. Only a small percentage (1–10%, depending on the plant species) of the incident solar energy is converted into chemical energy.

Much of the solar energy is lost due to factors such as reflection, absorption by non-photosynthetic pigments, heat dissipation, and inefficiencies in the photosynthetic process. This limited efficiency is primarily because photosynthesis is activated only by specific wavelengths of sunlight.

The most effective wavelengths fall in the blue and red regions of the spectrum, while green and yellow light are largely reflected, giving plants their characteristic green appearance. Chlorophyll, the natural catalyst central to this process, selectively absorbs radiation within these energetically useful frequencies.

To harness energy from biomass efficiently, two primary potential biomass sources are pursued:

- **Dedicated Biomass**
  Crops cultivated specifically for energy production, also known as energy crops.
- **Residual Biomass**
  Organic material derived from agricultural, urban, or industrial waste, repurposed after completing its primary use.

II.6.1.1 GHG Emissions of Biomass

So-called dedicated biomasses are plant species with high energy yields and significant mass productivity. Productivity depends on several factors: the species and type of plant used, climatic and meteorological conditions, and site-specific cultivation characteristics.

Unlike fossil fuels, which store concentrated energy accumulated over millions of years, biomass represents a more transient energy source. Overuse or mismanagement of biomass resources could upset the carbon cycle, potentially depleting organic material essential for other natural processes. A stark example would be indiscriminate deforestation for energy purposes, depriving trees of their role in CO_2 absorption.

Sustainability is at the heart of biomass use. Dedicated biomass is renewable, as it is cultivated for harvesting within a timeframe compatible with its consumption. Similarly, residual biomass must be sourced without diverting materials from other valuable applications. Only by adhering to these principles can biomass serve as a sustainable, integral component of the energy transition.

The biomass supply chain (from cultivation or collection to final use) intersects with CO_2 production at multiple points. Evaluating its emissions balance requires a comprehensive analysis of all processes involved, as the type of biomass (dedicated or residual) and its lifecycle significantly influence the final outcome.

Dedicated biomass, cultivated specifically for energy use, requires careful management of its entire supply chain, including the production and procurement phases.

During its growth phase, biomass absorbs CO_2 from the atmosphere, which helps improve its overall emissions balance. At first glance, this may suggest a near-neutral CO_2 balance.

However, other factors must be accounted for, such as the energy and emissions generated during cultivation, harvesting, transportation, processing, and conversion. These activities contribute additional CO_2 emissions, potentially reducing the perceived advantage of dedicated biomass. Thus, achieving a net benefit requires optimizing each step of the supply chain.

Residual biomass, derived from urban, agricultural, or industrial waste, operates under different constraints. Since the growth phase of the biomass is unrelated to its energy use, the CO_2 emitted during combustion cannot be offset by prior absorption. Consequently, it represents a net contribution to atmospheric CO_2, similar to fossil fuel combustion.

Despite this, waste incineration has economic and social benefits, such as reducing waste volumes and management costs. When used effectively, residual biomass can complement other renewable energy sources without requiring additional land or dedicated cultivation.

The CO_2 balance of biomass is heavily influenced by the scale of its supply chain. Large-scale operations often require long-distance transportation, which increases emissions and reduces efficiency. Biomass is more sustainable when sourced locally, minimizing transport-related energy expenditures.

II.6.1.2 Other Critical Issues about Biomass

Unlike other RESs, biomass combustion produces polluting emissions, including carbon monoxide (CO) and sulphur oxides (SO_x). Fortunately, these pollutants are not as persistent as CO_2 and can be managed with advanced mitigation technologies and strategic site selection, such as avoiding densely populated areas.

Unlike other RESs (like solar, wind, and hydro), biomass demands a continuous supply of material. Poor planning can result in a CO_2 balance even worse than fossil fuels. For example, neglecting transport emissions or overlooking lifecycle management could significantly underestimate its environmental impact.

Biomass also competes with other essential land uses. Allocating vast areas for biomass cultivation may conflict with food production, livestock farming, housing development, or industrial expansion. This competition can exacerbate food security concerns or limit urban growth.

To address these conflicts, innovative strategies have emerged like agrivoltaics (combining agriculture with photovoltaic installations optimizes land use, allowing simultaneous energy and food production) or advanced biofuels (utilizing raw materials like algae or organic residues that do not compete with food production reduces pressure on agricultural land).

These approaches help balance the need for energy with the preservation of critical land resources.

II.6.1.3 The Potential of CCS Technologies

The sustainability profile of biomass improves significantly with carbon capture and storage (CCS) technologies. By capturing and sequestering CO_2 during combustion, the CO_2 balance of biomass could become negative (removing more CO_2 from the atmosphere during plant growth than is emitted during use). This capability positions biomass as a powerful tool in reducing global emissions.

However, CCS technology is still in development. While the technical principles are well-understood, high implementation costs currently limit its widespread adoption.

II.6.1.4 Biofuels Potential in Hard-to-Abate Sectors

Biomass can be efficiently used to produce liquid fuels or serve as a direct substitute for fossil hydrocarbons, especially in the transportation sector. Its compatibility with existing infrastructure and technologies makes biomass an attractive transitional energy source, minimizing adaptation costs and timelines.

While public perception often favours other renewables as "greener", biomass has unique potential. Its ability to integrate CCS technologies and replace liquid fuels highlights its critical role in the energy transition.

Considering all potential advantages and drawbacks, biomass certainly offers significant opportunities but cannot form the sole foundation of a sustainable energy system. Its implementation must be carefully planned to minimize impacts on the carbon cycle, land use, and pollutant emissions. Relying exclusively on biomass would be neither practical nor sustainable, necessitating its integration with other renewable energy sources.

Despite these challenges, biomass remains essential and strategic in sectors like biofuel production. With the adoption of CCS technologies, it could play a transformative role in meeting ambitious emission reduction goals.

II.6.1.5 Main Biomass Constituents

Biomass primarily consists of carbon (C), oxygen (O), and hydrogen (H), which are present in average proportions of approximately 50%, 43%, and 6%, respectively. These proportions obviously may vary a lot depending on the specific source used. In addition to these major components, Biomass may contain trace amounts of mineral elements, including nitrogen (N), sulphur (S), potassium (K), phosphorus (P), silicon (Si), aluminium (Al), calcium (Ca), and sodium (Na).

From a biological-organic perspective, the primary components of biomass normally include cellulose, hemicellulose, pectin, lignin, and water.

For the purpose of studying biomass energy content (which could for example be freed during its direct combustion), only its key components involved in highly endothermic reactions are considered: commonly, carbon (C), hydrogen (H), and oxygen (O). Additionally, nitrogen (N) is considered as the most common inert component found in combustion products.

Among these, hydrogen, though present in relatively small amounts, provides a significant energetic contribution during combustion due to the high energy release associated with its oxidation.

Conversely, sulphur, while present only in small/trace amounts, requires careful attention due to its environmental implications.

To describe biomass composition concisely, a gross chemical formula is commonly used, which expresses the molar ratio of its main components without representing a specific molecule. This formula is: $CH_\alpha O_\beta N\gamma$ where α, β, and γ represent the molar ratios of hydrogen, oxygen, and nitrogen, respectively, relative to carbon.

These coefficients can be calculated from the mass percentages (x) of the constituent elements using their molar weights (M) through the following relations:

$$\alpha = \frac{x_H / M_H}{x_C / M_C}; \quad \beta = \frac{x_O / M_O}{x_C / M_C}; \quad \gamma = \frac{x_N / M_N}{x_C / M_C}$$

While hydrocarbons are typically represented by the general formula C_mH_n (or equivalently CH_α, where $\alpha = n/m$ with n and m being integers), biomass is chemically more complex. In addition to the primary elements (C, H, and O), biomass incorporates oxygen in its structure and various trace elements that influence its properties and behaviour during combustion.

II.6.1.6 Biomass Lower and Higher Calorific Values

Higher heating value (H_s) or lower calorific value (H_i) of a biomass may not only be determined by the mass percentage and chemical configuration of its primary elements: carbon (C), hydrogen (H), oxygen (O), and nitrogen (N). The estimation of calorific value in fact is more straightforward for hydrocarbons than for biomass. Hydrocarbons are simple compounds composed mainly of carbon and hydrogen with minimal impurities.

For instance, in methane (CH_4), the molecular structure and bonds are well-defined, enabling precise calculations. Biomass, instead, is chemically represented by a generic formula, $CH_\alpha O_\beta N\gamma$, which lacks detail regarding additional elements or specific chemical bonds. This complexity makes calorific value predictions less accurate for biomass.

Additionally, while hydrocarbons like gasoline exhibit predictable properties despite minor variations, biomass has greater variability in its composition. This variability, combined with the presence of inert water content, results in a lower calorific value for biomass compared to other fuels. Water further reduces the energy output by being non-combustible during the combustion process.

One of the more commonly used analytical formulas to estimate H_s is the so-called Dulong's formula:

$$H_s = 0.813\, x_C + \left(100\, x_H + \frac{x_N + x_O - 0.01}{0.08}\right) + 0.222\, x_S \quad \left[\frac{\text{kcal}}{\text{kg}}\right]$$

where: x_S is the mass percentage of sulphur (S). This formula estimates the H_s exclusively on a dry basis (denoted as $H_{s,D}$).

Therefore, the steps to obtain this value are as follows:

- Establish a Reference Cultivation: A representative sample of biomass is selected and harvested.
- Complete Drying: The biomass sample is fully dried to remove all moisture.
- Gas chromatographic analysis: The chemical composition, including the percentages of carbon (C), hydrogen (H), oxygen (O), and nitrogen (N), is determined.
- Application of Analytical Formulas like Dulong's, or determine $H_{i,D}$ experimentally using a so-called "calorimetric bomb".

The lower heating value on a dry basis ($H_{i,D}$) is derived by reducing the higher heating value $H_{s,D}$ by the energy associated with the water formed due to hydrogen oxidation. Then, $H_{i,D}$ is determinable by:

$$H_{i,D} = H_{s,D} - 9\, r_0\, x_H$$

where r_0 is the latent heat of vaporization of water.

Let us now come to the so-called heating value on a wet basis, H_W: this accounts for the moisture in biomass, which reduces the energy available during combustion. Both chemically bound water and retained moisture turn into steam during combustion, consuming energy and amplifying the difference between the higher calorific value ($H_{s,W}$) and the lower calorific value ($H_{i,W}$).

To minimize energy losses, reducing the moisture content of biomass before combustion is crucial.

Drying biomass lowers the weight and improves the calorific efficiency but incurs energy costs. If drying is performed at the combustion plant, the energy required reduces the net output. Alternatively, using another fuel to power the drying process adds operational costs. Drying before transport can reduce transportation expenses by lowering the biomass weight but may require additional resources and infrastructure.

Regardless of the drying method, knowing the moisture content at purchase is essential to avoid paying for non-usable mass. Natural drying is cost-free but slow and weather-dependent, while forced drying is faster but consumes energy. Even after drying, a minimal level of moisture is necessary for technological purposes, as achieving zero moisture is neither practical nor economical.

The calorific values on a wet basis ($H_{s,W}$ and $H_{i,W}$) are calculated by adjusting the dry basis values ($H_{s,D}$ and $H_{i,D}$) based on the moisture percentage ($U_{\%}$) using these formulas:

$$H_{i,W} = H_{s,D}\left(1 - \frac{U_{\%}}{100}\right) - r_0 \frac{U_{\%}}{100}$$

$$H_{s,W} = H_{s,D}\left(1 - \frac{U_{\%}}{100}\right)$$

Moisture significantly worsens the calorific value on a wet basis, with the reduction proportional to the water content. If biomass contained no moisture, the wet and dry calorific values would be identical. Conversely, if biomass were entirely water, the wet basis calorific value would drop to zero, as all energy would be consumed in vaporizing the water.

Wet biomass consistently shows lower HCV and LCV values compared to dry biomass. This reduction occurs because moisture in biomass negatively impacts combustion efficiency by consuming energy to evaporate water during the process. However, from a technological perspective, a small amount of moisture can be advantageous for the operation of specific systems, provided it remains within acceptable limits.

The calculation of calorific values requires careful attention, as the results are highly sensitive to the residual moisture content and chemical composition of the biomass. Even minor variations in these parameters can significantly influence the calculated values, emphasizing the importance of precise analyses of both the operating conditions and input data.

II.6.1.7 Biomass Types

This section focuses on the most commercially utilized raw materials for biomass energy production, distinguishing between:

- **Dedicated biomass** or **Energy crops**
  This biomass is cultivated specifically for energy purposes. The goal is to maximize energy yield, typically measured in terms of energy produced per hectare per year. Selecting the most suitable crops requires considering calorific values and various influencing factors, such as soil type, climate, and water availability. While general estimates exist, accurate evaluations depend on on-site measurements and analyses. Importantly, energy crops should be sourced locally to preserve environmental benefits and prevent emissions associated with long-distance transportation.

  Energy crops fall into two main categories: herbaceous crops and woody crops. Herbaceous crops, like sunflowers, have favourable chemical properties, including low nitrogen and no sulphur content, making them ideal for biofuel production. Woody crops, such as poplar, are primarily used for direct combustion in biomass power plants, where lower-quality biomass can be utilized due to the focus on energy content rather than chemical composition. Woody crops are more resilient to poor irrigation compared to herbaceous crops but still benefit from favourable climatic conditions and water availability.

  To avoid competition with food agriculture, energy crops are often grown on unused agricultural land. Most biomass plants operate within a radius of 50–100 km, as larger supply chains increase transportation costs and environmental impacts. Typical plant capacities range from a few megawatts to 30 MW. Very large plants, exceeding 200 MW, would in fact need a too high biomass consumption rate which could not come from a restricted territory without creating sustainability issues and land-use conflicts.

Rapid rotation techniques, where plants are harvested cyclically and replaced immediately, are commonly used to maximize yield. A common technique in energy crops is rapid rotation: the plants are harvested cyclically and immediately replaced, maximizing their early growth phase. Among woody crops, poplar is one of the most commonly used due to its high yield and rapid growth. Among herbaceous crops, oil crops such as sunflowers are important for the production of biofuels for transport.

- **Residual Biomass**

  Residual biomass refers to organic waste produced in sectors like urban areas, industry, agriculture, and forestry. These sources provide significant opportunities for energy generation while reducing environmental burdens.

  - Urban Waste

    Organic waste from cities, collected through proper separation, can supply biomass plants and help address waste disposal challenges. Locating such plants near cities must balance proximity for efficient energy use and distance to limit environmental impacts. Small cogeneration plants, which produce both heat and electricity, are particularly effective in urban contexts.
  - Industrial Waste

    Industries like furniture manufacturing and paper production generate large quantities of organic waste, which can be repurposed for energy production. Many companies establish small-scale energy plants to convert their waste into power, improving efficiency and reducing operational costs.
  - Agricultural and Agro-Livestock Residues

    Agricultural operations produce organic waste, including plant residues and animal carcasses. These can be transformed into biofuels or used in biomass plants, reducing reliance on landfills and minimizing harmful emissions. Animal fats, similar to vegetable oils, can also be converted into biofuels, offering another sustainable energy pathway.
  - Forest Residues

    Natural forest cycles produce biomass as vegetation dies or through pruning operations necessary for forest health. Pruning prevents overgrowth, or "crowding," and promotes regeneration, enhancing the forest's capacity to capture CO_2. Biomass derived from forest management is considered sustainable and beneficial for the ecosystem. It is essential to note that pruning differs from deforestation, as it supports forest conservation and resource management.

II.6.1.8 Biofuels Types

Biofuels are categorized into four "generations" based on the type of biomass raw material used (see Figure II.6.1) and the associated production methods:

FIGURE II.6.1 Classification of biofuels based on source types.

- **First generation**
  These biofuels are derived from **edible biomass**: food sources, including starch, cereals, vegetable oils, sugarcane, and animal fats. They rely on microorganisms to ferment the raw materials, which are then processed into the desired fuel. However, their reliance on food crops has raised concerns about competition with food production and potential impacts on food security.
- **Second generation**
  These biofuels are produced from **non-edible biomass**: wood, straw, and crop residues left after harvesting the edible parts. They are also known as cellulosic biofuels, utilize high-cellulose materials, and avoid the food-versus-fuel debate.
- **Third generation**
  These biofuels come from **algae biomass**, where oil is extracted for fuel production. Algal biofuels are cost-effective but remain in the developmental stage for large-scale production.
- **Fourth generation**
  These biofuels represent the most advanced category, utilizing **sophisticated biochemical techniques** to address limitations and environmental concerns of earlier generations. These biofuels aim to mitigate challenges such as land use conflicts and sustainability issues. Currently, fourth-generation biofuels remain experimental, with ongoing research and development.

II.6.1.9 Biomass Conversion Technologies

The processes used to achieve the energy conversion of biomass are diverse in type and purpose. This variety is justified by:

- The significant heterogeneity of biomass sources.
- Biomass utilization processes are not necessarily linked to the final use; they may also involve the transformation of biomass into energy carriers.

Figure II.6.2 depicts a possible categorization of biomass conversion processes.

Possible conversion technologies are grouped into two broad categories: thermochemical processes and biochemical processes.

The processes discussed here are those currently established and widely utilized, primarily referring to first- and second-generation biofuels.

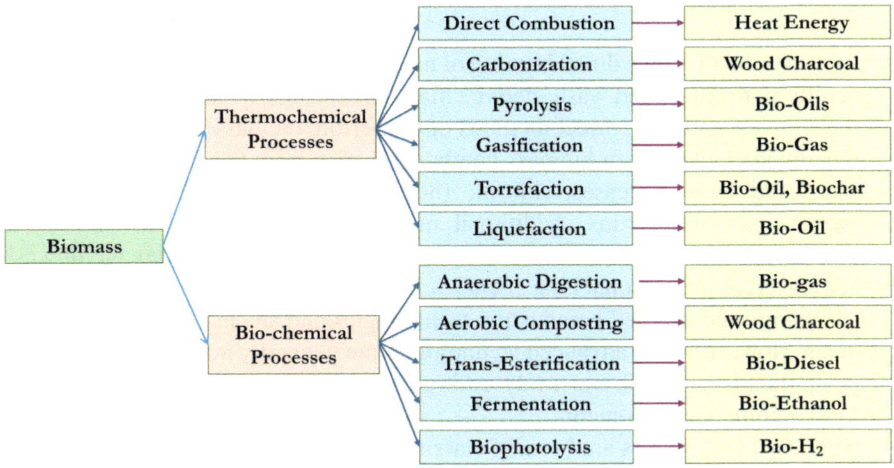

FIGURE II.6.2 Biomass conversion processes.

Thermochemical processes are well-established technologies with limited potential for groundbreaking innovations. Current research efforts focus on optimizing existing methods rather than developing entirely new approaches.

These processes treat biomass as a combustible fuel, leveraging its capability to release thermal energy by breaking molecular bonds. Interestingly, the organic nature of biomass is not a strict requirement for thermochemical processes. The same principles can apply to non-organic combustible substances. In fact, all these processes have been used for decades for solid and liquid fossil hydrocarbons also.

In contrast, **biochemical processes** are highly diverse and continue to evolve rapidly, driven by ongoing research and innovation in the field of "green chemistry". These processes depend on biochemical reactions that require the organic nature of the reagents to activate. Each technology is tailored to specific types of biomass, enabling a wide variety of applications.

Among the most prominent biochemical processes are anaerobic digestion (also called composting) and aerobic digestion, both of which are well-suited to processing a broad spectrum of organic residual biomasses.

These include solid waste, industrial by-products, and municipal effluents. Anaerobic and aerobic digestion technologies are widely implemented in landfills and organic waste treatment plants, where biomass is transformed into biogas or compost.

Other key biochemical processes, such as **alcoholic fermentation** and **transesterification**, are primarily aimed at producing **transportation fuels** (respectively for spark-ignition and compression ignition ICEs), to replace gasoline and diesel fossil-based fuel.

These processes are described in Part III of this book, which is dedicated to sustainable mobility, with Chapter III.3 devoted to alternative fuels for ICEs.

II.6.1.10 Direct Biomass Combustion

Combustion involves the direct burning of materials, primarily of forest origin, though agricultural, residual, and waste-derived materials can also be utilized. It remains one of the simplest and most direct methods of converting biomass into energy.

The main challenges in biomass combustion arise from the presence of nitrogen, sulphur, and, in some cases, chlorine (notably in municipal solid waste). These substances generate toxins. Additional metal (phosphorus, silicon, sodium, aluminium, …) content of the biomass can instead create low-temperature melting ashes which can create operational issues. These challenges may be mitigated in larger-scale plants using fluidized bed combustion technologies and advanced flue gas treatment systems. However, such advanced technologies are generally not feasible for smaller-scale applications, which must rely on high-quality biomass with very low sulphur and other components content.

Solid fuel combustors are broadly classified into two main categories:

- Grate Combustors (Small-Scale Plants)
  In grate combustors, the fuel and oxidizer are fed perpendicularly to one another. The fuel rests on a perforated grate, with air introduced from below to burn the fuel while keeping it suspended. This design supports combustion along the flow path of the hot gases.
- Bed Combustors (Larger-Scale Plants)
  In bed combustors, both the fuel and oxidizer flow parallel to one another, which promotes more complete and uniform combustion. There are different types of bed combustors. The best-available technology (BAT) is the fluidized "bubbling" bed, which is widely used in large-scale plants with capacities in the tens of megawatts.

  The bubbling fluidized bed combustor operates on an artificially created bed formed by a flow of oxidizer, which automatically draws pulverized fuel into the combustion chamber and carries it along an extended path. The process is often improved by the way the fuel is fed into the chamber, with helicoidal motions generated by fans to enhance the system's efficiency. This design promotes better contact between the combusted gases and the chamber walls, which are equipped with economizer tubes, vaporizer tubes, and, where applicable, superheater tubes. The "bed" is essentially composed of the moving oxidizing fluid and the fluidized fuel.

  This configuration increases both the space and time available for combustion, enabling multi-zone combustion to occur while maintaining a low combustion temperature of approximately 800–900°C, thereby reducing the formation of nitrogen oxides (NO_x) and sulphur oxides (SO_x).

  The process begins with the oxidizer drawing pulverized fuel into the chamber. During this initial stage, a limited amount of oxidizer is supplied to maintain a low temperature, creating a fuel-rich combustion environment. In the second stage, additional air is introduced through supplementary inlets to provide the remaining oxidizer, surpassing the stoichiometric amount required for the residual calorific value of the combustion gases.

This excess air absorbs the calorific energy while maintaining a low temperature throughout the system. Although the total amount of air introduced into the chamber remains stoichiometric, staging the combustion process ensures that the temperature is consistently kept low.

This approach not only limits the formation of NO_x but also addresses the issue of low-melting-point ashes, a common problem in biomass combustion caused by the mineral content of the fuel. Minerals such as silicon, phosphorus, magnesium, and calcium form oxides during combustion, which tend to liquefy at temperatures exceeding 800°C, causing them to adhere to the chamber walls.

By distributing the air unevenly across both space and time, the combustion process prevents ash from sticking to the walls while maintaining optimal efficiency and performance. This strategy also minimizes heat losses through the exhaust gases, further enhancing the system's effectiveness.

For domestic-scale boilers (ranging from approximately 50 kW to 1 MW, suitable for individual homes or apartment buildings), high-quality biomass is essential.

Preferred fuels include wood chips and pellets, which are both derived from waste materials like furniture production scraps or pruning residues from trees such as oak, fir, or pine.

Wood chips are small, high-quality pieces of wood that provide favourable chemical properties and high fluidity, which are crucial for continuous boiler operation.

Pellets are produced by compressing fine wood powder into small cylindrical shapes. Their uniform size and powdered consistency ensure smooth, consistent combustion, addressing one of the critical challenges of direct combustion.

These characteristics make wood chips, and especially pellets, ideal for smaller systems, as they promote uniform combustion and minimize the presence of harmful oxides.

Large-scale plants employ more diverse types of biomass, which are pre-treated on-site. This pre-treatment typically includes shredding, pulverizing, and feeding the biomass into the combustion chamber as a fine powder. Advanced combustion systems help ensure uniform combustion and minimize the production of harmful oxides and atmospheric pollutants. These facilities generally integrate sophisticated technologies to handle the operational complexities associated with larger capacities.

II.6.1.11 Carbonization

Carbonization is the process of converting woody biomass into charcoal (different from mining-derived carbon coke), which differs from biomass primarily due to its lack of oxygen. In this transformation, the oxygen bonds in the biomass (specifically the weaker C–O bonds) are broken, making carbonization essentially an accelerated form of fossilization.

Fossilization in nature occurs over geological timescales, during which organic matter is gradually transformed into inorganic substances, starting with coal (the earliest fossil product) and eventually into oil and natural gas.

The carbonization process begins with the anaerobic heating of wood (heating in the absence of oxygen) at a temperature of approximately 165°C. During this phase, the breakdown of oxygen in lignin and cellulose produces carbon dioxide, water vapour, methane, and heat. Additionally, the process yields by-products such as acetic acid, acetone, and tar.

Wood charcoal has several advantages over raw biomass. It is easier to transport, has a higher calorific value per unit of both mass and volume, and is more manageable in combustion processes. These benefits make charcoal an efficient and practical fuel for various energy applications.

II.6.1.12 Pyrolysis

Pyrolysis is a process involving the thermal decomposition of combustible substances in the complete absence, or with a very limited presence, of oxygen or other oxidizers. It occurs at elevated temperatures ranging from 300 to 800°C and is primarily applied to lignocellulosic plant materials.

The products of pyrolysis are diverse and include high-calorific-value gases (containing tar in vapour form), liquids (commonly referred to as bio-oil; which have fossil fuel origins), and solids (known as "char"; which consist of carbon, ash, sulphur compounds, and volatile hydrocarbons).

The specific outcome of the process depends largely on the duration of the pyrolysis, which can be classified as fast, slow, or conventional. Pyrolysis can be considered an extension of carbonization, as it operates under anaerobic conditions to break weak oxygen bonds and release oxygen from the biomass.

Unlike carbonization, which leaves biomass in a solid state, pyrolysis introduces oxygen to degrade the material further. This breakdown shifts the biomass towards a liquid form by breaking additional molecular bonds. The controlled increase in temperature facilitates this transformation while ensuring the process remains stable.

By carefully managing the oxygen supply and process temperature, pyrolysis enables the conversion of solid biomass into a liquid state, primarily used for the production of bio-oils. Depending on the specific conditions (such as the amount of oxygen and the temperature), the final products can vary, ranging from solids to liquids or gases. When the objective is bio-oil production, the focus is on achieving the liquid state. However, pyrolysis can also act as a preparatory step for gasification, setting the stage for further processing.

II.6.1.13 Gasification

Gasification is a process initially developed for coal combustion and electricity generation, aiming to achieve low environmental impact and high efficiency in combined-cycle plants, known as IGCC (integrated gasification combined cycle) plants. It involves partial oxidation under oxygen-deficient conditions and high temperatures (800–1000°C), resulting in a gas mixture primarily composed of hydrogen (H_2), carbon monoxide (CO), carbon dioxide (CO_2), water vapour (H_2O), methane (CH_4), higher hydrocarbons (C_nH_m), and nitrogen (N_2).

The composition and quality of the resulting gas depend on the source of oxygen used in the process. Gasification using air yields a gas with high nitrogen content and low calorific value, termed "producer gas". Conversely, if pure oxygen is used, the process produces "synthesis gas" or "syngas", which has a higher calorific value (although still only about one-third that of methane). To conserve thermal energy and reduce energy losses during transport, gasification is generally performed close to the end-user site, often located downstream of the gasification plant at the final combustion facility.

The gasification process unfolds in three distinct stages, often separated physically within the plant.

The first stage involves biomass drying, where temperatures exceeding 100°C remove moisture.

The second stage, pyrolysis, occurs at temperatures between 300–500°C, breaking down the biomass into simpler components.

The final stage, termed "low-temperature pre-combustion", takes place at 800–1100°C. In this phase, the molecular bonds of the fuel are reformed, yielding a lighter, energy-dense chemical compound dominated by C–H bonds. Steam is introduced to moderate the temperature and facilitate the production of hydrogen (H_2). The composition and energy content of the resulting gas are heavily influenced by the duration of the pyrolysis and pre-combustion phases. Once this new compound is formed, it undergoes secondary combustion.

Gasification serves as a transformative process that reformulates biomass into a new energy carrier compatible with conventional power generation technologies. However, implementing this process poses challenges due to the scale mismatch between combined-cycle plants and biomass plants. For instance, biomass plants with a capacity of 300 MW are not economically feasible due to supply chain limitations, making the large-scale application of gasification technologies for biomass particularly challenging.

II.6.1.14 Anaerobic and Aerobic Digestion

Digestion is a biological process that allows us to release the bond energy stored using solar energy. Digestion breaks the C, H, and O bonds present in materials. All living organisms perform this process, but some carry out anaerobic digestion, while others perform aerobic digestion. Generally, more complex organisms carry out a more advanced and efficient process (aerobic digestion) which involves the use of oxygen and leads to greater energy release (breaking most biomass bonds).

- **Anaerobic digestion**
  Anaerobic digestion is a process performed by biological and non-biological entities, particularly bacteria, but also by microalgae and other substances. These entities are capable of digesting organic matter in the absence of oxygen. This type of digestion does not break all bonds but only the weaker ones (the oxygen available is only that contained in the organic material

itself). It is therefore a process (pushed to the gasification of solid organic matter) of partial degradation of organic matter that releases **biogas** rich in methane (CH_4), CO_2, and other impurities. Since only a small amount of bond energy is released, the resulting gas is warm but not hot enough to harness its thermal energy.

This process is of energy interest because the resulting biogas has a very high calorific value. However, a challenge with synthesis gas is that, besides CH_4, it also contains a significant amount of CO_2, which is energetically inert. To address this, CO_2 is separated from the synthesis gas, resulting in a gas that is almost entirely CH_4 – pure enough to be injected into natural gas pipelines.

Practically, this technology is applied in landfills dedicated to organic waste, where the presence of bacteria or organisms capable of digesting the waste is promoted. The landfill is covered with tarps to prevent the release of methane-rich biogas into the atmosphere (as methane is a greenhouse gas). Instead, the gas is captured in extraction wells to be either utilized on-site or transported through pipelines. For example, synthesis gas may be burnt in a small ICE for co-generative electric and thermal local final uses.

It is important to note that organisms capable of anaerobic digestion include pathogens such as *Salmonella*, *Escherichia coli*, and *Staphylococcus*, which are all bacteria causing diseases. When present in excessive quantities (small amounts are tolerated by the human body), these bacteria digest organic matter in our bodies instead of us, posing a real risk of disease. These bacteria are also abundant in number, though food industries mitigate the risk through cleaning and food treatment processes.

- **Aerobic digestion**

  In aerobic digestion (or composting), more complex organisms capable of aerobic digestion are utilized to digest the biomass. This process is safer as it does not release methane, is faster, and is generally more efficient than anaerobic digestion. However, it has limited energy applications since biomass degradation does not release methane (which can be easily collected) but instead releases thermal energy (in the form of water vapour and CO_2).

  Collecting this thermal energy is challenging as it is widely dispersed. Consequently, this type of process is used primarily for the rapid disposal of organic waste rather than for energy purposes. It could, however, serve energy applications when organic matter exists in liquid form, allowing the thermal energy to be collected via a heat exchanger to heat a process fluid. This approach is indeed applied in industries that produce biomass, where part of the biomass is used to sustain the process itself (e.g., in paper mills). Nonetheless, aerobic digestion is primarily employed to address the issue of organic waste disposal.

  Aerobic digestion requires greater attention to leachate management. In landfill sites, adding organic matter alters the site's organic nature. While this modification is not overly impactful when dealing solely with organic matter, it can become problematic if other materials, such as plastics or rubber, are present. In such cases, the risk of groundwater contamination becomes more significant. Moreover, aerobic digestion produces H_2O, which naturally humidifies the soil, regardless of isolation measures, climate, or other factors. (In contrast, anaerobic processes do not produce H_2O.)

II.6.2 ON E-FUELS

e-Fuels, also known as "electrofuels", are synthetic fuels typically produced through chemical reactions involving hydrogen (H_2) and carbon sources like carbon dioxide (CO_2) or carbon monoxide (CO). Another variant, e-ammonia, is synthesized using hydrogen and nitrogen (N_2) as the primary components.

Figure II.6.3 offers an overview of renewable and sustainable technology pathways, including feedstock requirements, production processes such as Fischer–Tropsch synthesis and methanol dehydration, transportation options, and potential applications.

These fuels, which can exist in gaseous or liquid forms, exhibit properties similar to those of fossil fuels. Their production relies on electricity, with sustainability hinging on the energy source and the method used to produce hydrogen. The use of hydrogen derived from renewable electrolysis, also known as green hydrogen, ensures a low-carbon production process. In contrast, fossil-based hydrogen, whether grey or blue, diminishes sustainability.

When powered by renewable electricity and produced with sustainable hydrogen, e-fuels emerge as a highly sustainable alternative to conventional fuels. Utilizing CO_2 as a feedstock ensures that combustion emissions do not contribute additional greenhouse gases to the atmosphere. However, significant challenges remain, including high production costs, limited availability of critical materials for large-scale production, and the localized emissions generated during combustion.

e-Fuels hold particular promise as energy carriers in sectors such as transportation, where they facilitate energy storage derived from intermittent renewable sources.

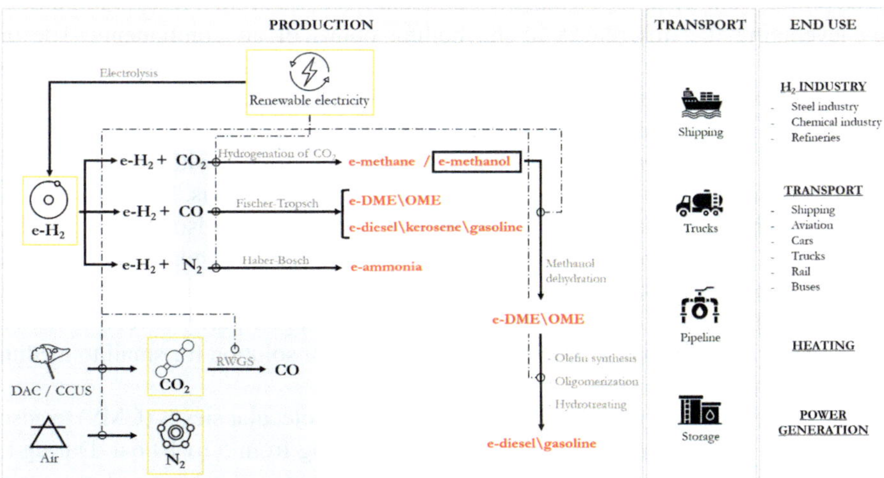

FIGURE II.6.3 Overview of technology pathways for renewable e-fuels production.

Importantly, they do not necessitate changes to existing fuel distribution networks or infrastructure.

Their high energy density and compatibility with conventional engines make them especially well-suited for applications in aviation, shipping, and heavy transportation, where electrification is less feasible.

Despite these advantages, uncertainties related to production efficiency, economic feasibility, and environmental impacts complicate their large-scale deployment. Active research efforts and pilot projects are addressing these issues by enhancing production technologies and reducing costs, offering hope for future scalability.

The terminology associated with e-fuels, such as power-to-X (PtX), power-to-liquid (PtL), power-to-gas (PtG), gas-to-liquid (GtL), and renewable fuels of non-biological origin (RFNBO), has caused some confusion within the scientific community. Policymakers and researchers are working to clarify these terms and standardize definitions.

Policy incentives and financial subsidies will play a crucial role in bridging the cost gap between e-fuels and conventional fossil fuels, enabling their widespread adoption. While many experts regard e-fuels as essential to achieving a zero-emission energy future, others argue that alternative solutions may offer more immediate and practical pathways to a clean energy system.

II.6.2.1 Feedstocks for e-Fuels Production

II.6.2.1.1 Nitrogen

Nitrogen is a critical feedstock in the production of e-ammonia, where it combines with hydrogen to form the final product. As the main component of air, nitrogen is extracted commercially through three primary methods: cryogenic distillation, membrane separation, and pressure swing adsorption (PSA).

Cryogenic distillation is the most commonly used method for large-scale nitrogen production. Renowned for its maturity, efficiency, and flexibility, it separates nitrogen by leveraging the differences in the boiling points of air components. The process begins by compressing ambient air, cooling it to around 10°C, and filtering it to remove impurities such as moisture and dust. The purified air is then cooled to below –195.8°C, causing liquefaction. Through high-pressure distillation, nitrogen is separated from oxygen and other gases, yielding nitrogen with adjustable purity levels that are tailored for a wide range of industrial applications.

Membrane separation employs hollow-fibre membranes to isolate nitrogen from compressed and filtered air. In this method, oxygen, water vapour, and other impurities permeate the membrane walls, while nitrogen passes through the membrane's centre and is collected. This technique achieves nitrogen purity levels of between 95% and 99.5%, offering a compact and cost-effective solution for small to medium-sized industrial applications.

Pressure swing adsorption (PSA) utilizes carbon molecular sieves (CMS) to adsorb oxygen and other impurities at high pressures ranging from 5 to 10 bar. During this process, compressed air is purified and directed into two synchronized pressurized vessels. One vessel adsorbs impurities while allowing nitrogen to flow into storage tanks, while the second vessel regenerates the sieves by depressurizing and releasing the trapped oxygen. PSA systems can produce nitrogen with purity levels as high as 99.999%, making them an excellent choice for applications requiring ultra-high-purity nitrogen.

II.6.2.1.2 Carbon Dioxide

Captured CO_2 plays a crucial role in e-fuel synthesis, serving as a key feedstock to produce synthetic hydrocarbons or methanol. This approach facilitates a circular carbon economy by recycling emissions into low-carbon fuels, reducing reliance on fossil fuels, and enabling decarbonization in hard-to-abate sectors such as aviation, maritime, heavy-duty transport, and heavy industry.

Two main approaches exist for CO_2 capture: carbon capture, utilization, and storage (CCUS) and direct air capture (DAC). CCUS focuses on capturing CO_2 from industrial point sources, where concentrations range between 10–50%. This method is especially effective for industries with high-volume emissions and relatively pure CO_2 streams, making it more cost-effective than alternatives. DAC, by contrast, extracts CO_2 directly from the atmosphere, where its concentration is approximately 400 parts per million (ppm). Although DAC addresses diffuse emissions, it is typically more expensive due to the thermodynamic challenges posed by processing large air volumes.

II.6.2.1.2.1 Point Sources Technologies

Technologies for capturing CO_2 from industrial processes can be classified into separation capture and non-separation capture, based on whether CO_2 needs to be separated from other gases in the stream.

- **Separation capture** involves isolating CO_2 from gas effluents. This approach is further divided into pre-combustion and post-combustion methods:
  - Pre-combustion: This process involves converting fuel into syngas (a mixture of CO and H_2) via steam reforming or partial oxidation. CO is then converted to CO_2 through the water–gas shift (WGS) reaction, resulting in a high-pressure CO_2 stream that facilitates separation. Commonly applied in integrated gasification combined cycle (IGCC) systems, this method captures CO_2 for storage or utilization, while the resulting H_2 is used as fuel.
  - Post-combustion: CO_2 is captured from exhaust gases after combustion, which typically have low CO_2 concentrations (3–14%) and may contain contaminants like SO_x and NO_x. Large gas volumes and high exhaust temperatures necessitate advanced separation techniques, making this method suitable for retrofitting existing plants.
- **Non-separation capture** eliminates the need for CO_2 separation by using pure oxygen for combustion instead of air, producing a gas stream primarily composed of CO_2 and water vapour. Two key methods include:
  - Oxyfuel combustion: Combustion with pure oxygen creates a highly concentrated CO_2 stream (85–95%), but the process is energy-intensive due to the need for oxygen separation from air.

- Chemical looping combustion: Metal oxides act as oxygen carriers, alternating between oxidized and reduced states in air and fuel reactors. This method produces CO_2 and water vapour without direct air contact, resulting in a highly efficient, concentrated CO_2 stream.

For processes requiring the separation of CO_2 from gas effluents, various technologies have been developed to concentrate CO_2 and remove impurities efficiently. Below is an overview of the main methods:

- **Absorption**: This widely used method treats gases post-combustion and is divided into chemical and physical absorption, both involving solvent regeneration.
    - Chemical Absorption: CO_2 is absorbed into a liquid solvent, typically aqueous amines like monoethanolamine (MEA) or diethanolamine (DEA), in an absorber column. The CO_2-rich solution is then regenerated in a desorber column, releasing pure CO_2 for capture and reuse. This process operates efficiently under high pressure and low temperature for absorption, and low pressure and high temperature for regeneration. However, challenges include amine degradation, corrosion, and high energy costs for regeneration. Additionally, impurities like H_2S in the gas stream can lead to salt formation (e.g., carbamates), which further complicates the process.
    - Physical Absorption: Organic or inorganic solvents (e.g., Selexol, Rectisol, or Purisol) absorb CO_2 based on Henry's law. This method requires less energy for regeneration but is effective only at low temperatures, necessitating cooling for hot gas streams.
- **Adsorption** employs solid adsorbents (e.g., activated carbon, zeolites, silica gel) with micropores that bind CO_2 molecules through physical or chemical interactions. The process typically occurs in adsorption beds and is followed by desorption to regenerate the adsorbent and recover CO_2.
  Techniques such as pressure swing adsorption (PSA) and temperature swing adsorption (TSA) enable continuous operation by alternating between adsorption and regeneration cycles. While effective for concentrated CO_2 streams, adsorption faces limitations in low CO_2 concentrations and high temperatures, which reduce adsorbent efficiency.
- **Membrane Capture** separation utilizes semi-permeable barriers to isolate CO_2 based on pressure differentials and membrane selectivity.
    - Gas Separation Membranes: These operate at high pressure, allowing CO_2 to pass through the membrane while retaining other gases.
    - Gas Absorption Membranes: These combine microporous membranes with chemical absorbents for enhanced separation.
    - Polymeric Membranes: These dominate this field but degrade under high temperatures and pressures, reducing their effectiveness. Research into advanced materials such as ceramic composites aims to address these issues. Membranes are best suited for smaller-scale applications or as part of hybrid systems.

- **Cryogenic Capture** methods involve cooling gas streams to liquefy CO_2 at temperatures and pressures above its triple point (–56.6°C, 5.18 bar). This process is particularly suitable for streams with CO_2 concentrations above 75%, where it becomes economically viable. However, the method is energy-intensive due to the cooling requirements and may require additional distillation for high-purity CO_2 recovery.
- **Calcium Looping** process utilizes calcium oxide (CaO) as a sorbent to capture CO_2 by forming calcium carbonate ($CaCO_3$) in a carbonator reactor at 650°C.

  $CaCO_3$ is then decomposed back into CaO and CO_2 in a calciner at 900°C. The reaction cycle is highly efficient for industries like cement manufacturing, where CaO is readily available and cost-effective. Limitations include the gradual loss of reactivity in CaO over multiple cycles, requiring regeneration techniques to maintain efficiency.

Carbon capture technologies vary significantly in energy demands, scalability, environmental impact, and cost-effectiveness, necessitating careful alignment with industrial needs. Energy-intensive methods like cryogenic capture and oxyfuel combustion excel in concentrated CO_2 streams, while energy-efficient options such as chemical looping and membrane systems suit cost-sensitive applications.

Scalability depends on context: pre-combustion capture is ideal for large-scale facilities with high CO_2 concentrations, whereas post-combustion capture offers versatility for retrofitting and use in diverse industrial settings.

Environmental considerations are critical. While chemical absorption is mature and widely used, amine solvents may degrade, generate waste, and cause corrosion, requiring diligent management. Chemical looping, though efficient, raises concerns about recycling or disposing of spent materials. Addressing such issues is essential for sustainable deployment.

Technology readiness also influences adoption. Absorption methods are commercially established, while adsorption and membrane technologies are still evolving, with research aiming to enhance durability and cost-efficiency. Emerging techniques like oxyfuel combustion and chemical looping, though less mature, show promise for scalability and cost reduction.

Selecting a CO_2 capture method requires balancing energy efficiency, scalability, environmental impact, maturity, and cost against specific application needs. Chemical absorption remains widely implemented due to moderate costs and proven performance, but emerging methods like chemical looping offer potential for more sustainable, cost-effective solutions. Each approach brings distinct opportunities and challenges, shaping its role in industrial decarbonization and broader sustainability goals.

II.6.2.1.2.2 Direct Air Capture (DAC) Technologies

Direct air capture (DAC) offers an alternative to capturing CO_2 from industrial sources by removing CO_2 directly from the atmosphere, where its concentration is approximately 400 ppm. This low concentration presents thermodynamic challenges, requiring the processing of large air volumes, but it enables the achievement of negative emissions. The goal of DAC is to produce a concentrated CO_2 stream suitable for storage or utilization. Commercial DAC processes primarily rely on liquid solvents or solid adsorbents for CO_2 capture.

In the **liquid solvent process**, CO_2 is absorbed into a basic solution, typically potassium hydroxide (KOH), forming potassium carbonate (K_2CO_3). This reacts with calcium hydroxide ($Ca(OH)_2$), precipitating calcium carbonate ($CaCO_3$) pellets, which are calcined at high temperatures (~900°C) to release CO_2 and regenerate calcium oxide (CaO) for reuse. The system includes a contactor loop for air absorption and a calciner loop for CO_2 recovery, but its reliance on natural gas and oxygen for calciner heating compromises renewable credentials.

In the solid adsorbent process, CO_2 is captured on the surface of porous materials such as metal-organic frameworks (MOFs), zeolites, activated carbons, silica materials, or molecular sieves. The strength of the interaction between CO_2 and the adsorbent depends on the nature of the adsorption, which may involve weak intermolecular forces (physisorption) or strong covalent bonds (chemisorption). The system operates through a cycle of adsorption at ambient conditions and desorption under reduced pressure and elevated temperatures using temperature-vacuum swing adsorption (TVSA). During adsorption, air is drawn into a contactor where CO_2 adheres to the adsorbent. In the desorption phase, the contactor is sealed, and residual air is evacuated to prevent CO_2 dilution. Heat and water vapour are introduced to release CO_2, which is separated from the vapour through condensation and then compressed for storage or use.

Both DAC methods are energy-intensive, with about 80% thermal energy (primarily for regeneration) and 20% electrical energy (for fans, compressors, and pumps). Solid adsorbent systems require ~6 GJ/t thermal energy and 1.5 GJ/t electrical energy, while liquid solvent systems need 5.25–8 GJ/t and 1.2–1.8 GJ/t, respectively. The modularity of DAC systems, where components are divided into scalable units, enables flexibility and adaptability to various operational needs.

II.6.2.1.3 Hydrogen

Hydrogen is a vital energy carrier with a high energy density by mass (120 MJ/kg), making it a powerful fuel. However, its low energy density by volume (10.8 MJ/Nm^3) creates challenges for transportation and storage. Currently, hydrogen is predominantly produced from fossil sources like natural gas, oil, and coal. Electrolysis offers a sustainable alternative by using electricity to split water into hydrogen and oxygen, enabling low-carbon hydrogen production when powered by renewable energy sources.

The main electrolysis technologies currently employed (see Figure II.6.4) are alkaline electrolysis (AEL), proton exchange membrane electrolysis (PEM), and solid oxide electrolysis (SOEC). Each technology varies in efficiency, operating conditions, and energy requirements, as explored in the following sections.

- **Alkaline Electrolysis (AEL)**
  Alkaline electrolysis is a well-established technology used for large-scale hydrogen production. It operates with electrodes immersed in a 25–30% aqueous solution of potassium hydroxide (KOH) or sodium hydroxide

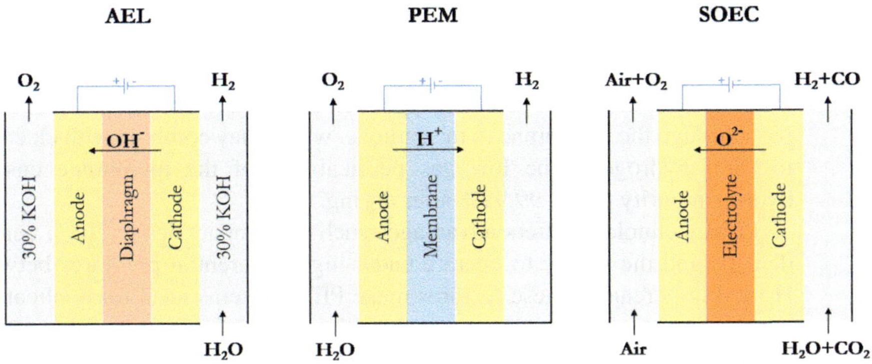

FIGURE II.6.4 Overview of the main electrolysis technologies in current use.

(NaOH), which facilitates the transfer of hydroxide ions (OH^-) between the cathode and anode. A diaphragm separates the electrodes to prevent the mixing of hydrogen and oxygen gases while allowing the passage of the electrolyte.

The electrolyte is continuously recirculated to remove gas bubbles and dissipate heat, maintaining optimal operating conditions. Liquid–vapour separators for H_2 and O_2 ensure the recovery of the residual electrolyte and gas purification. After drying, hydrogen purity ranges from 99.5% to 99.9%, which can exceed 99.999% with catalytic purification.

The electrochemical reactions at the electrodes are:

- Cathode reaction: $2H_2O + 2\ e^- \rightarrow H_2 + 2\ OH^-$
- Anode reaction: $2\ OH^- \rightarrow \frac{1}{2}O_2 + H_2O + 2\ e^-$
- Overall reaction: $2\ H_2O \rightarrow 2\ H_2 + O_2$
  Water is consumed at the cathode and regenerated at the anode, requiring the KOH currents to mix and maintain electrolyte concentration.

  AEL systems are cost-effective, simple, and capable of producing high-purity hydrogen. However, they are prone to corrosion, require long startup times, and perform poorly under variable energy inputs, making them better suited for steady-state operations.

- **Proton Exchange Membrane Electrolysis (PEM)**
  Proton exchange membrane electrolysis (PEM) relies on a polymer electrolyte membrane that separates two half-cells. Electrodes are mounted directly onto the membrane, forming a compact membrane electrode assembly (MEA). This acidic system uses noble metal catalysts, with iridium at the anode and platinum at the cathode. The electrochemical reactions are as follows:

  - Cathode reaction: $H_2O \rightarrow \frac{1}{2}O_2 + 2\ H^+ + 2\ e^-$

- Anode reaction: $2\ H^+ + 2\ e^- \rightarrow H_2$
- Overall reaction: $H_2O \rightarrow H_2 + \frac{1}{2}O_2$

Water is split at the anode into oxygen, electrons, and protons (H^+). Protons pass through the membrane to the cathode, where they combine with electrons to form hydrogen. The low gas permeability of the membrane ensures hydrogen purity above 99.99% after drying.

PEM technology offers advantages such as compact design, high current density, and the ability to operate under high differential pressures between H_2 and O_2 streams. These features make PEM systems ideal for applications with intermittent energy inputs, such as those powered by renewable energy sources.

However, PEM systems require expensive materials for catalysts and membranes, resulting in higher costs. Despite this, their efficiency, adaptability to fluctuating loads, and high-purity output make them a valuable option for low-carbon hydrogen production.

- **Solid Oxide Electrolysis Cell (SOEC)**

Solid oxide electrolysis cells (SOEC), also known as high-temperature electrolysis (HTE), operate using steam instead of liquid water, enabling higher electrical efficiency than AEL and PEM electrolysis. Operating at temperatures of 700–900°C, SOEC achieves superior efficiency due to enhanced reaction kinetics and thermodynamic advantages, as well as the utilization of internal heat for steam conversion. However, the high operating temperatures can compromise material stability, presenting challenges in durability.

The electrochemical reactions are as follows:

- Cathode reaction: $H_2O + 2\ e^- \rightarrow H_2 + O^=$
- Anode reaction: $O^= \rightarrow \frac{1}{2}\ O_2 + 2\ e^-$
- Overall reaction: $H_2O \rightarrow H_2 + \frac{1}{2}O_2$

The SOEC process begins with preheated steam, supplied to the cathode along with recycled hydrogen, where water is reduced to hydrogen and oxygen ions ($O^=$). These ions migrate through the solid oxide electrolyte to the anode, where oxygen is formed. The reaction's efficiency benefits from heat recovery within the system, as steam is preheated using the hot gas exiting the apparatus. Integrating low-temperature heat sources further avoids the need for electric heating to supply the vaporization energy, enhancing overall efficiency.

The SOEC stack consists of multiple electrically connected cells in series. Steam utilization efficiency ranges between 40–80%, depending on the electrical current density. After the reaction, the hydrogen and remaining steam are separated by cooling, with water condensed for reuse. To maintain the high input temperature (700–900°C), steam must be superheated, typically using external heat sources such as solar thermal systems, industrial waste heat, or exothermic chemical processes. Electric heating is generally avoided to preserve energy efficiency.

SOEC systems often include co-electrolysis, where steam and CO_2 are converted into syngas (H_2 and CO), a key feedstock for e-fuel production. This integration with exothermic reactions like Fischer–Tropsch synthesis enhances efficiency.

Recent advancements in materials science have addressed durability and cost issues, making SOEC systems more viable. Modern designs incorporate cost-effective materials for electrodes and electrolytes, reducing capital costs and improving system longevity. When integrated with high-temperature heat sources, SOEC offers a highly efficient solution for hydrogen production and e-fuel applications.

II.6.2.1.4 Electricity Consumption

Electricity plays a vital role in every stage of e-fuel production, including hydrogen generation via electrolysis, nitrogen separation, CO_2 capture, and the synthesis of final fuels. Among these processes, hydrogen production through electrolysis is the most energy-intensive, requiring 45–73 kWh per kilogram of hydrogen, depending on the electrolyser efficiency.

Nitrogen separation, typically achieved via cryogenic distillation, consumes only 0.11 kWh per kilogram of nitrogen, whereas CO_2 capture methods, such as direct air capture (DAC), require around 1.45 kWh per kilogram of CO_2. If carbon capture utilization and storage (CCUS) is added, energy demand can increase by 10–40%. The final synthesis stage, such as Fischer–Tropsch (FT) synthesis, consumes a comparatively modest 0.50 kWh per kilogram of product. Despite variation in energy use across these processes, the overall energy demand for e-fuel production remains significant.

The lifecycle emissions of e-diesel production through methanol synthesis depend heavily on the electricity source, as shown in Figure II.6.5. Renewable electricity results in emissions below the liquid fossil fuel comparator threshold (approximately 100 gCO_2eq/MJ).

In contrast, fossil-based electricity, particularly from coal, leads to emissions exceeding 500 gCO_2eq/MJ. Natural gas-powered electricity results in emissions ranging between 150–300 gCO_2eq/MJ, depending on whether carbon capture is utilized.

This underscores the critical need for decarbonizing the electricity grid to ensure sustainable e-fuel production.

Producing e-fuels sustainably relies on abundant low-carbon or renewable electricity. However, renewable energy currently accounts for only 20% of global electricity generation, with the average global carbon intensity of electricity at 438 gCO_2/kWh in 2022. If e-fuels were produced using grid electricity at this intensity, their greenhouse gas (GHG) emissions would be nearly three times higher than fossil fuels, making renewables indispensable.

Economic feasibility requires cost-effective renewable electricity and high operational hours (3000–4000 full-load hours per year) to sustain e-fuel plants. However,

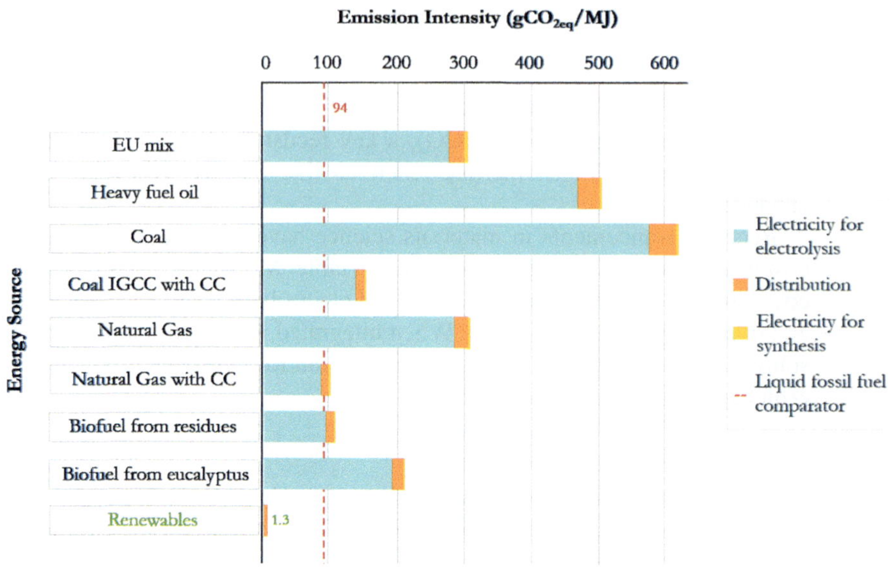

FIGURE II.6.5 Emission intensity comparison across electricity generation sources.

renewable energy availability often falls short; for example, in Germany, average availability is just 1500 hours per year.

Even regions with surplus renewable energy, such as European offshore wind farms, typically generate only 10% excess electricity annually (insufficient to meet large-scale e-fuel demand). Substantial investments in renewable energy infrastructure, such as offshore wind farms in Europe or solar and wind energy installations in North Africa or the Middle East, are therefore essential.

Nuclear energy offers an alternative for steady, low-carbon electricity supply. Coupled with high-temperature electrolysis (HTE), nuclear-powered hydrogen production achieves up to 90% efficiency, with lifecycle emissions as low as 0.28 $kgCO_2$ per kilogram of hydrogen. Renewable sources like solar and wind minimize emissions but face challenges from seasonal variability and the need for energy storage.

Geographic factors also play a significant role in e-fuel production economics. Regions with abundant renewable resources, such as North Africa, can produce e-fuels more cost-effectively than energy-scarce areas, potentially shifting global energy dynamics by reducing fossil fuel exports and increasing renewable energy exports.

Meeting ambitious e-fuel targets requires vast investments in renewable electricity. For example:

- Supplying 50% of Europe's aviation fuel with e-fuels by 2050 would require renewable electricity equivalent to 25% of current EU electricity generation.
- Replacing 50% of truck fuel by 2050 would demand renewable electricity surpassing one-third of the EU's current output.

These estimates highlight the need for substantial renewable energy deployment to scale e-fuel production as a major contributor to GHG reduction targets.

The economic viability of e-fuels hinges on the availability of cost-competitive renewable electricity and advancements in production technology. Currently, the high capital investment and reliance on abundant renewable electricity make e-fuels suitable for niche markets with relatively small demands.

However, progress in energy storage, grid stability, and renewable integration could accelerate adoption, enabling e-fuels to transition from niche applications to broad climate solutions in the future.

II.6.2.2 Production Technologies

II.6.2.2.1 e-Methane

Methane, the primary component of natural gas, is a cleaner-burning fuel with lower carbon dioxide emissions and fewer local pollutants compared to other hydrocarbon-based fuels. However, methane's global warming potential (GWP100) is significant: 1 kg of methane (CH_4) has a warming effect equivalent to 28 kg of CO_2 over a 100-year period. Therefore, managing methane emissions is critical to mitigating its environmental impact.

Most methane today is fossil-derived, sourced from natural deposits or fracking technologies. Fracking, especially in sub-sea contexts, presents risks of volatile methane emissions over time, further contributing to greenhouse gas (GHG) emissions.

In contrast, e-methane production involves combining electrolysis with methanation. Hydrogen is produced from water electrolysis, and this hydrogen reacts with CO_2 in a reactor, following the Sabatier reaction:

$$CO_2 + 4\ H_2\ \ CH_4 + 2\ H_2O$$

During this process, CO_2 is compressed to 5–50 bar and introduced into a methanation reactor at temperatures between 250–400◦C, converting it into CH_4.

This process, while straightforward, requires significant amounts of CO_2 (about 5.5 kg for every kg of H_2), which is often challenging to obtain due to the distance between CCUS systems and renewable plants, increasing CO_2 transport costs.

Despite these challenges, e-methane production is an effective way to reduce GHG emissions. When renewable electricity powers hydrogen production and synthesis, e-methane offers a sustainable path to utilize CO_2 while supporting renewable energy integration.

The produced e-methane can be injected into the existing natural gas grid, used as fuel for natural gas vehicles, or utilized in industrial applications and electricity generation. Moreover, e-methane production allows for the storage of renewable electricity in chemical form, making it available when needed.

II.6.2.2.2 e-Methanol

Methanol (CH_3OH) is the simplest alcohol, primarily produced from fossil fuels – 85% from natural gas and about 15% from coal. Global production reaches 98 million tons annually, emitting 0.3 gigatons of CO_2, which accounts for roughly 10% of the chemical sector's total emissions. Despite its widespread use, only 0.2% of methanol is derived from renewable sources.

e-Methanol, a sustainable alternative, is synthesized through CO_2 catalytic hydrogenation in an adiabatic fixed-bed reactor. The process uses catalysts based on copper (Cu) and zinc (Zn), with additives like aluminium (Al), zirconium (Zr), chromium (Cr), silicon (Si), boron (B), and gallium (Ga). Pre-heated hydrogen and carbon dioxide are converted into methanol via the following three chemical reactions:

- CO_2 hydrogenation: $CO_2 + 3\ H_2 \rightarrow CH_3OH + H_2O$
- Reverse water gas shift: $CO + H_2O \rightarrow H_2 + CO_2$
- CO hydrogenation: $CO + 2\ H_2 \rightarrow CH_3OH$

This process operates at temperatures of 250–300°C and pressures of 50–100 bar, using Cu/Zn/Al-based catalysts.

An alternative method for e-methanol production involves the direct conversion of e-methane. In this pathway, water acts as an oxidizing agent in the presence of a copper bed containing zeolite as a catalyst, producing hydrogen as a by-product. However, direct conversion processes are still in early stages of development and are estimated to require 5–20 years to achieve industrial feasibility.

Methanol has approximately half the volumetric energy density of gasoline but features a high octane number and low cetane number, making it suitable for spark-ignition engines with minimal modifications. It is widely used in the production of chemicals such as formaldehyde, dimethyl ether (DME), and methyl tertiary-butyl ether (MTBE).

Methanol also serves as a solvent, clean synthetic fuel, and energy source for transportation, industrial boilers, wastewater treatment, and electricity generation. Furthermore, methanol derivatives, including DME, synthetic gasoline, and kerosene, can be synthesized for broader applications.

China leads the world in methanol production and consumption, with national standards for methanol blends such as M15 (15% methanol) and M85 (85% methanol) fuels.

II.6.2.2.3 e-DME/OME

Dimethyl ether (DME), also known as methoxymethane (CH_3OCH_3), is the simplest form of ether. It was originally a by-product of high-pressure methanol synthesis but is now primarily produced via methanol dehydration in a two-stage synthesis process or directly from syngas. The direct hydrogenation of CO_2 into DME has been studied, but its stage of technological development remains limited.

Methanol dehydration involves converting CO_2 into DME using hydrogen through the following steps:

- CO_2 hydrogenation to methanol: $CO_2 + 3\ H_2 \rightarrow CH_3OH + H_2O$
- Methanol dehydration to DME: $2\ CH_3OH \rightarrow CH_3OCH_3 + H_2O$

The overall reaction is:$2\ CO_2 + 6\ H_2 \rightarrow CH_3OCH_3 + 3\ H_2O$

An undesired side reaction, the reverse water–gas shift (RWGS), produces CO and H_2O:

$$CO_2 + H_2 \rightarrow CO + H_2O$$

For direct synthesis from syngas, two possible pathways are described by the following reactions:

$$3\ CO + 3\ H_2 \rightarrow CH_3OCH_3 + CO_2$$

$$2\ CO + 4\ H_2 \rightarrow CH_3OCH_3 + H_2O$$

Both reactions are employed in commercial plants. In this process, in addition to hydrogen, carbon monoxide is used, which is easily obtained from CO_2 through the RWGS reaction.

DME oligomers, or oxymethylene ethers (OME_x), consist of a chain of DME-derived ester units. For example: OME_0 is equivalent to DME (there are no additional ester units); OME_1 presents one additional ester block added to DME; OME_2 presents two additional ester blocks, and so on.

OME_x has a higher molecular mass than DME, resulting in altered physical properties such as higher boiling points.

In principle, DME can be blended with LPG for spark-ignition engines, but its low octane number presents challenges. It is more commonly explored as a diesel alternative for compression-ignition engines, where it offers significantly lower CO_2 emissions and reduced exhaust pollutants such as NO_x, SO_x, and particulate matter.

Moderate modifications to the engine and injection system, along with dedicated recalibration, are necessary for its use. To date, DME has been primarily used in small commercial fleets, including buses and heavy-duty vehicles, particularly in urban areas to address air quality concerns.

Beyond transportation, DME has diverse applications, including combustors for heating, thermoelectric power plants, and the production of chemicals such as dimethyl sulphate (DMS), used in detergents and household products. Its direct use in fuel cell devices remains in early laboratory stages.

OME_x fuels, derived from DME, lack carbon–carbon bonds and have a high oxygen content. While their volumetric energy density is higher than for methanol, it is still lower than diesel. OME_x fuels are not compatible with current infrastructure and fall outside European diesel specifications (EN 590, EN 15940).

Consequently, their use is limited to small quantities unless engine and fuel systems are entirely adapted. Approval for OME_x use in existing vehicles is also needed, as their compatibility with current engine materials is uncertain.

Both DME and OME_x are easy to transport and store. Despite their potential – particularly in the heavy-duty transport sector – most studies do not consider e-DME and e-OME as viable large-scale alternatives to existing fuels.

II.6.2.2.4 e-Diesel/Kerosene/Gasoline

Technologies for producing e-diesel, e-kerosene, and e-gasoline fall under the category of power-to-liquid (PtL) applications. These technologies utilize hydrogen (H_2) and carbon monoxide (CO) as feedstocks to produce liquid fuels via Fischer–Tropsch (FT) synthesis. Alternatively, methanol can serve as an intermediate, which is converted into gasoline and diesel through processes such as methanol-to-gasoline (MTG) or methanol-to-olefin-to-gasoline and diesel (MOGD).
In the MTG process, the following reactions occur:

- DME synthesis: $2\ CH_3OH \rightarrow CH_3OCH_3 + H_2O$
- Olefin synthesis: $CH_3OCH_3 \rightarrow (CH_2)_2 + H_2O$
- Oligomerization: $\frac{1}{2}n(CH_2)_2 \rightarrow C_nH_{2n}$
- Hydrotreating: $C_nH_{2n} + H_2 \rightarrow C_nH_{2n+2}$

Methanol is first converted into dimethyl ether (DME), followed by the synthesis of olefins. These olefins undergo oligomerization to form hydrocarbons, which are then hydrotreated to produce saturated hydrocarbons suitable for use as fuels. The development of the MTG process dates back to the 1970s, when methanol was first converted to olefins and gasoline through these chemical pathways.

The Fischer–Tropsch (FT) process, originally developed in Germany as an alternative to petroleum, synthesizes hydrocarbons by reacting CO and H_2. These reactants can be supplied as syngas or separately. When CO_2 is the carbon source, it must first be reduced to CO via the reverse water–gas shift (RWGS) reaction. Alternatively, syngas can be generated directly from CO_2 and water through co-electrolysis in solid oxide electrolysis cells (SOECs). The FT synthesis then converts CO and H_2 into hydrocarbons according to the reaction:

$$(2n+1)\ H_2 + nCO \rightarrow C_nH_{2n+2} + H_2O$$

This reaction occurs at temperatures between 190–350°C and pressures of 20–50 bar, with cobalt- or iron-based catalysts being commonly used.

Longer-chain hydrocarbons, such as waxes, can be cracked to increase yields of middle distillates like kerosene and diesel.

The resulting hydrocarbons are classified by their chain lengths. Fuel gases (C_1–C_5), naphtha (C_5–C_{10}), kerosene (C_{10}–C_{15}), diesel (C_{15}–C_{18}), and base oils and waxes (C_{18+}) are obtained, as shown in Figure II.6.6.

The types of hydrocarbons produced during FT synthesis depend on the operating conditions, such as temperature and catalyst type. High temperatures (320–350°C) favour the production of low-molecular-weight hydrocarbons, whereas lower temperatures (190–270°C) promote the formation of high-molecular-weight hydrocarbons.

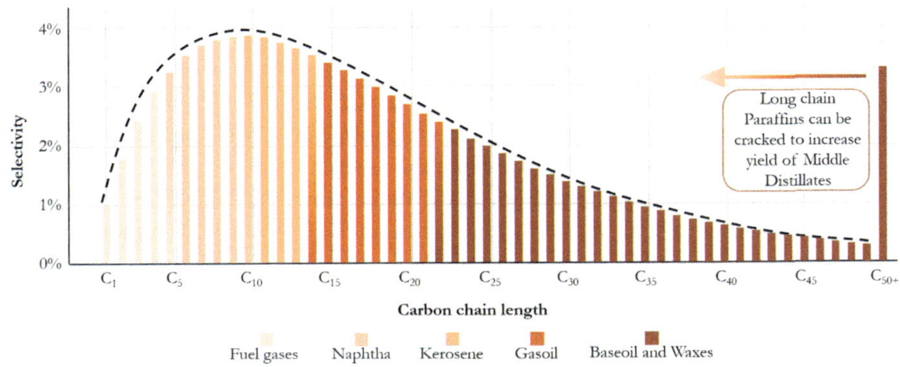

FIGURE II.6.6 Hydrocarbon chain length distribution from FT synthesis.

Catalyst composition also plays a critical role. Cobalt-based catalysts are preferred for diesel production due to their selectivity and ability to operate at low temperatures (190–240°C), though they require a narrow H_2/CO ratio of 2.0–2.3. Iron-based catalysts are more versatile, accommodating a broader H_2/CO ratio (0.5–2.5) and operating across a wider temperature range (200–360°C).

These catalysts are also more cost-effective, although cobalt catalysts offer longer lifespans. Both catalyst types are typically supported by materials such as aluminium oxide (Al_2O_3) or silica (SiO_2), which provide structural stability. Promoters like potassium (K), manganese (Mn), or platinum (Pt) are often added to enhance activity and selectivity.

Recent advancements in FT synthesis have focused on laboratory-scale catalysts, such as Fe–Mn–K systems, which demonstrate high conversion rates and selectivity for sustainable aviation fuel (SAF).

Once hydrocarbons are synthesized in the FT reactor, refining processes are necessary to maximize desired fuel yields. Fractional distillation separates components based on their boiling points, enabling the production of e-diesel, e-kerosene, and e-gasoline.

These technologies, through either the MTG or FT processes, play an essential role in enabling the production of liquid fuels, providing sustainable alternatives for the transportation and aviation industries.

II.6.2.2.5 e-Ammonia

Ammonia (NH_3) has emerged as a potential alternative fuel and a carrier for hydrogen energy. Traditionally, about 80% of global ammonia production is allocated to the agricultural fertilizer industry. Beyond this, it is a critical raw material for producing plastics, synthetic fibres, resins, refrigerants, and explosives. Currently, approximately 98% of ammonia is produced from fossil fuels, including natural gas, naphtha, and coal, contributing to about 1.8% of global CO_2 emissions.

The most common method of production is the Haber–Bosch (HB) process, which combines nitrogen (N_2) and hydrogen (H_2) under high temperatures (400–450°C) and pressures (150–200 bar) in the presence of an iron-based catalyst. The reaction:

$$N_2 + 3\,H_2 \rightarrow 2\,NH_3$$

achieves a hydrogen-to-nitrogen ratio of 2:1 to 3:1, with a conversion efficiency of 25–35% per cycle. Ammonia is separated via cooling at –25°C, causing it to liquefy, and unreacted gases are recycled back into the reactor.

When ammonia is produced from fossil fuels combined with CCS technologies and blue hydrogen, it is termed "blue ammonia". Conversely, ammonia derived entirely from renewable energy sources (RES) and green hydrogen is classified as "green ammonia" or e-ammonia.

Ammonia is the simplest nitrogen hydride but has a lower specific energy compared to most hydrocarbon-based fuels. Its environmental impact requires careful management; ammonia leakage into soil, water, or air can lead to biodiversity loss, eutrophication, air pollution, and greenhouse gas emissions. Additionally, ammonia combustion generates significant levels of nitrogen oxides (NO_x), necessitating measures to mitigate these risks.

Ammonia can be used directly as a fuel in various applications, including gas turbines, industrial furnaces, or internal combustion engines (ICEs). Its use often requires partial or complete thermal cracking (at 400–600°C and 30–100 atm) to overcome its high ignition temperature.

Alternatively, it can be burned with combustion promoters such as diesel or hydrogen, which can be generated via onboard reforming of ammonia. Ammonia also shows potential as a hydrogen transport vector, but complete dehydrogenation to produce high-purity hydrogen is currently associated with considerable efficiency losses.

The Commonwealth Scientific and Industrial Research Organization (CSIRO) of Australia has explored ammonia's use in fuel cells (FCs) by testing two specially designed cars. Their approach employs membrane technology integrated into modular units at refuelling stations.

These membranes selectively allow hydrogen molecules to pass through while blocking larger nitrogen molecules. Ammonia can also be directly used in alkaline FCs, which have already been commercialized.

Ammonia storage is relatively simple, as it liquefies at low pressures (approximately 1 MPa), avoiding the need for high-pressure tanks. However, the infrastructure for its widespread use as a fuel is still under development.

Ammonia is considered a promising energy carrier for hydrogen transport and storage. Its moderate liquefaction conditions and high hydrogen content make it an attractive option.

While the production cost of ammonia is slightly higher than hydrogen, transportation and storage costs are significantly lower. This cost efficiency makes ammonia preferable for energy storage and distribution.

In summary, biomass, biofuels and e-fuels could increasingly contribute in the near future to a sustainable energy mix. Ensuring their sustainability, however, involves a number of crucial activities to be realized, optimizing feedstocks, reducing land-use conflicts, and improving conversion technologies.

Further Readings

1. Sustainable Energy Sources' Characteristics

Boyle, G., *Renewable Energy: Power for a Sustainable Future*. Oxford University Press, 2017.

Ellabban, O., Abu-Rub, H., & Blaabjerg, F., Renewable Energy Resources: Current Status, Future Prospects and Their Enabling Technologies. *Renewable and Sustainable Energy Reviews*, *35*, 748–764, 2014.

Kalogirou, S. A., *Solar Energy Engineering : Processes and Systems*. Elsevier. 2023.

Letcher, T. M., *Future Energy: Improved, Sustainable and Clean Options for Our Planet*. Elsevier, 2008.

Sørensen, B., *Renewable Energy: Physics, Engineering, Environmental Impacts, Economics & Planning*. Academic Press, 2017.

2. Solar Energy

Green, M. A., *Solar Cells: Operating Principles, Technology, and System Applications*. Prentice Hall, 2016.

Kalogirou, S. A., *Solar Photovoltaic Technology Production, Design, and Efficiency Considerations*. Elsevier, 2016.

Luque, A., & Hegedus, S., *Handbook of Photovoltaic Science and Engineering*. Wiley, 2019.

Parida, B., Iniyan, S., & Goic, R., A Review of Solar Photovoltaic Technologies. *Renewable and Sustainable Energy Reviews*, *15*, 1625–1636, 2016.

Smil, V., *Energy and Civilization: A History*. MIT Press, 2018.

3. Wind Energy

Ackermann, T., *Wind Power in Power Systems*. Wiley, 2017.

Burton, T., Jenkins, N., Sharpe, D., & Bossanyi, E., *Wind Energy Handbook*. Wiley, 2015.

Gipe, P., *Wind Power: Renewable Energy for Home, Farm, and Business*. Chelsea Green, 2016.

Manwell, J. F., McGowan, J. G., & Rogers, A. L., *Wind Energy Explained: Theory, Design and Application*. Wiley, 2019.

Veers, P., Dykes, K., Lantz, E., et al, *Grand Challenges in the Science of Wind Energy*. *Science*, *366*(6464), eaau2027, 2019.

4. Hydro Energy

Kibler, K. M., & Tullos, D. D., Cumulative Biophysical Impact of Small and Large Hydropower Development in Nu River, China. *Water Resources Research*, *51*(7), 5214–5230, 2015.

Moran, E. F., Lopez, M. C., Moore, N., Müller, N., & Hyndman, D. W., Sustainable Hydropower in the 21st Century. *Proceedings of the National Academy of Sciences*, *115*(47), 11891–11898, 2018.

Paish, O., Small Hydropower: Technology and Current Status. *Renewable and Sustainable Energy Reviews*, *6*(6), 537–556, 2015.

Twidell, J., & Weir, T., *Renewable Energy Resources*. Routledge, 2015.

5. Geothermal Energy

Di Pippo, R., *Geothermal Power Plants: Principles, Applications, Case Studies and Environmental Impact*. Butterworth-Heinemann, 2015.

Dickson, M. H., & Fanelli, M., *Geothermal Energy: Utilization and Technology*. Earthscan, 2015.

Glassley, W. E., *Geothermal Energy: Renewable Energy and the Environment*. CRC Press, 2018.

Gutierrez Negrin, L. C. A., G. W. *Evolution of Worldwide Geothermal Power 2020–2023*. Geothermal Energy, 2024.

Lund, J. W., Freeston, D. H., & Boyd, T. L., Direct Utilization of Geothermal Energy 2015 Worldwide Review. *Geothermics*, *60*, 29–68, 2018.

6. Biomass, Biofuels, and E-Fuels

Cherubini, F., The Biorefinery Concept: Using Biomass Instead of Oil for Producing Energy and Chemicals. *Energy Conversion and Management*, *51*, 1412–1421, 2017.

Dahiya, A., *Bioenergy: Biomass to Biofuels and Waste to Energy*. Springer, 2020.

Dell'Aversano, S., & Villante, C., et al., *E-Fuels: A Comprehensive Review of the Most Promising Technological Alternatives towards an Energy Transition*. Energies, MDPI, 2024.

Demirbas, A., *Biorefineries: For Biomass Upgrading Facilities*. Springer, 2017.

Drapcho, C. M., et al., *Biofuels Engineering Process Technology.* Mc Graw Hill, 2009.

Mousdale, D. M., *Introduction to Biofuels*. Taylor & Francis, 2010.

Ramesh, C. R, *Sustainable Biofuels Opportunities and Challenges*. Springer, 2021.

Sims, R. E., Mabee, W., Saddler, J. N., & Taylor, M., An Overview of Second-Generation Biofuel Technologies. *Bioresource Technology*, *101*(6), 1570–1580, 2015.

Soler, A. et al. *E-Fuels: A Technoeconomic Assessment of European Domestic Production and Imports towards 2050*. Concawe, 2024.

Detailed Contents

PART III Sustainable Mobility: Sector Evolution, Propulsion Systems

Part III

Sustainable Mobility

Sector Evolution, Propulsion Systems

INTRODUCTION

In Part III of this book, the focus shifts to sustainable mobility, one of the most hard-to-abate energy sectors. This section discusses the evolution of transportation systems, the environmental impact of internal combustion engines, and the potential of alternative fuels such as hydrogen and biofuels. It also highlights innovations in electric vehicles, fuel cell technologies, and Mobility-as-a-Service (MaaS) paradigm, illustrating how these solutions are reshaping urban mobility and reducing emissions, especially within smart urban environments.

- **Chapter III.1: Sustainable Mobility Issues**
  This chapter identifies the challenges of decarbonizing transportation, including dependency on fossil fuels, emissions, and urban congestion. It introduces strategies for promoting sustainable mobility, such as public transport and Mobility-as-a-Service (MaaS).
- **Chapter III.2: Vehicular Energy and Power Requirements**
  This chapter analyses the energy and power demands of vehicles, focusing on efficiency improvements and renewable integration. It provides insights into optimizing energy use in the transportation sector.
- **Chapter III.3: ICEs and Their Local and Global Impacts**

DOI: 10.1201/9781003631125-22

This chapter discusses the environmental impacts of internal combustion engines (ICEs) and the evolution of emission regulations. It highlights the need for alternative propulsion technologies to reduce ICE dependency.

- **Chapter III.4: Alternative Fuels for ICEs**
  This chapter explores alternative fuels like natural gas, LPG, hydrogen, biofuels, and e-fuels, emphasizing their potential to reduce emissions and dependence on fossil fuels.
- **Chapter III.5: Vehicle Electrification**
  The chapter highlights the benefits and challenges of electrifying transportation systems, focusing on battery technologies, charging infrastructure, and grid integration.
- **Chapter III.6: Fuel Cell Electric Vehicles (FCEVs)**
  This chapter examines hydrogen fuel cell technologies and their applications in transportation. It discusses their efficiency, cost trends, and potential for reducing emissions.
- **Chapter III.7: Battery Electric Vehicles (BEVs)**
  This chapter discusses battery electric vehicles, battery technologies, vehicle-to-grid (V2G) systems, and charging infrastructure. It highlights BEVs' role in achieving sustainable mobility.
- **Chapter III.8: Hybrid Electric Vehicles (HEVs)**
  This chapter examines motivations for the introduction of hybrid electric vehicles. It focuses on main HEV typologies, focusing on power optimization, control strategies, and highlighting their contribution to reducing emissions in the transportation sector.

III.1 Sustainable Mobility Issues

As already explained in Part I, global energy systems are far from being sustainable. In fact, they determine significant issues in terms of energy dependency, exhaustibility of energy sources, and GHG emissions.

The cause of about one-third of these issues can be attributed to the transportation sector and its demand for mechanical energy. The sector's impact is even more pronounced when considering local air pollution, as internal combustion engines (ICEs) used in vehicles are the primary sources of pollution in urban areas.

Addressing these problems is complex and requires a comprehensive approach that aims to revolutionize the entire transportation sector within the next decade. This ambitious transformation is often referred to as the "Transition to Sustainable Mobility", which encompasses several key interventions broadly categorized as follows:

- **Actions on Transportation Modalities and Services**
  Transportation energy demand arises from the balance between transportation demand and supply for both passengers and freight. The sector is undergoing significant evolution:
  - **Freight Transportation**
    Freight logistics increasingly leverage ICT technologies to trace and optimize goods movement globally. This includes reducing energy consumption through enhanced intermodal transport systems, prioritizing smaller, low-impact vehicles for last-mile delivery – vehicles that will likely become the only ones permitted in urban areas in the near future.
  - **Passenger Transportation**
    A paradigm shift, known as **Mobility as a Service (MaaS)**, is redefining how people approach their transportation needs. MaaS offers collective and private mobility services provided by both public and private entities, not only for long-distance travel (already common) but also for short, everyday journeys. This flexibility addresses any kind of personal transportation demands. With MaaS, individuals are expected to increasingly prefer purchasing mobility services, such as car-sharing, rather than owning private vehicles. This shift would significantly reduce the number of vehicles sold and operated, while increasing their usage cycles, particularly in urban areas.

    For example, the average privately owned vehicle in urban areas is currently used for only ~5000 km/year. This low usage results in

DOI: 10.1201/9781003631125-23

unsustainable outcomes: too many vehicles are concentrated in urban areas, and these vehicles tend to be older and more polluting.

Under the MaaS paradigm, the lower number of vehicles in urban areas would alleviate challenges like parking, charging infrastructure availability, and traffic congestion, thereby making urban living more sustainable.

Long-distance travel demands would similarly benefit from integrated intermodal infrastructures, including dedicated parking facilities. These services are now feasible within a Smart City framework, supported by existing technologies enabling keyless, geo-referenced, and instantly billable vehicles.

- **Actions on Transportation Means**
  With the MaaS paradigm, vehicles could potentially be utilized for at least 30,000 km/year, enabling faster recovery of investment costs and allowing for their frequent replacement with models featuring the latest technologies. This frequent turnover would also help ensure compliance with advanced environmental standards.

  For instance, the penetration of electric mobility would accelerate if dominated by private MaaS providers instead of individual owners who must bear the costs of vehicle purchases, infrastructure installation, and management.

  MaaS also enables the use of vehicles designed specifically for urban environments – smaller, lighter, and more energy-efficient compared to multipurpose vehicles (so-called MPVs or SUVs) that have dominated passenger car markets over the last two decades.

  MaaS could drive a fundamental change in vehicle homologation, manufacturing, and design processes. Currently, vehicles are designed and homologated for multipurpose applications, including both urban and extra-urban uses. This results in heavy, over-engineered designs to meet structural and crash test standards, which could be significantly relaxed for urban-only vehicles.

  By prioritizing specialized designs for urban vehicles, energy consumption for urban transport could be reduced by half. Similarly, emissions regulations could be strengthened for urban vehicles while being relaxed for vehicles used exclusively for extra-urban or long-distance travel, provided these vehicles utilize designated intermodal hubs.

This evolution highlights the transportation sector's move towards adopting technologies best suited to specific applications and contexts. Long-distance road and air transportation, for instance, are unlikely to forgo chemical energy carriers for onboard energy storage due to their demanding requirements for flexibility and range.

These energy carriers could be derived from biomass or renewable electricity, with the continued use of ICEs potentially acceptable in remote areas far from residential zones where emissions would cause less harm.

In urban environments, however, emissions are nearly intolerable, making electric mobility the only viable and fully sustainable solution. For urban applications, as we will clearly show, energy storage requirements are lower and less critical, particularly if vehicles are designed exclusively for urban use.

Therefore, keeping all these considerations in mind, the subsequent chapters will explore how the transportation sector can transition to sustainability through various solutions:

- **Chapter III.2**: Examines vehicular energy and power requirements to understand how to reduce overall transportation energy demand and identifies the key influencing factors.
- **Chapter III.3**: Focuses on ICEs, analysing their local and global impacts, pollutant formation mechanisms, and how emission directives and ICE technologies have evolved to meet environmental challenges.
- **Chapter III.4**: Discusses alternative fuels for ICEs, such as liquefied petroleum gas, compressed/liquefied natural gas, hydrogen, hydro-methane, biofuels, and e-fuels, to enhance environmental performance.
- **Chapter III.5**: Introduces vehicle electrification concepts, categorizing vehicle configurations and evaluating their environmental performance.
- **Chapter III.6**: Delves into fuel cell electric vehicles (FCEVs), exploring their potential as a sustainable long-term technology despite current challenges in technological readiness.
- **Chapter III.7**: Focuses on battery electric vehicles (BEVs), analysing their design, charging infrastructure, and potential role in stabilizing the electric grid through vehicle-to-grid (V2G) technologies within a smart-grid framework.
- **Chapter III.8**: Explores hybrid electric vehicles (HEVs), highlighting their configurations, environmental performance, and optimization techniques.

This comprehensive approach will outline a roadmap for transitioning to a more sustainable transportation sector, addressing its energy, environmental, and social challenges.

III.2 Vehicular Energy and Power Requirements

This chapter focuses on vehicular energy and power requirements, emphasizing the longitudinal dynamics of vehicles.

Vehicle longitudinal dynamics refers to the study of forces and motions occurring along the vehicle's longitudinal axis, primarily governing acceleration, braking, and resistance during motion.

Understanding these dynamics is essential for optimizing performance, safety, and efficiency, particularly as they dictate how vehicles respond to varying road and environmental conditions.

III.2.1 TRACTION FORCE EVALUATION

Traction force is the force generated at the contact point between a vehicle's tyres and the road surface, propelling the vehicle forward (or backward, in reverse). This force overcomes resistive forces, including aerodynamic drag, rolling resistance, gravitational components on inclines, and inertial forces due to acceleration. The powertrain transmits traction force through the tyres, influenced by the tyre material properties and road surface conditions.

Mathematically, making reference to the rigid body diagram for the vehicle depicted in Figure III.2.1, the traction force (F_t) can be expressed as the algebraic sum (with appropriate signs) of four contributions: aerodynamic drag (F_a), rolling resistance (F_r), gravitational components on inclines (F_g), and inertial forces generated by vehicle acceleration (F_i).

Then, for a known vehicle speed (v), both traction power (P_t) and energy (E_t) can be evaluated according to the following set of governing equations:

$$F_t = F_a + F_r + F_g + F_i, \qquad P_t = F_t \cdot v, \qquad E_t = \int_0^t P_t \cdot dt$$

For a given vehicle mass M_v, and road climbing angle α, F_g can be easily calculated as:

$$F_g = M_v \cdot g \cdot cos(\alpha)$$

DOI: 10.1201/9781003631125-24

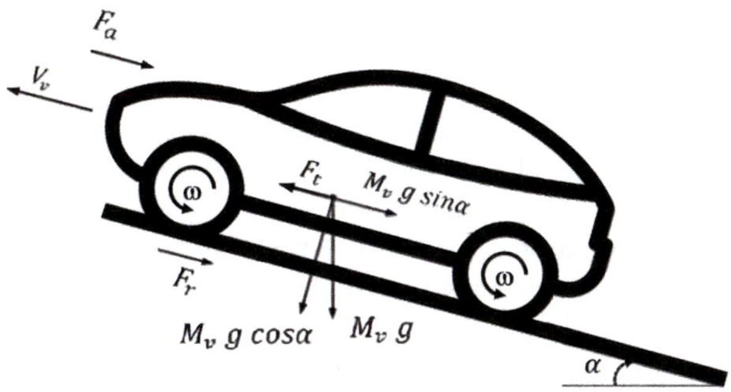

FIGURE III.2.1 Rigid body diagram for a vehicle identifying the main acting forces.

The road slope ($S_{\%}$, or grade) describes how steep a road is. It is calculated as the ratio of vertical rise (elevation change) to horizontal run (distance travelled horizontally), expressed as a percentage: $\alpha = arctan(S_{\%}\ /100)$.

For example, a 10% grade means the road rises 10 units vertically for every 100 units of horizontal distance.

Road slope can locally reach up to ±15% for road transport, while it is normally below ±3.5% for rural low-speed railways, and below ±1.5% for high-speed railways.

Inertial forces generated by vehicle acceleration (F_i) are normally simply calculated as:

$$F_i = M_v \cdot \frac{dv}{dt}$$

However, this calculation must account for the rotational inertia of vehicle components like wheels and engine shafts.

This is achieved by adjusting the vehicle's equivalent mass (M_v^*).

In the following equation, R_w is the wheel radius, while for each i-th rotating part, I_i is its inertia momentum, and λ_i is the "transmission ratio" on i-th rotating axis with respect to the wheels ($\lambda_i = \omega_i\ /\ \omega_w$):

$$M_v^* = M_v + \sum_i I_i \cdot \frac{\lambda_i^2}{R_w^2}$$

The relevance of this additional inertia, required to obtain a desired speed acceleration (and deceleration), depends on vehicle type, but also on the instantaneous gear ratio at the gearbox (which is typically higher for lower gear numbers):

- In passenger cars it has a moderate impact. The equivalent mass typically adds a small percentage (e.g., 2–5%) to the vehicle's total mass. Its effect is more pronounced during rapid acceleration or deceleration.

- In heavy trucks and buses, rotational inertia is more significant due to the larger and heavier rotating components (e.g., wheels and driveshafts). Accounting for this is crucial for accurate performance modelling and brake design.
- In motorcycles, rotational inertia is critical due to the lighter overall mass of the vehicle. The wheels' rotational inertia can significantly affect acceleration, braking, and handling dynamics.
- In trains, the rotational inertia of wheels and axles is a significant factor because of the large masses involved. It influences energy efficiency and braking systems design.
- In aviation, rotational inertia is less relevant in overall performance calculations due to the high total mass of the aircraft. However, it is critical in engine dynamics and landing gear design.
- In electric vehicles (EVs), rotational inertia is often less significant due to the instant torque provided by electric motors. However, lightweight designs amplify its relative impact.

Now, coming to aerodynamic forces, these are commonly evaluated as:

$$F_a = C_D \cdot A \cdot \rho \cdot \frac{1}{2} \cdot v^2$$

where A is vehicle frontal area, ρ is air density, and C_D is vehicle drag coefficient. Typical drag coefficients for different types of vehicles are:

- Passenger Cars: 0.25–0.35 for modern sedans and hatchback; 0.20–0.30 for sports cars; 0.35–0.45 for SUVs and trucks.
- Heavy-Duty Vehicles: 0.50–0.70 for semi-trucks.
- Buses: 0.60–0.80.
- Motorcycles: normally 0.50–0.70; 0.20–0.30 for racing motorcycles.
- Bicycles: 0.90–1.10 for standard upright bicycles; 0.70–0.90 for racing bicycles with aerodynamic posture.
- Rail Vehicles: 0.15–0.30 for high-speed trains; 0.40–0.60 for normal trains.
- Aerospace Vehicles: 0.02–0.1 for commercial aircraft in cruise conditions.

Rolling resistances are very complex to be completely described. They certainly depend linearly on the vehicle weight component which is perpendicular to the road surface ground ($M_v \cdot g \cdot sin(\alpha) \approx M_v \cdot g$ for low ± values).

This linearity is expressed by a rolling resistance coefficient, C_r, by the following equation:

$$F_r = C_r \cdot M_v \cdot g \cdot sin(\alpha)$$

However, the rolling resistance coefficient is not constant, being dependent on road (or rail) and wheel materials, as well as on environmental conditions, tyre pressure, and vehicle speed. Vehicle speed, in fact, can alter the contact conditions modifying resistance coefficient.

All this considered, C_r is commonly modelled as:

$$C_r = C_{r,0} + C_{r,1} \cdot v$$

with the constant $C_{r,0}$ being the most important, especially in urban contexts (effect of speed is practically negligible for speed lower than 60 km/h).

Summarizing, typical mean values for C_r are:

- Passenger cars: 0.008–0.015, standard tyres running on asphalt or concrete.
- Heavy-Duty Vehicles: 0.006–0.010.
- Buses: 0.60–0.80.
- Motorcycles: normally 0.50–0.70; 0.20–0.30 for racing motorcycles.
- Rail Vehicles: 0.001–0.002 for steel wheels on steel tracks. They may increase during curves.

By understanding these fundamental forces and their interactions, we can comprehensively evaluate and optimize vehicle performance across various applications.

III.2.2 TYPICAL VEHICLE MISSIONS

According to the parameters required in vehicle traction forces calculation, a vehicle mission is defined when its expected speed and slope variations with time are known, together with the possible variation of its weight during the mission (more typically happening for freight or public transportation means).

For some vehicles, the mission to accomplish is almost known or foreseeable with a certain approximation (e.g. rail systems, airplanes), while, for other transportation means, it may vary a lot and it's not at all programmable (e.g. passenger cars).

In this latter case, calculations are made based on standard vehicle testing profiles defined by homologation procedures for environmental emission regulations and commonly used also for vehicle dynamic performances evaluation.

The most relevant limitation in these missions is the complete absence of a slope road profile. As examples of standard and real vehicle mission profiles, consider the three vehicular missions depicted in Figure III.2.2:

- The left side of the figure depicts the so-called NEDC Test Cycle, used in EU homologation procedures from the 1990s up to 2017.
- The central part depicts the WLTP Test Cycle substituting NEDC since 2017.
- On the right side, a typical urban cycle experimentally acquired by the authors in the city of L'Aquila (Italy) is reported.

Table III.2.1 reports the main overall characteristics of both NEDC and WLTP test cycles.

TABLE III.2.1
Main Characteristics of EU Test Driving Cycles

Feature	NEDC	WLTP
Duration	~20 minutes	~30 minutes
Distance	~11 km (6.8 miles)	~23.3 km (14.5 miles)
Average speed	~33.6 km/h (21 mph)	~46.5 km/h (29 mph)
Top speed	~120 km/h (74.5 mph)	~131 km/h (81 mph)
Driving conditions	Simplistic, less dynamic	Real-world-like
Focus	Laboratory-based	Realistic driving conditions

III.2.3 TYPICAL VEHICULAR ENERGY AND POWER REQUIREMENTS

Figure III.2.3 depicts the result of power terms calculation on the three identified vehicular missions, for a mid-size sedan car (weighing about 1500 kg).

Figure III.2.4, instead, reports data about energy terms for the same missions and vehicle.

Some considerations can be drawn from the date presented in Figures III.2.3 and III.2.4:

- Power requirements due to rolling resistances are normally highly dominant with respect to those due to aerodynamic drag. Aerodynamic drag is the only one amongst the four power terms which is not proportional to vehicle weight. Therefore, in most cases, overall traction energy requirements are only proportional to vehicle weight.
- Traction energy requirements are of the order of around 100 Wh/km/ton. Therefore, a mid-sized sedan car (weighing about 1500 kg) averagely needs about 15 kWh of energy for each 100 km mileage.
- All simulated missions are closed ones starting from the stopped vehicle in the same position. Therefore, the energy terms due to integration vs. time of both inertial and gravitational terms are null.

  This means that, from a theoretical point of view, all the energy required to accelerate the vehicle is accumulated in terms of vehicle inertia and may be recovered during braking phases.

  At the same time, the energy used by the vehicle for climbing a hill may be recovered during hill descent.

  With these aims, a conservative (so-called "regenerative") braking device must be used on-board during all vehicle deceleration phases. A regenerative braking device should subtract this mechanical energy from the vehicle, convert it, and store it in some form on-board.

  The whole chain of components used to this aim should be sized considering both the maximum mechanical power to be subtracted during these phases and the maximum energy to be stored during the most energy-demanding deceleration transients.

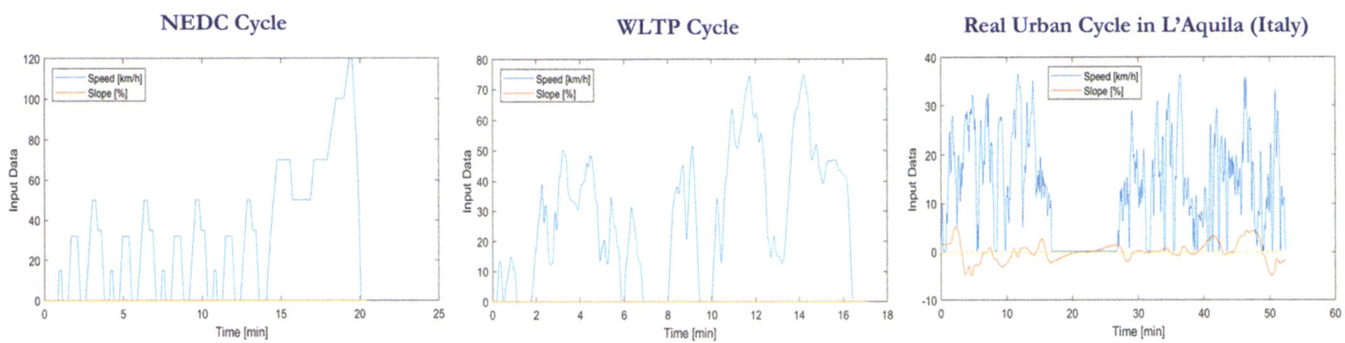

FIGURE III.2.2 Typical vehicle missions for road vehicles.

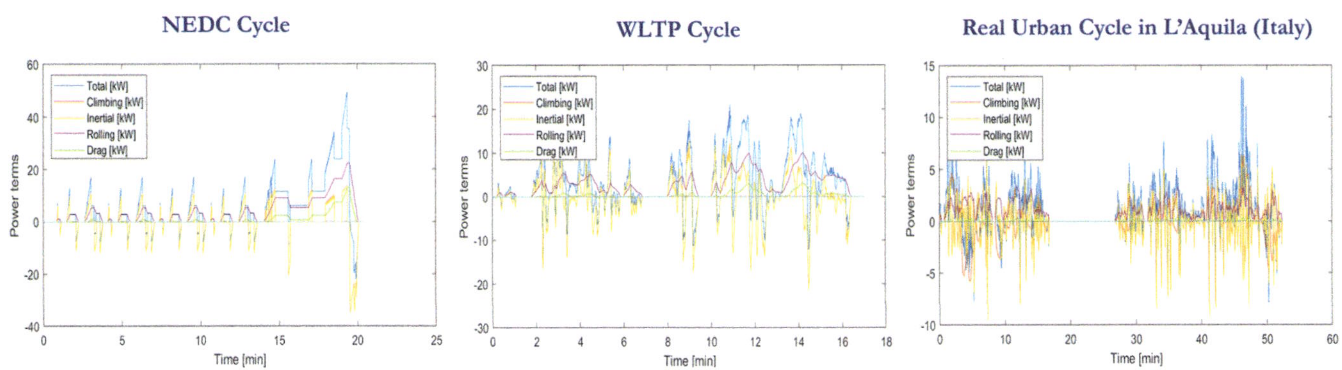

FIGURE III.2.3 Power terms evaluations for the vehicle missions depicted in Figure III.2.2 and a mid-sized sedan car.

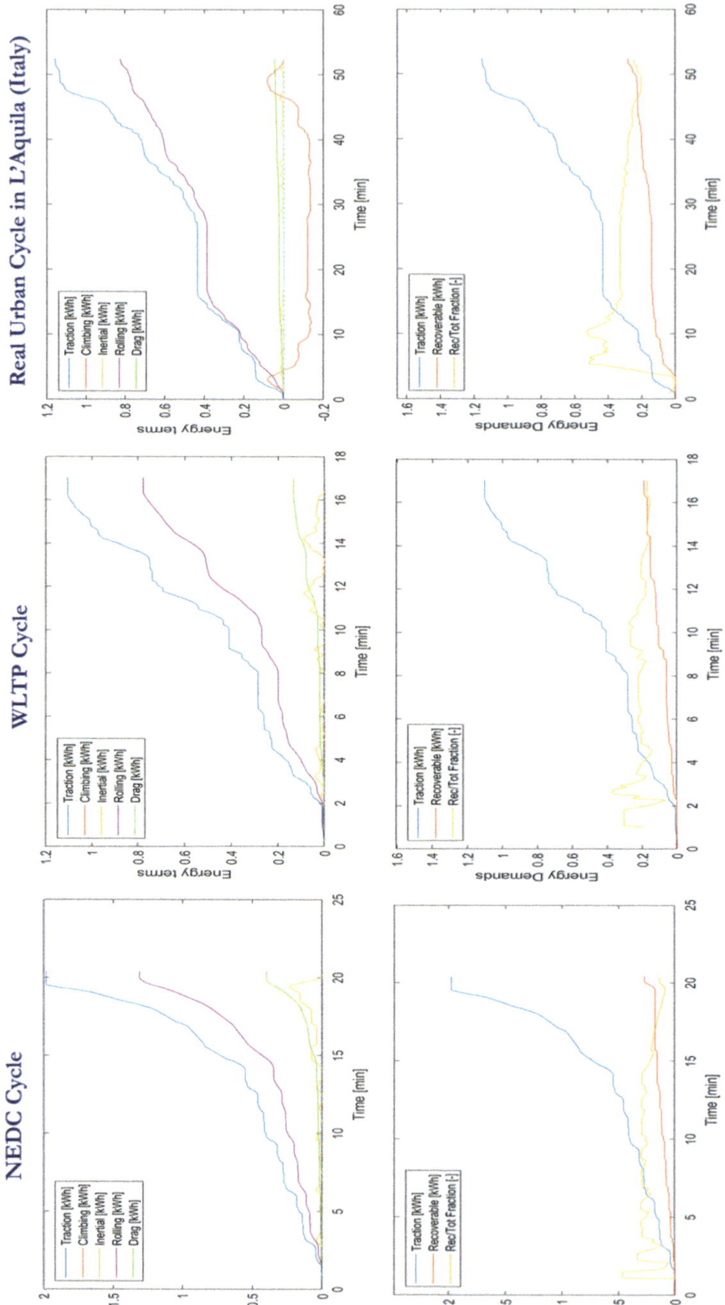

FIGURE III.2.4 Energy terms evaluations for the vehicle missions depicted in Figure III.2.2 and a mid-sized sedan car.

Data show that to regenerate energy due to inertial transients, a minimum of 10–20 kW/ton (for a unit ton of vehicle weight) of regenerative braking power has to be installed, while about 100 Wh/ton are sufficient to store energy in between for subsequent inertial phases. About the same power, but a much higher energy, is instead required to store energy during a hill descent (up to 1 kW/ton or even more).

- Both rolling resistances and aerodynamic drag are always positive power terms contributing to traction power requirements. Therefore, those energy contributions to vehicle "energization" cannot be recovered in any way. Unfortunately, in typical driving cycles, those are the most contributing factor, so that typically, a maximum of 15–25% of total traction energy may be recoverable through properly designed devices.

  Considering also the expected efficiencies of possible regenerative devices (around 70% at most), this means that only a maximum additional weight of the vehicle of around 10% of original vehicle weight can be acceptable for those devices.

  In fact, if this condition is not satisfied, the weight of the device would produce an additional energy consumption which would be higher than the energy it could recover during regeneration phases, resulting in an increase in vehicle overall energy consumption.
- Similar considerations have to be made when choosing the energy carrier to feed a vehicle. If the weight of the energy carrier is high, including the weight of its storage system ("fuel tank"), vehicle energy consumption could increase significantly, so that a significant part of the fuel is consumed only to transport itself.
- This may be the case of electric vehicles, where most of the energy carrier specific weight (for unit energy stored/storable on board) is in fact due to the weight of the energy storage system (commonly invertible electrochemical batteries).

  If a very high mileage is required for a typical MPV (up to 700 km in some vehicles currently on the market), vehicle weight could more than double with respect to an equivalent gasoline model, so that vehicle consumption is doubled as well because half of the energy stored on board has to be consumed to transport the energy storage system, and not the passenger, nor the vehicle.

III.2.4 IMPLICATIONS ON ENERGY CARRIERS CHOICE AND VEHICULAR DESIGN

To generalize the implications of energy carrier choice on vehicle design, the following critical vehicle parameters must be considered:

- L, maximum mileage in km required for the vehicle without recharging the ESS.
- η_v, mean propulsion system efficiency, expressing the ratio between the energy stored at the tank, E_{ESS}, and that made available at the wheels for traction purposes, E_t:

$$E_t = \eta_v \cdot E_{ESS}$$

Vehicle efficiency is much lower for conventional ICE vehicles (which store energy at the tank in chemical form, and do not recover any energy during braking phases), being around 0.3, and higher for electric vehicles (which store energy at the tank in electric form and recover a part of traction energy during braking phases), being around 0.9.

- $\bar{E}_t$, mean specific mechanical traction energy required for a unit km of vehicle mileage at the wheels and unit ton of vehicle weight. Typical values of this parameter, as described, are in the range 70–130 Wh/(km*ton): a value of 0.1 kWh/(km*ton), or also Wh/(km*kg) can be used as reference. If the total mass of the vehicle $M_{v,tot}$ and its maximum mileage L are defined, total maximum energy to be consumed at the wheels (E_t) can be evaluated as:

$$E_t = \bar{E}_t \cdot M_{v,tot} \cdot L$$

- M_v, vehicle mass without its ESS (energy storage system, having a total mass M_{ESS}) in kg. This ranges from 500 kg for small vehicles for restricted urban use, to 1500 kg for typical mid-size MPVs, to around 10 tons for a mid-sized bus or truck. Total vehicle weight $M_{v,tot}$ would be:

$$M_{v,tot} = M_v + M_{ESS}$$

- $\bar{M}_{ESS}$, mean specific weight of the ESS: this includes both the weight of the energy carrier and of the needed devices to store and use it on-board. Values are expressed for unit maximum energy storable on-board. Typical values for this parameter range from around 1 kg/kWh for traditional ICE vehicles using gasoline or diesel fuel, to around 10 kg/kWh of innovative BEVs using Li-ion batteries, to 50–100 kg/kWh of previous-generation BEVs using lead-acid batteries. For a given on-board maximum ESS energy capacity installed, E_{ESS}, it would be:

$$M_{ESS} = \bar{M}_{ESS} \cdot E_{ESS}$$

Gathering together the equations stated above leads to:

$$E_t = \bar{E}_t \cdot L \cdot \left(M_v + \bar{M}_{ESS} \cdot \frac{E_t}{\eta_v} \right)$$

Which could be solved for E_t, leading to:

$$E_t = \frac{\bar{E}_t \cdot L \cdot M_v \cdot \eta_v}{\eta_v - \bar{E}_t \cdot L \cdot \bar{M}_{ESS}}$$

Some interesting considerations are derived introducing a parameter describing the percentage of additional mass vehicle weight due to its ESS, defining that as:

$$\Delta W_{Ess,\%} = \frac{M_{Ess}}{M_v} \cdot 100$$

Manipulating the equations above leads to the following relation:

$$\frac{\Delta W_{Ess,\%}}{100} + 1 = \frac{\eta_v}{\eta_v - \bar{E}_t \cdot L \cdot \bar{M}_{ESS}}$$

Which could be solved depending on the unknown parameters and be used for various design purposes. Let us make some examples.

- For a conventional ICE-based mid-sized MPV: $\eta_v = 0.3$; $\bar{E}_t$=100 Wh/(km*ton); M_v = 1500 kg; $\bar{M}_{ESS}$ = 1 kg/kWh. A request of L=500 km is therefore easily obtained, only increasing the vehicle weight of the $\Delta W_{Ess,\%}$=20%.
- For a BEV-based mid-sized MPV: η_v=0.9; $\bar{E}_t$ = 100 Wh/(km*ton); M_v = 1500 kg; $\bar{M}_{ESS}$ = 10 kg/kWh, a similar request of L = 500 km is obtained, more than doubling its weight $\Delta W_{Ess,\%}$ = 125%, and overall energy consumption. As a result, the costs of the propulsion system are also relevantly increased, since a huge BESS of 187.5 kWh has to be installed on-board.
- For a small-sized BEV vehicle for restricted urban use: η_v = 0.9; $\bar{E}_t$ = 100 Wh/(km*ton); M_v = 500 kg; $\bar{M}_{ESS}$ = 10 kg/kWh, so that a request of L = 100 (which is a mileage more than compatible with urban applications) can be obtained with an increase in vehicle weight of $\Delta W_{Ess,\%}$ = 12.5%, so limiting the energy consumption increase due to electric mobility and also the cost of the propulsion system, being only 6.25 kWh of BESS installed on-board.

Figure III.2.5 depicts the so-called WTW comparative analysis approach, describing graphically the different WTT paths for conventional ICE-based vehicles and BEVs.

A question arises: is electric mobility sustainable in all cases and for all applications? In current conditions, probably this is not the case, both for costs and environmental constraints, due to the huge prospective increase in energy consumption which would derive from the utilization of electric mobility solutions in MPVs or in other long-distance heavy-duty applications.

CO_2 emissions due to the mobility sector, in fact, are not only the result of local emissions at the energy carrier final use (the so-called Tank-To-Wheels, TTW, impacts), but also of these produced in all carrier production and distribution phases (so-called Well-To-Tank, WTT, impacts) up to its final utilization site.

The whole WTT process has to be considered when comparing different propulsion systems using different energy carriers.

Figure III.2.5 depicts this comparative analysis approach, describing graphically the different WTT paths for conventional ICE-based vehicles and BEVs.

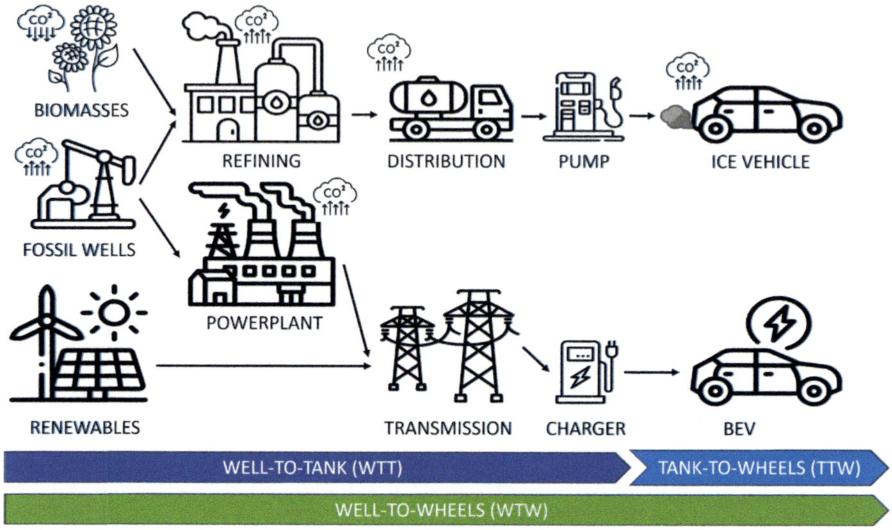

FIGURE III.2.5 WTW comparative analysis approach of conventional ICE-based and BEVs.

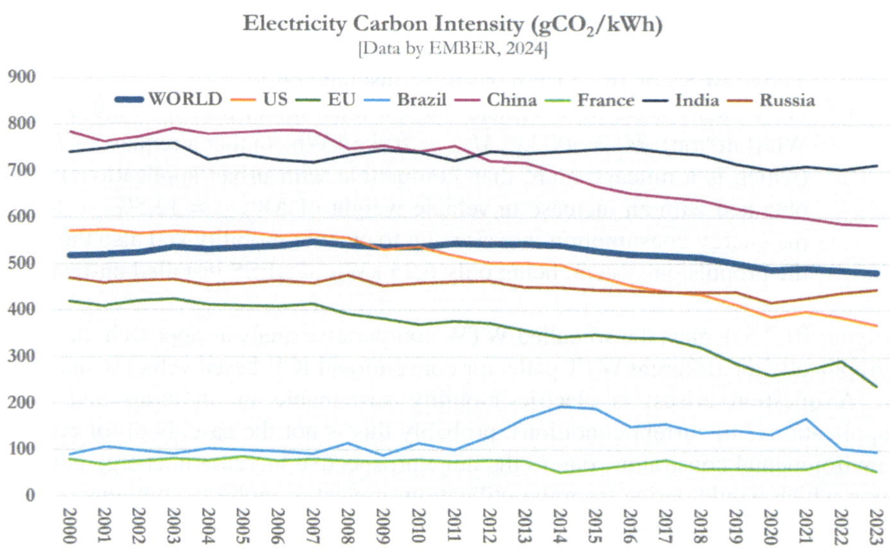

FIGURE III.2.6 Time variations in electricity carbon intensity, both at the global level and for individual countries or groups.

It is clear how, from this point of view, the overall impact of electric mobility solutions strictly depends, as fully explained in Part I of this book, on the specific emissivity of the electric sector (e.g. expressed in grams of CO_2 emissions for unit energy of electricity made available for final uses, which is also the “D” parameter of Kaya identity, when restricted to the electric sector).

Values for emissivity, however, are very site-dependent, and are strictly correlated to the penetration of RESs in the electric sector. As a reference, Figure III.2.6 shows the time variations in electricity carbon intensity, both at the global level and for individual countries or groups.

Therefore, electric mobility has to penetrate the transportation sector only if, where, and when RESs will sufficiently penetrate the electricity sector. If not, their massive introduction would mainly produce global disadvantages.

As explained, this would not be the case if electric mobility only addresses small vehicles for restricted urban uses, , with in this case the expected increase in overall energy consumption being much lower and this drawback being much more than compensated for by the elimination of toxic pollutant sources from very critical areas.

These are the main reasons why electric mobility still is not the most suitable option for long-distance transportation systems: to this aim, BESS specific weight should at least decrease to 2–4 kg/kWh.

On the contrary, electric vehicles already are the most viable and sustainable transportation solution for urban applications, especially in small vehicles for restricted urban use (weighing at most 500 kg, typically hosting two or three passengers and having to cover short travel distances).

III.3 ICEs and Their Local and Global Impacts

One of the most pressing issues associated with internal combustion engine (ICE)-based passenger vehicles is their emission of toxic pollutants into the atmosphere. This issue was briefly discussed in Part I, Chapter I.9, where we defined primary pollutants and examined air quality standards that establish acceptable concentration levels for various polluting species. These standards are particularly critical in urban areas, where the majority of the population spends most of their time.

The situation is dire, as a significant portion of the global population lives in areas where air quality standards are not adequately met. The most straightforward and effective strategy to address this issue is to reduce the specific emissions of pollution sources. Among these sources, passenger vehicles bear a substantial share of the responsibility for poor air quality.

The key question, therefore, is: how rapidly and effectively are we acting to mitigate these critical environmental and health challenges? To address this, we will primarily examine the situation from a European Union (EU) perspective, where stringent regulations and policies are already shaping the evolution of vehicle emissions. However, as global car markets are deeply interconnected, we will also show that similar efforts are underway in other parts of the world.

III.3.1 EMISSION DIRECTIVES EVOLUTION

The evolution of EU homologation procedures for passenger cars reflects significant advancements in technology, environmental awareness, and road safety.

This process, which began with the EU harmonization efforts in the 1970s, was formally established with the introduction of the Euro-1 emission standards in 1992.

Since then, new emission standards have been introduced approximately every four years, culminating in the current Euro-6 standards. The next phase, Euro-7, is expected to be set in 2025.

Tables III.3.1 and III.3.2 illustrate the evolution of emission limits established by Euro Directives for light-duty vehicles, respectively, for spark ignition (SI) and compression ignition (CI) internal combustion engines.

The reduction of vehicular emissions over the years has been rapid and challenging. While the tables above summarize emission limits, the testing procedures

DOI: 10.1201/9781003631125-25

TABLE III.3.1
EU EURO Directives Emission Limits Evolution for SI ICE Vehicles

Spark Ignition (SI) ICEs						
[g/km]	Year	CO	NO_x	HC	PM	Notes
Euro 1	1992	2.72	–	0.97	–	
Euro 2	1996	2.2	–	0.5	–	Lower HC
Euro 3	2000	2.3	0.15	0.2	0.05	NO_x/PM limits
Euro 4	2005	1.0	0.08	0.1	0.025	Tightening
Euro 5	2009	1.0	0.06	0.1	0.005	Focus on PM
Euro 6	2014	1.0	0.06	0.1	0.005	RDE tests

TABLE III.3.2
EU EURO Directives Emission Limits Evolution for CI ICE Vehicles

Compression Ignition (CI) ICEs						
[g/km]	Year	CO	NOx	HC	PM	Notes
Euro 1	1992	2.72	–	0.97	–	
Euro 2	1996	1.0	0.7	0.5	–	NO_x limits
Euro 3	2000	0.64	0.50	0.56	0.05	PM limits
Euro 4	2005	0.50	0.25	0.30	0.025	Focus on PM
Euro 5	2009	0.50	0.18	0.23	0.005	Focus on PM
Euro 6	2014	0.50	0.08	0.17	0.005	RDE tests

have also evolved significantly. The modern requirements involve stricter test cycles, shorter warm-up times, and more realistic operational scenarios, making compliance with the limits far more difficult than in the past.

Similar Directives are stated for heavy-duty vehicles: these are normally referred to using Roman instead of Arabic numbers (Euro-I to Euro-VI) and normally are slightly more permissive in terms of pollutant emissions.

Limits for light-duty vehicles are imposed in g/km and are measured on vehicles running a lab test-cycle in a rolling test-bench. The test-driving cycle was the so-called NEDC cycle up to 2015, and since then it has been substituted by the WLTP cycle: the main characteristics of these two test cycles were already discussed in Chapter III.2.

Since 2017, Real Driving Emission (RDE) tests were introduced, during a WLTP cycle realized on a real road. The actual limits are evaluated using a more permissive "conformity factor" (CF) with respect to lab environment (CF=2.1 in 2017, CF=1.5 from 2021 onwards), meaning that in the RDE test the vehicle can emit 50% more than the limit values (CF=2.1).

This shift to more stringent testing was a direct response to the "Dieselgate" scandal, which exposed emissions test manipulations by several manufacturers. The

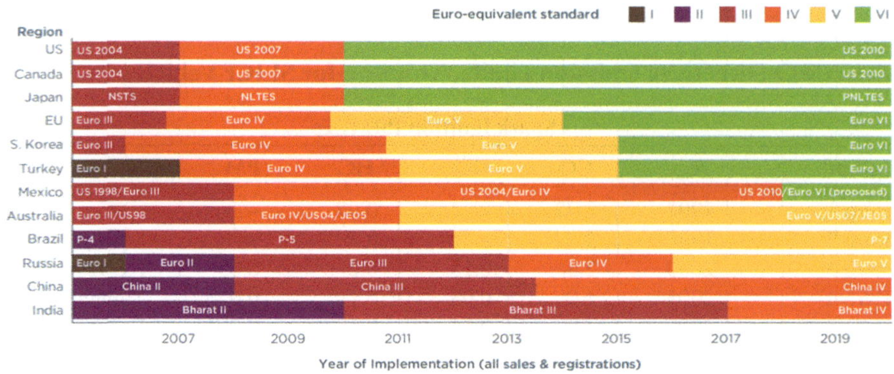

FIGURE III.3.1 Passenger car toxic emission standards comparison for various countries around the world.

scandal highlighted weaknesses in the EU's homologation system, prompting reforms to restore public trust and enforce stricter standards.

The EU's approach to emissions reduction has influenced policies globally. Many OECD countries have adopted similar standards, while others are still in the earlier stages of enforcement. As shown in Figure III.3.1, a clear alignment exists among most industrialized nations, while developing regions often permit higher emission levels due to economic and infrastructural challenges.

Traditionally, homologation Directives only act on main primary toxic pollutants (HC, CO, NO_x, PM). However, since Euro-6's introduction in 2014, the EU's commitment to achieving climate neutrality by 2050 has spurred the introduction of ambitious CO_2 emission targets for passenger cars.

Manufacturers are now required to meet fleet-wide average emissions limits, with significant penalties for non-compliance. These measures have accelerated the adoption of low-emission technologies and reshaped the automotive industry.

In more detail, the EU imposes TTW CO_2 emission limits for passenger cars primarily through fleet-wide average targets, designed to reduce greenhouse gas emissions from vehicles. Key milestones in this evolution include:

- **2015**: A fleet-wide average target of **130 g CO_2/km** was introduced for new passenger cars.
- **2020/21**: The target was reduced to **95 g CO_2/km,** calculated using the NEDC (New European Driving Cycle) test procedure. For vehicles tested under the WLTP (Worldwide Harmonized Light Vehicles Test Procedure), this value is adjusted to reflect real-world conditions.
- **2025**: A further 15% reduction from the 2021 baseline is required for new fleets: The target was reduced to **81 g CO_2/km**.
- **2030**: A 37.5% reduction compared to the 2021 baseline is mandated, effectively targeting around **60 g CO_2/km**.

Manufacturers exceeding these limits face financial penalties. For each gram of CO_2/km that exceeds the target, the penalty is **€95** per vehicle sold. For example, if a manufacturer exceeds its target by **1 g CO_2/km** and sells **1,000,000 vehicles** in a given year, the penalty would be **95 M€**.

This penalty structure strongly incentivizes manufacturers to develop and sell low-emission or zero-emission vehicles to offset higher-emission models. Basically speaking, a device costing less than €95 can be introduced in a vehicle if it determines energy saving, and a corresponding CO_2 emission saving greater than 1 g/km.

III.3.1.1 Vehicular CO_2 Emissions

Are the limits imposed in terms of CO_2 emissions met by conventional vehicles? How can vehicular CO_2 emissions be estimated?

Let us start with the evaluation of vehicle traction energy E_t (as defined in Chapter III.2), for a single average mission km. This is simply given by the multiplication of two factors we already defined: $\bar{E}_t \approx 100$ Wh/(km*ton); $M_v \approx 1500$ kg, for a typical MPV.

CO_2 emissions are proportional to vehicle energy consumption (which we named as E_{ESS} in Chapter III.2, being $E_{ESS} = E_t / \eta_v$, where η_v is the mean propulsion system efficiency: 0.32 for CI ICEs and 0.25 for SI ICEs). Summarizing:

$$E_{ESS} = \frac{E_t}{\eta_v} = \frac{\bar{E}_t \cdot M_v}{\eta_v}$$

$$E_{ESS} \approx 600\frac{Wh}{km}, \text{ for SI ICEs}$$

$$E_{ESS} \approx 469\frac{Wh}{km}, \text{ for CI ICEs}$$

Which estimates the chemical energy (in kWh) stored at the tank needed to run an average mission km. Therefore, vehicle fuel consumption, FC, in kg/km may be evaluated as:

$$FC = \frac{E_{ESS}}{H_i} = \frac{\bar{E}_t \cdot M_v}{\eta_v \cdot H_i}$$

$$FC \approx 49.5\frac{g_{Gasoline}}{km}, \text{ for SI ICEs}$$

$$FC \approx 38.0\frac{g_{Diesel}}{km}, \text{ for CI ICEs}$$

where H_i is the lower heating value of the fuel (about 12.1 kWh/kg for gasoline, and 13.9 kWh/kg for diesel fuel).

The CO_2 emissions ratio of single fuels (R_f) has now to be used, which depends on the empirical formula of the fuel (C_nH_m) and may be calculated as:

$$R_F = \frac{44}{12+\frac{m}{n}} \frac{g_{CO_2}}{g_{C_nH_m}}$$

$R_F \approx 3.05$, for SI ICES, gasoline $\approx C_7H_{17}$, then $m/n \approx 17/7$
$R_F \approx 3.18$, for CI ICES, diesel $\approx C_{12}H_{22}$, then $m/n \approx 22/12$

recalling what was already explained in Part I Chapter I.7.
Consequently, vehicular CO_2 emissions can be finally evaluated by:

$$q_{CO_2} = FC \cdot R_F = \frac{\bar{E}_t \cdot M_v \cdot R_F}{\eta_v \cdot H_i}$$

$q_{CO_2} \approx 151 \frac{g_{CO_2}}{km}$, for a typical SI ICE MPV

$q_{CO_2} \approx 120 \frac{g_{CO_2}}{km}$, for a typical CI ICE MPV

Which shows how difficult it is for a medium-sized MPV to meet CO_2 emissions thresholds. However, the threshold imposed is not a homologation limit.

These vehicles can therefore freely circulate but will determine a taxation for the manufacturer if its fleet-wide average emission is above the target of 95 g/km, with the fleet being composed of both conventional vehicles (which normally exceed the threshold) and lower TTW CO_2-emitting vehicles: all BEVs emit 0 g/km; plug-in HEVs are normally around 20 g/km, while "full" HEVs are normally around the threshold value.

III.3.2 MAIN GASOLINE AND DIESEL ICES EVOLUTIONS IN RECENT DECADES

Addressing both ICE's thermodynamic and thermochemistry to explain its pollutant formation and abatement dynamics is far beyond the objectives of this text and should be treated in separate academic courses. Here, instead, we only briefly list the main technical evolutions regarding ICEs in the last two to three decades, explaining how they made it possible to reduce their emissions so rapidly, following the very fast reduction path imposed by the homologation Directives.

This process has been key to meeting a rapidly increasing demand for mobility without worsening the level of air quality in urban environments, however, unfortunately, it is not expected to continue with the same speed and effectiveness in the near future.

On the other hand, we now fortunately have reliable BEVs, which are emitting zero local toxic pollutants and are therefore the best solution to be implemented in urban contexts. As already explained, this does not necessarily mean that ICE-based vehicles will not be sold anymore for other long-distance and extra-urban applications, although EU regulations for the private passenger mobility sector currently impose a complete thermal vehicles phase-out by 2035.

This measure, in fact, is part of the EU's broader climate goals under the **European Green Deal**, which targets climate neutrality by 2050.

III.3.2.1 SI ICEs

In SI ICEs, a pre-mixed air–fuel mixture is inlet in the cylinder and burnt as fast as possible (in a few milliseconds) at the end of the compression stroke. The fuel must have high anti-detonating properties (indicated by a sufficiently high octane number).

The mixture ratio must be at least stoichiometric to avoid leaving a portion of the fuel unburnt, but it should not significantly exceed this value, as an overly rich mixture can slow the flame front within the combustion chamber, promoting the formation of HC and CO, and reducing engine efficiency. Therefore, these engines are almost always stoichiometric, and they are forced to regulate the load by closing the inlet duct through a throttling valve (losing efficiency with respect to CI ICEs in partialization).

Additionally, after the 1990s, operating at this ratio became particularly critical for the optimal functioning of **three-way catalytic converters**, which simultaneously have to abate HC, CO, and nitrogen oxides (NO_x).

These converters only perform reasonably well when the exhaust gases result from very precise stoichiometric combustion (much less than 1% error can be tolerated). This is because NO_x has to be chemically reduced (to N_2), while HC and CO have to be oxidized.

In fact, these three chemical reactions can only happen if no residual oxygen is present at the engine exhaust with the exception of those due to the failure of oxidation of HC and CO in the combustion chamber, and to that which has to be recovered from the reduction of the NO_x (which takes oxygen needed for the combustion as a consequence of the chemical activation of N_2 at flame temperatures above 1400 °C).

Controlling the air-to-fuel ratio around the stoichiometric value is achieved using an electronically controlled closed-loop integrated feedback system. the key components include:

- Oxygen Sensors (Lambda Sensors): These sensors, placed in the exhaust system, measure the oxygen content of exhaust gases. The data indicates whether the burning mixture was too rich (excess fuel) or too lean (excess air).
- Electronic Control Unit (ECU): The ECU uses feedback from the oxygen sensors to adjust fuel injections in real time, ensuring the mixture remains stoichiometric under varying operating conditions (e.g., idling, acceleration, or cruising).
- Fuel Injection Systems: Modern engines use precise, electronically controlled fuel injectors (which replaced carburettors) that meter fuel delivery with high

accuracy. Paired with sensors measuring airflow, throttle position, and engine load, this ensures precise air–fuel mixing. These injectors are placed very near the inlet valves and directed so that the injection can be phased with valves opening. This limits the possibility for the fuel to create a liquid film on the walls on the inlet manifolds, therefore partially losing control of the fuel quantity inlet in the cylinder at each engine cycle.

This dynamic control system allows SI engines to maintain almost perfect stoichiometric combustion across a wide range of conditions, and during very fast transients both in load and rotation speed, ensuring both optimal performance and compliance with stringent emissions regulations.

III.3.2.2 CI ICEs

In CI ICEs, the cylinder is only "breathing" fresh air, while the fuel is injected directly into the combustion chamber just before its ignition. In a compression-ignition engine, fuel ignition happens only because of the thermodynamic conditions reached in the cylinder at the end of its compression stroke. The fuel must therefore have high auto-igniting properties (indicated by a sufficiently high cetane number).

Combustion dynamics are completely defined by fuel injection laws. Engine load is controlled through the injected fuel quantity as well, so that the load control is not dissipative (leading to a much greater overall efficiency with respect to SI ICEs).

However, CI ICEs typically have chronic problems regarding PM formation and emissions, due to the fact that the fuel cannot completely vaporize before ignition.

To reduce these problems, the injection pressure has to be increased as much as possible. This process led to very advanced and costly injection systems known as **common-rail systems**, which are able to inject the fuel at more than 2500 bars.

These components are fully electronically controlled and can realize three or even more injection phases inside each engine cycle. These are needed to optimize both fuel consumption and engine environmental performance.

Another relevant upgrade in CI ICEs was due to the introduction of **variable-geometry over-charging devices**: a turbine takes part of the exhaust gases extracting their mechanical energy and using that to increase the air inlet pressure in the cylinder.

Consider, in fact, that CI ICEs may be highly overcharged while SI ICEs can not (due to the possible desaturation of fuel detonation conditions at the end of compression).

Overcharging permits "dilution" of the charge, mixing fresh air (containing oxygen) with exhaust recirculated gases (not containing oxygen), lowering flame temperatures and reducing NO_x formation.

Coupling these technologies gave CI ICEs a certain number of freedom degrees which can be used to optimize their control, both from an environmental and a motor performance point of view, so that, at the start of 2000s, CI ICEs were largely prevailing over SI ICEs.

Unfortunately, most original equipment manufacturers (OEMs) used this freedom to manipulate emission tests, writing control codes for the ECU which recognized test conditions optimizing engine environmental performances only during homologation procedures, while on-road the ECU was set to optimize vehicle performance resulting in emission up to 10 times higher than the homologation limits.

This was the so-called “Dieselgate scandal”, which unfortunately almost resulted in CI ICEs being banned, even though they were the most promising technology within the ICE-based vehicles framework.

In summary, internal combustion engines (ICEs) have been central to mobility for decades but continue to present significant environmental challenges. Stricter regulations are therefore necessary to govern their use.

One possible approach to extending their viability is the adoption of alternative, more environmentally friendly, energy carriers, a topic explored in the next chapter.

However, more substantial changes are required in urban contexts, necessitating deeper modifications to vehicle propulsion systems. These alternative solutions will be discussed in Chapters III.5–III.8.

III.4 Alternative Fuels for ICEs

As already explained, ICE-based vehicles could continue to play a relevant role in the next decades only if they evolve in ways which permit at least a maintenance of their toxic emission levels (which are acceptable for restricted extra-urban applications), while drastically reducing their GHG emissions.

To this end, a number of alternative chemical energy carriers could be used to feed them. In this chapter we focus on these alternatives, highlighting the advantages and drawbacks of each one.

III.4.1 COMPRESSED AND LIQUIFIED NATURAL GAS (CNG AND LNG)

Methane (CH_4), the primary component of natural gas, has emerged as a viable and promising fuel for internal combustion engines (ICEs). Its high hydrogen-to-carbon ratio, abundance, and favourable combustion properties make it an attractive alternative for ICEs.

Additionally, methane can be derived from both fossil fuel sources and renewable processes such as anaerobic digestion, contributing to its versatility as a fuel for sustainable energy systems.

Methane has several physical and chemical properties that make it particularly suitable for SI ICEs:

- **High Octane Number**
  Methane has an octane number of approximately 120, significantly higher than conventional gasoline. This allows methane-fuelled engines to operate at higher compression ratios without the risk of pre-ignition or knocking phenomena, enhancing thermal efficiency.
- **Clean Combustion**
  Methane's simple molecular structure facilitates cleaner combustion, resulting in lower emissions of toxic pollutants compared to liquid fuels. It should be emphasized, however, that this depends on calibration activities. These are usually stopped when the same homologation limits are obtained as for gasoline engines, but better results in terms of limited pollutant emissions could be obtained with more detailed calibration.

  Moreover, the use of methane, due to its high hydrogen content, results in around 20% lower carbon dioxide (CO_2) emissions per unit of energy produced.

DOI: 10.1201/9781003631125-26

Using the same calculation scheme derived in the previous chapter, being $H_i \approx 13.9$ kWh/kg, and $R_f = 2.75$ for methane, we can derive:

$$q_{CO_2} \approx 119 \frac{g_{CO_2}}{\text{km}}$$

Which shows that the CO_2 emissions per unit of energy produced of a NG SI MPV are approximately 21% lower than those of gasoline-driven SI ICE.

- **Gaseous State**
  Methane exists as a gas at standard atmospheric conditions, enhancing its mixing properties in the air–fuel mixture. However, its gaseous state presents challenges for storage and handling. Methane, in fact. has a much lower energy density (measured in MJ/l, per unit litre of fuel storage system) compared to liquid fuels, requiring pressurization or liquefaction to enable efficient on-board storage.

  Gasoline, with a density of approximately 746 g/L, has a much greater volumetric energy density compared to methane gas, which has a density of about 0.78 g/L under standard conditions.

  Moreover, the injection of the fuel in a gaseous state determines a significant reduction in engine power, due to the absence of the charge refrigeration (and consequent density increase) effect which is normally obtained through gasoline vaporization during the inlet process. This loss in power may be balanced by increasing the compression ratio (which can be done due to CH4's high octane number) but this is not normally done because most current CH4 ICEs are dual-fuel engines and are commonly started using gasoline to avoid cold start issues.

Methane can be supplied and stored in tanks in either compressed gas or liquid form, depending on the application and infrastructure:

- **Compressed Natural Gas (CNG)**
  CNG involves compressing natural gas to around 300–350 bar, allowing it to be stored in high-pressure tanks in vehicles, while having a very low mileage (less than 30% of a similar gasoline vehicle). This solution is widely used in light-duty vehicles, buses, and fleet applications due to its cost-effectiveness and reduced emissions. New vehicular tanks are being standardized with a storage pressure of 750 bar in composite materials tanks. These are normally used only in heavy-duty applications.

  Refuelling requires time to compress the fuel from the off-board storage level to final tank pressure on-board. This compression also heats the fuel, so that if the refuelling is not intercooled and/or is too fast, the filling of the tank cannot be full, especially if the tank is pressurized at 750 bar.
- **Liquefied Natural Gas (LNG)**
  LNG is produced by cooling methane at least below −162°C (its critical temperature), converting it into a liquid state for higher energy density and easier

transportation. LNG is particularly suitable for heavy-duty vehicles, such as trucks and ships, which require longer driving ranges. LNG is stored in cryogenic tanks: these are not particularly pressurized but have to be thermally insulated to prevent fuel boiling.

Even with advanced insulation (like vacuum-jacketed tanks), some heat transfer inevitably occurs, especially for automotive applications, due to weight and volume constraints which limit the use of insulating materials.

When heat enters the cryogenic tank, the stored liquid begins to absorb this heat and evaporates, forming vapor. This causes the pressure inside the tank to increase since the vaporized gas takes up more volume. To prevent excessive pressure buildup, which could damage the tank, "venting" occurs through pressure-relief valves.

The valves automatically release small amounts of vaporized gas into the atmosphere to maintain safe pressure levels. Even if limited, these emissions have to be taken into account to define vehicle environmental performance, because of the high GWP of methane as a GHG (see Part I, Chapter I.5).

The venting phenomenon can be very significant for light-duty vehicles where the utilization time of the fuel at the tank can be very long, and this is one of the main reasons why LNG is only used in heavy-duty vehicles.

Methane can also be produced by renewable sources, referred to as **biomethane**. Biomethane is a renewable form of methane produced through the anaerobic digestion of organic waste. It is purified to meet natural gas specifications and can be used interchangeably with fossil-derived methane.

Utilizing biomethane in ICEs significantly reduces lifecycle greenhouse gas (GHG) emissions, contributing to decarbonization efforts.

III.4.1.1 Methane Injection Systems for Spark-Ignition ICEs

Methane injection systems are pivotal in optimizing combustion and enhancing engine efficiency in spark-ignition (SI) ICEs. These systems are meticulously designed to regulate and deliver methane into the combustion chamber, ensuring the maintenance of a stoichiometric air–fuel ratio across varying operating conditions.

Methane can be introduced into the engine using two primary methods:

- **Port Fuel Injection (PFI)**
  In PFI systems, methane is injected into the intake port, upstream of the combustion chamber. This allows the methane to mix with incoming air, forming a homogeneous air–fuel mixture before entering the cylinder. PFI systems are particularly well-suited for retrofitting existing gasoline engines.
  The main components of PFI systems include (see Figure III.4.1):
  - High-Pressure Fuel Tank
    Typically cylindrical and made of steel, these tanks are designed to store methane at high pressures. However, they impose significant weight, space, and safety constraints, particularly in small-sized vehicles.

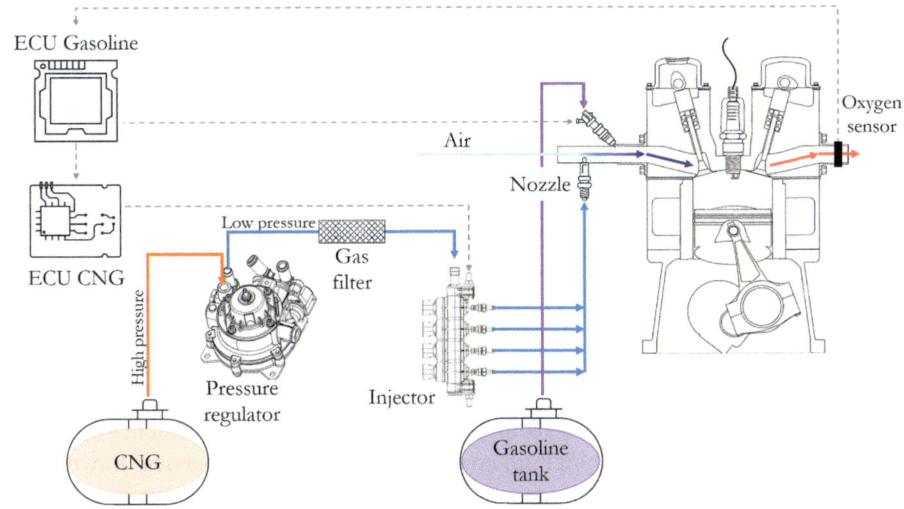

FIGURE III.4.1 Schematic of a typical methane dual-fuel light-duty SI ICE layout.

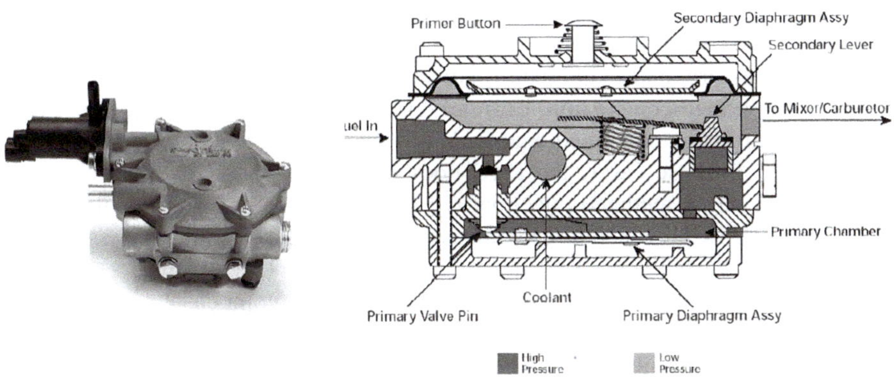

FIGURE III.4.2 Photo and schematic of a typical pressure regulator for methane/LPG SI ICEs layout.

- Pressure Regulator
  The pressure regulator reduces methane pressure from storage levels to injection pressures suitable for PFI systems, which range between 3 and 10 bars. The regulator's dynamic performance is essential for maintaining a stable injection flow rate, precise air-to-fuel ratios, and compliance with ICE emission standards.

  Two self-regulating pressure control valves are introduced in the pressure regulator (see Figure III.4.2): both of them react to a pressure drop in the fuel lines reducing fuel passage sections, using two leveraged-actuated valves based on both spring and pressurized membrane actions.

The pressure value acting on the membrane may be external atmospheric pressure, or engine inlet pressure, depending on the application.

Methane pressure at the tank varies widely depending on its filling level (from 300 to 30 bars). Fuel has to be injected at a stable pressure (ranging from 3 to 10 bars depending on the injector design). Considering that, the first of the two valves is sized and calibrated to have a rougher action, disposing of most of the pressure drop, while the second is more precisely acting to stabilize injection pressure during injection phases.

The overall expansion ratio is very high, resulting in a significant cooling of the fuel. It removes much energy from the metal parts of the pressure regulator, also making its functioning critical.

To avoid problems, the temperature inside the regulator has to be sustained: to this aim, the engine refrigerant fluid is used, making it circulate in the regulator before being eventually directed to the vehicle's radiator.

Engine refrigerant fluid normally is at about 90°C (having drawn energy from the ICE) but during cold starts, when the refrigerant is at ambient temperature, methane ICE operation can become more critical.

- Fuel Injectors
  Fuel injectors precisely meter and deliver methane into the intake port.
- Electronic Control Unit (ECU)
  The ECU monitors engine parameters (e.g., air–fuel ratio, cylinder pressure, and temperature) adjusting the injection timing and duration to optimize combustion. Currently, as already said, most methane ICEs are equipping retrofitted dual-fuel light-duty vehicles.

  In these cases, the ECU does not control the ICE directly, but only manages methane injection, relying on communication with the gasoline ICE ECU, which already measures and processes all the necessary values to control the gasoline injectors. The methane ECU, therefore, only modifies the results obtained by the gasoline ECU, compensating for variations in fuel density and inhibiting gasoline injection.
- Fuel Lines and Rails
  High-pressure fuel lines and rails distribute methane from the tank to the injectors. Proper insulation and material selection are critical to prevent leaks and pressure drops.

- **Direct Injection (DI)**
  In DI systems, methane is injected directly into the combustion chamber at high pressures, during either the intake or compression stroke. This approach allows for precise control of the air–fuel mixture, enabling stratified combustion and improving efficiency. The DI concept parallels that of modern gasoline DI ICEs, but its implementation is somewhat simpler due to methane's gaseous state.

  These engines are much more complex and costly than PFI systems, requiring specialized high-pressure injectors, fuel management systems, and long-lasting design activities on a single engine. Their primary applications are in the heavy-duty transportation sector, where efficiency gains justify the higher costs and complexity.

III.4.1.2 Dual-Fuel Compression-Ignition ICEs

As mentioned earlier, methane-fuelled internal combustion engines (ICEs) typically operate using spark-ignition (SI) systems. However, methane can also be utilized in dual-fuel compression-ignition (CI) ICEs, where it is mixed with intake air and ignited by a small quantity of diesel fuel, acting as a pilot for combustion initiation.

Dual-fuel natural gas engines represent a viable technology for retrofitting existing diesel vehicles into methane-powered systems. However, the use of natural gas in diesel engines faces challenges due to methane's poor ignition characteristics, including its high auto-ignition temperature and low cetane number compared to diesel. Consequently, a reliable ignition source is essential to initiate combustion within the cylinder.

In dual-fuel engines, natural gas is typically injected into the intake manifold, where it forms a premixed air–gas blend. This mixture is subsequently introduced into the cylinder. During the compression stroke, a small quantity of diesel fuel is injected, which ignites due to the compression effect. The ignition of the diesel triggers combustion of the air–gas mixture at multiple points within the cylinder.

The quantity of diesel injected, known as the diesel replacement rate, can range from 20% to 100%, depending on engine load conditions. However, dual-fuel engines must operate exclusively on diesel under specific circumstances:

- At low loads: Combustion stability issues necessitate using only diesel.
- During startup and full load: Methane use is limited to prevent detonation, which could occur due to the high compression ratios inherent in diesel engines.

Dual-fuel technology offers several benefits, particularly in heavy-duty applications, where it maintains high power output and thermal efficiency. These advantages include:

- **High Compression Ratio Utilization**
  Dual-fuel engines can fully leverage the high compression ratios typical of diesel engines, resulting in lower fuel consumption while staying within detonation limits.
- **Flexibility in Fuel Usage**
  Dual-fuel engines can operate entirely on diesel when natural gas is unavailable, thereby enhancing vehicle autonomy, particularly in areas without natural gas refuelling stations.
- **Lower NO_x Emissions**
  Compared to diesel engines without abatement devices, dual-fuel engines produce fewer nitrogen oxides due to lower end-of-combustion temperatures.
- **Reduced Particulate Emissions**
  Depending on the diesel replacement rate, these engines exhibit a marked reduction in solid particulate emissions relative to conventional diesel engines.

Despite their advantages, dual-fuel engines encounter several issues, primarily linked to the slow combustion speed of lean gaseous mixtures. These challenges include:

- **Unburned Hydrocarbon Emissions**
  Methane slip occurs during valve overlap when a portion of the unburned natural gas–air mixture escapes directly into the exhaust. This unburned methane contributes to hydrocarbon emissions.
- **High Carbon Monoxide (CO) Emissions**
  In certain areas of the cylinder, the air–gas mixture may be excessively lean, preventing the flame front from propagating and halting CO oxidation reactions.
- **Combustion Instability**
  Lean mixture combustion can result in misfiring and instability, reducing engine reliability.
- **Power Reduction**
  The lower flame propagation speed of methane–air mixtures leads to a reduction in power output compared to diesel-only engines.

These limitations, particularly regarding methane slip and lean mixture challenges, underline the importance of optimizing combustion and calibration techniques to enhance dual-fuel engine performance and environmental compatibility. However, the inherent flexibility, improved emissions profile, and potential fuel cost savings make dual-fuel engines a promising alternative, especially for heavy-duty transportation applications.

III.4.1.3 Methane in ICEs: Benefits, Drawbacks, Potentiality

Summarizing, using methane in ICEs gives some relevant advantages:

- **Environmental Benefits**
  Methane offers notable advantages in terms of emissions. It produces approximately 20–25% less CO_2 compared to gasoline or diesel per unit of energy, primarily due to its higher hydrogen-to-carbon ratio. Additionally, methane-powered engines can potentially reduce particulate matter (PM), carbon monoxide (CO), and hydrocarbon (HC) emissions. However, nitrogen oxide (NO_x) emissions remain a concern due to the high local flame temperatures during combustion.

  Achieving reductions in toxic emissions requires extensive and costly engine calibration efforts. These calibrations often stop once the engine meets Euro-6 emission standards, but better environmental performance could likely be achieved with engines specifically designed for methane, rather than gasoline-based engines adapted for dual-fuel use. This is typically seen in heavy-duty methane-only engines designed explicitly for this fuel. Furthermore, the use of biomethane can enable near-carbon-neutral operations, depending on the feedstock and production methods used.

- **Economic and Resource Advantages**
  Methane presents economic and resource-related benefits. Natural gas reserves remain abundant, ensuring a stable and medium-term supply. Methane fuels, particularly compressed natural gas (CNG), are often cheaper than gasoline or diesel, resulting in reduced operational costs for vehicle owners and fleet operators.
  Despite its advantages, several challenges limit methane's broader application in ICEs. These challenges must be addressed to maximize its potential as a sustainable fuel source.
- **Storage and Infrastructure**
  Methane's gaseous nature requires compression (CNG) or liquefaction (LNG) for on-board storage, adding complexity to vehicle design and refuelling infrastructure.
  The development of a robust network of CNG and LNG refuelling stations is essential for widespread adoption.
  CNG and LNG have lower energy densities than gasoline or diesel, requiring larger storage tanks to achieve equivalent driving ranges. This can be a limiting factor for long-haul transport applications.
- **Methane Slip**
  Incomplete combustion of methane can result in the emission of unburned methane, a phenomenon known as methane slip. Methane is a potent greenhouse gas with a high global warming potential (GWP). Reducing methane slip requires advancements in engine design, combustion control, and aftertreatment technologies to minimize its environmental impact.

As technology evolves and the global focus on reducing greenhouse gas emissions intensifies, the use of methane in ICEs is expected to grow. Key developments shaping the future of methane-fuelled ICEs include:

- **Advanced Combustion Technologies**
  Innovations such as high-pressure direct injection (HPDI), homogeneous charge compression ignition (HCCI), and lean-burn strategies are improving methane engine performance while simultaneously reducing emissions.
- **Biomethane and Power-to-Gas**
  The production of renewable methane, such as biomethane derived from organic waste or synthetic methane through power-to-gas processes, significantly enhances methane's sustainability as a fuel. Governments and industries are actively promoting renewable methane adoption as part of broader climate targets.
- **Hybridization and Methane ICEs**
  Combining methane ICEs with hybrid electric drivetrains could improve fuel efficiency and reduce toxic emissions, making these systems viable for urban and restricted-area applications.
- **Infrastructure Development**
  Expanding CNG and LNG refuelling networks will facilitate the adoption of methane-fuelled vehicles, particularly in heavy-duty and marine transport sectors.

Methane represents one of the most promising alternative fuels for ICEs, offering significant environmental, economic, and performance benefits. While challenges such as storage, infrastructure, and methane slip remain, ongoing technological advancements and the growth of renewable methane sources will continue to drive its adoption. As part of a broader strategy to decarbonize transportation, methane-fuelled ICEs have the potential to play a pivotal role in achieving cleaner, more efficient energy systems.

III.4.2 LIQUIFIED PETROLEUM GASES (LPG)

Liquified petroleum gases (LPG) are a mixture of very light gaseous hydrocarbons, primarily butane (C_4H_{10}) and propane (C_3H_8), and occasionally small amounts of ethane (C_2H_6), typically derived from petroleum refining.

Unlike methane, these hydrocarbons have critical temperatures above normal environmental conditions, enabling liquefaction with simple compression. However, their respective saturation pressures differ significantly: Butane requires approximately 2 bars, propane around 20 bars, and ethane about 70 bars at 25°C.

This means that an LPG mixture, particularly with minimal ethane content, can remain in liquid form under moderate pressurization in uninsulated tanks, ensuring safety and usability.

Both butane and propane exhibit octane numbers higher than conventional gasoline but lower than methane, making them compatible with spark-ignition (SI) ICEs. Their use as a fuel shares many of methane's advantages and challenges, with notable distinctions:

- **High Octane Number**
  Depending on its composition, LPG has an octane number of approximately 100–105, slightly higher than conventional gasoline. "This allows LPG operating SI ICEs, but there is no margin to increase its compression ratio, like in Methane-fed ICEs." While this enables LPG to be used in standard SI ICEs, it does not offer the same margin for increasing the compression ratio as methane does.
- **Clean Combustion**
  Due to the simple chemical structure of the fuel, LPG ICEs may generate fewer toxic emissions compared to conventional liquid fuels. However, the actual environmental performance of LPG-powered engines depends significantly on calibration. As with methane engines, calibration activities are typically discontinued once the engine meets homologation standards, even though further environmental gains might be achievable with more advanced calibrations.

  The high hydrogen content in LPG reduces CO_2 emissions per unit of energy produced by approximately 7% compared to gasoline. Using the calculation scheme established in the previous chapter (with LPG having a heating value H_i of about 12.8 kWh/kg and a CO_2 emission factor $R_f \approx 3.0$), the CO_2 emissions for an LPG-powered medium passenger vehicle (MPV) can be estimated as:

$$q_{CO_2} \approx 141 \frac{g_{CO_2}}{km}, \text{ for an LPG CI MPV 7\% reduction.}$$

- **Liquified State and Tank Design**
  In their liquid states, butane and propane have densities lower than gasoline. Gasoline, under typical conditions (25°C, 1 bar), has a density of approximately 746 g/L, whereas propane and butane exhibit densities of around 500 g/L and 580 g/L, respectively.
  For safety, LPG tanks are filled to a maximum of 80% of their capacity with liquid fuel, leaving the remaining volume for vapour to maintain the mixture in saturated conditions.
  This requirement, combined with LPG's lower density, reduces the vehicle's mileage by approximately 30–35%, a penalty much smaller than the 75% reduction observed in methane-powered vehicles.

LPG can be injected into the engine in either liquid or gaseous form. Gaseous-phase injection is the most common method, employing systems similar to those used for methane. However, gaseous injection eliminates the charge refrigeration effect that gasoline engines benefit from during fuel vaporization, resulting in a slight reduction in engine power.

Alternatively, injecting LPG in its liquid phase exploits its high vaporization enthalpy, significantly cooling the intake charge. This cooling increases air density, potentially boosting engine power by 5–10% compared to the original gasoline engine configuration. Despite this performance benefit, liquid-phase LPG injection systems are uncommon due to their higher complexity and cost. Details about these systems are provided in the subsequent section.

III.4.2.1 LPG Injection Systems for Spark-Ignition ICEs

As already said, LPG injection of the fuel in the engine may happen both in the gaseous and liquid phases. In both cases, the fuel is introduced from the bottom of the tank, in its liquid state, at saturation conditions.

- **Gaseous-phase injection systems**
  These systems are very similar to those used in methane SI ICE injection setups. One of the most critical components in such systems is the pressure regulator (see Figure III.4.2). For LPG systems, the first valve in the regulator not only reduces the pressure but also vaporizes the fuel. This phase change significantly alters the fuel's properties, potentially causing instabilities in the system, particularly during cold-start conditions.
  The quantity of fuel injected must account for its properties (e.g., stoichiometric value), which can be challenging as LPG composition varies. The ratio of butane to propane in LPG depends on the season and the country of refuelling. Additionally, the composition of the liquid and gaseous phases in the tank is not identical: the vapour phase is richer in propane (the more volatile component), while the liquid phase, which feeds the engine, is richer in butane.
  As the tank empties, the liquid phase gradually enriches in propane, requiring real-time compensation by the electronic control unit (ECU).

To address these challenges, the ECU estimates the fuel composition by monitoring tank pressure and temperature. The estimation relies on:

- Saturation curves of pure butane and propane.
- Dalton's law, which states that the total pressure of a vapour phase is the weighted sum of the saturation pressures of its components, based on their volumetric fractions.
- Mass conservation equations for the two components.

- **Liquid-phase injection systems**

Since LPG in the tank is at saturation conditions, even a slight drop in pressure or increase in temperature can cause it to vaporize.

To address this, the system includes a fuel pump that increases the pressure by approximately 3 bars, creating a safety margin against vaporization. This pump is submerged at the bottom of the tank to prevent cavitation at its inlet.

The fuel circulates continuously in a closed-loop circuit between the tank and the common rail volume, which feeds the injectors. This setup is analogous to gasoline injection systems. A pressure regulator positioned after the common rail stabilizes the pressure during injection.

The regulator's membrane-leveraged valve is controlled via connections to the tank and the engine's inlet manifold, ensuring the pressure difference between the two remains consistent and stabilizing the pressure drop across the injectors regardless of variations in tank pressure.

The fuel line going from the tank to the common rail has to be insulated. In some systems (like that sketched in Figure III.4.3) the in-going and out-going pipes connecting the tank to the common rail are concentrically realized, so that the out-going pipe is insulated using the returning one.

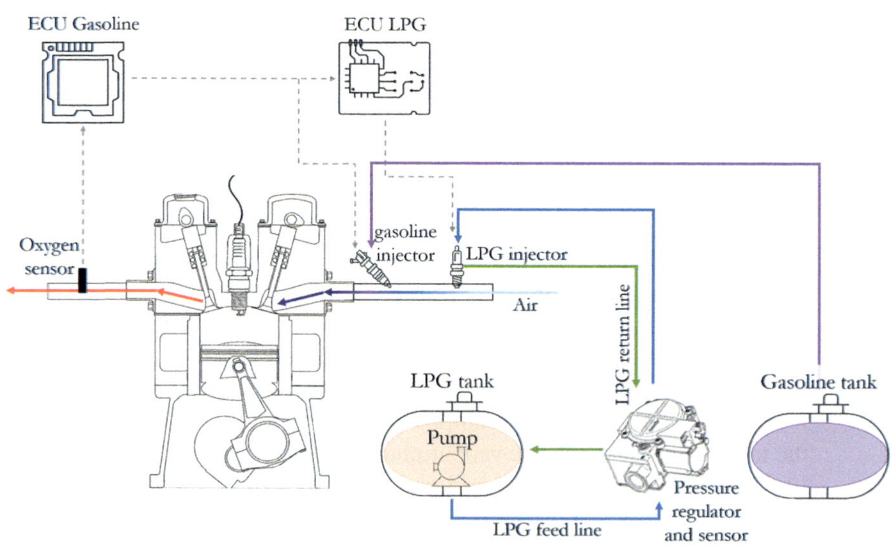

FIGURE III.4.3 Schematic of a liquid-phase LPG injection system.

As already explained, liquid-phase LPG have many perspective advantages:

- They significantly increase engine power output.
- They could lead to much easier air-to-fuel ratio control, and therefore to better engine environmental performance.
- They do not have any cold-start issues.

However, they have two additional problems to be faced:

- They may have some hot-start issues. If the fuel pump stops after the engine is turned off, the LPG in the fuel lines can vaporize. This results in a delay during the next engine start, as the system must purge the vaporized fuel before ignition.
- Some ice can form in the inlet manifold. Rapid fuel vaporization can cause water vapour to frost, potentially leading to engine instability. This issue is more pronounced in port-fuel-injected engines where injectors are not positioned close enough to the intake valves.

III.4.2.2 LPG in ICEs: Benefits, Drawbacks, Potentiality

Summarizing, using LPG in ICEs gives some relevant advantages:

- **Environmental Benefits**
  Lower CO_2 emissions: LPG produces approximately 7–10% less CO_2 than gasoline or diesel per unit of energy. Potentially reduced PM, CO, HC, and NO_x emissions. This reduction in toxic emissions is only obtained by long-lasting and costly engine calibration activities which are normally stopped when Eure-6 limits are obtained, while better performances could probably be reached in ICE engines which are not derived from a gasoline-based approach but dedicated to the alternative fuel.

  Like methane, butane and propane also can be produced by renewable sources. In this case it is referred to as bio-LPG.
- **Economic and Resource Advantages**
  Possible LPG sources are both natural gas and oil reserves, which are still abundant. Butane and propane are sub-products of refining operations. LPG is often cheaper than gasoline or diesel, reducing operational costs.

However, there are still some relevant and unsolved challenges limiting LPG use in ICEs:

- **Storage and Infrastructure**
  LPG vehicles have a 30–35% reduced mileage. Development of LPG refuelling stations is necessary to support widespread adoption, but is much easier than that of methane, with the refuelling station being very similar to gasoline ones. Refuelling times are limited.

III.4.3 HYDROGEN

Hydrogen is emerging as a promising alternative fuel for decarbonizing the transportation sector. Hydrogen-powered ICEs leverage existing engine technology with modifications to accommodate the unique properties of hydrogen as a fuel.

Hydrogen, in fact, has unique characteristics that make it both advantageous and challenging as a fuel for internal combustion engines. It has an energy content, H_i, of approximately 120 MJ/kg, nearly three times that of gasoline, offering significant energy density by weight (but unfortunately not per unit volume at environmental conditions, as its density very low).

When combusted, hydrogen produces water vapour as the primary by-product, eliminating CO_2 emissions. It also eliminates most of the toxic pollutants coming from fossil fuel combustion, with the sole exception of NO_x.

Hydrogen can combust over a broad range of air–fuel ratios, allowing for lean-burn operation, which improves thermal efficiency and reduces nitrogen oxide (NO_x) emissions. It burns faster than traditional fossil fuels, enabling higher engine efficiency and power output.

However, hydrogen also presents various challenges in its use as an energy carrier for transportation purposes in ICEs.

It has a very low volumetric energy density, requiring high-pressure storage (e.g., at 350–700 bar) or cryogenic conditions. Therefore, hydrogen storage and fuelling infrastructure remain underdeveloped compared to conventional fuels.

Its high flammability and low ignition energy increase the risk of pre-ignition and knocking in SI ICEs.

Another issue is related to NO_x emissions: while hydrogen combustion produces no carbon emissions, high combustion temperatures can lead to the formation of NO_x if not properly managed.

III.4.3.1 Modifications Required for Hydrogen ICEs

Adapting ICEs to use hydrogen as a fuel involves several modifications to address its unique properties. Due to its high octane number, hydrogen is normally used as fuel in SI ICEs. Those are more commonly port fuel injection (PFI) engines, so that the simpler option is to inject hydrogen into the inlet manifold.

However, backfiring issues may arise, so that more commonly, hydrogen SI ICEs are direct injection (DI) engines. Hydrogen is injected directly into the combustion chamber, also reducing the risk of detonation and pre-ignition. Precise control of ignition timing has to be implemented to prevent knocking and ensure stable operation.

EGR is widely used to lower combustion temperatures, reducing NO_x emissions without compromising engine performance.

Enhanced cooling systems are necessary to manage the higher combustion temperatures associated with hydrogen.

III.4.3.2 Combustion Strategies for Hydrogen ICEs

As in other DI SI ICEs, different combustion strategies are used to optimize the performance and efficiency of hydrogen-powered ICEs:

- Lean-Burn Operation: Hydrogen's wide flammability range enables lean-burn operation, which is a very limited option with gasoline, improving thermal efficiency and reducing NO_x emissions.

- Stoichiometric Combustion: Combustion at a stoichiometric air–fuel ratio simplifies the use of three-way catalytic converters for emission control.
- Dual-Fuel Operation: Hydrogen (like methane) can be used also as a supplementary fuel alongside diesel in CI ICEs, reducing overall emissions and improving efficiency.

III.4.3.2.1 Main Perspective Applications of Hydrogen ICEs

Hydrogen ICEs are particularly suited for specific applications where their advantages more easily outweigh their limitations:

- **Heavy-Duty Vehicles**
  Trucks, buses, and off-road vehicles benefit from hydrogen ICEs due to their ability to store large amounts of fuel and operate over long distances.
- **Power Generation**
  Hydrogen ICEs can be used in stationary power generation applications, providing clean and reliable energy.
- **Marine and Aviation**
  Hydrogen ICEs offer a potential solution for decarbonizing marine vessels and small aircraft, where fuel cells face size and weight constraints.

Hydrogen ICEs represent a viable pathway for reducing carbon emissions in sectors that are difficult to electrify. However, their success will depend on addressing technical challenges, reducing costs, and developing a robust hydrogen infrastructure. As hydrogen production becomes greener and more cost-effective, hydrogen ICEs could play a critical role in the transition to a sustainable energy future.

The use of hydrogen in ICEs offers a promising, albeit transitional, solution for decarbonizing the transportation and energy sectors. By leveraging existing ICE technology and infrastructure, hydrogen ICEs provide a practical means to reduce emissions while paving the way for broader adoption of hydrogen-based systems. With continued advancements in engine design, combustion strategies, and hydrogen infrastructure, hydrogen ICEs could become an integral part of the global energy landscape.

III.4.4 BIOFUELS

In nature, alcohols possess a high octane number, making them comparable to gasoline, while vegetable oils exhibit a favourable cetane number, aligning them more closely with diesel fuel. This correlation is not coincidental, as diesel and Otto cycles were initially designed to operate on organic substances. For instance, early diesel engines ran on seed oil.

However, the evolution of modern engine technologies has optimized their performance for fossil fuels, creating a need for alternatives when using vegetable oils or alcohols. This adaptation can take two forms: modifying engine technologies or altering the characteristics of oils and alcohols to mimic fossil-based diesel or gasoline.

III.4.4.1 Alcoholic Fermentation: Biogasoline

Alcoholic fermentation is a biochemical process that converts sugars, such as glucose ($C_6H_{12}O_6$), into ethyl alcohol (C_2H_5OH), commonly known as **bioethanol**. This process produces an inorganic compound derived from organic materials. It is primarily applied to sugar-rich crops like sugarcane, sugar beet, sweet sorghum, and certain fruits.

In addition, **lignocellulosic materials**, such as crop residues or wood, can serve as feedstocks for bioethanol production. These materials must first undergo hydrolysis with sulphuric acid to break them into fermentable sugars. These sugars are subsequently processed by genetically modified bacteria or yeast to produce ethanol.

To be a viable fuel, bioethanol must demonstrate resistance to premature detonation, measured by the **octane number**, ensuring combustion occurs only when ignited by a spark. Ethyl alcohol naturally possesses a high octane number, making it compatible with gasoline in spark-ignition engines. However, its performance is not ideal when used alone. As a result, bioethanol is often blended with gasoline in ratios of up to 15% (e.g., E10 or E15), providing a practical and widely accepted fuel mix.

Another application involves combining bioethanol with isobutene to create **ETBE** (ethyl tertiary-butyl ether), a gasoline substitute. However, ETBE production is less common due to its reliance on isobutene, a petroleum-derived compound, which ties its supply chain to fossil fuels. Moreover, blending ethanol directly with gasoline is both more cost-effective and compatible with a broader range of vehicles.

Biogasoline production in Europe is limited due to the region's unfavourable soil and climate for growing sugar-rich crops. Instead, large-scale production occurs in regions with suitable agricultural conditions, such as South America, Asia, and Africa. However, transporting raw materials or finished bioethanol to Europe results in significant CO_2 emissions, undermining the environmental advantages of biogasoline.

Currently, the primary production method for bioethanol involves fermenting corn or sugarcane using yeast. This process includes several key stages:

- **Harvesting** the biomass.
- **Grinding** the material into fine particles.
- **Mixing** the ground material with water and yeast to form a mash.
- **Fermenting** the mash to convert sugars into alcohol.
- **Separating and purifying** the alcohol.

The overall chemical reaction for alcoholic fermentation is:

$$C_6H_{12}O_6 \text{ (glucose in biomass)} + \text{yeast} \rightarrow C_2H_5OH \text{ (ethanol)} + CO_2$$

Similarly, **biomethanol** can be produced by fermenting wood or agricultural waste using specialized microorganisms capable of converting these materials into methanol. This process follows analogous stages, demonstrating how fermentation can produce various biofuels to complement existing energy sources.

III.4.4.2 Transesterification and Hydrogenation: Biodiesel and HVO

Diesel-like fuels can theoretically be derived from a wide range of fatty substances, including vegetable oils, animal fats, acidic oils, and cooking oils. However, several factors limit the range of usable raw materials:

- **Cost**: The material must be economically viable, aligning with the market price of the final product.
- **Availability and Quality**: Sufficient quantities of consistent quality are essential to ensure steady production.
- **Compatibility**: The chemical and physical properties of the biodiesel produced must meet the specifications required for its intended application.

Currently, **rapeseed and soybean oils** are the predominant raw materials for biodiesel production, as they are affordable, readily available in consistent quality and quantity, and yield methyl esters with properties closely resembling those of mineral diesel.

When comparing fuels derived from these raw materials to conventional diesel, notable differences emerge. Organic fuels, for instance, contain lower sulphur residues than mineral diesel. Additionally, rapeseed bioester produces fewer carbonaceous residues than unprocessed rapeseed oil, although it still generates more residues than standard diesel fuel. Rapeseed oil has approximately 10% less calorific value than mineral diesel, which slightly reduces performance. However, this difference is negligible under partial load conditions.

The cetane numbers of oils are comparable to those of diesel fuel, typically exceeding 45. Rapeseed bioester further improves upon this, with cetane numbers ranging from 52 to 56. The density of vegetable oils is slightly higher than that of diesel fuel, though their lower energy content offsets this difference.

One critical distinction between vegetable oils and diesel fuels lies in viscosity. Vegetable oils have a viscosity nearly ten times higher than that of mineral diesel. While viscosity does not directly impact the thermodynamic cycle, it significantly affects the design and performance of machinery. In diesel engines, high viscosity creates particular challenges for injectors. Over the past 10–15 years, advances in diesel technology, such as increasing injection pressures, have improved engine efficiency and reduced emissions. However, these systems cannot tolerate the high viscosity levels of vegetable oils.

This limitation also affects the combustion process. High viscosity leads to larger fuel droplets, which burn unevenly, generating more particulates. While technologies like common rail systems and turbocharging have mitigated this issue in engines using conventional fuels, the high viscosity of vegetable oils prevents their full utilization, resulting in injector malfunctions and increased particulate emissions.

To make vegetable oils suitable for modern engines, their chemical properties must be modified. This is achieved by converting them into bioesters or products similar to mineral diesel through processes such as transesterification or hydrogenation. These processes drastically reduce viscosity while retaining the oils' desirable properties, making them blendable with mineral diesel.

Bioesters have an additional benefit: their unique solvent capacity improves the solubility of mineral diesel, reducing the frequency of fuel filter replacements. However, this property can also pose a challenge, as biodiesel may dissolve plastic engine components, requiring these parts to be replaced with metal ones when high biodiesel concentrations are used.

III.4.4.3 Transesterification

Biodiesel is a renewable fuel composed of fatty acid methyl esters (FAME) and is regarded as a low-carbon alternative to fossil diesel, particularly in the transportation sector. Among the four primary methods of biodiesel production (oil blending, micro-emulsion, pyrolysis, and transesterification) the transesterification process is the most widely used due to the superior quality of the fuel it produces.

The transesterification process reduces the viscosity of vegetable oils by converting them into esters through a chemical reaction. Esters retain a structure similar to oils but exhibit significantly lower viscosity, making them compatible with modern engines.

Esterification is an alcohol condensation reaction in which a carboxylic acid (fatty acid) reacts with an alcohol. When vegetable oil or animal fat are used, the reaction involves triglycerides.

A triglyceride (see Figure III.4.4) consists of a three-dimensional structure composed of three carbon atoms bonded to each other. Each carbon atom is linked to two hydrogen atoms and a carboxyl group. The radicals (denoted as R) represent variable-length chains organized in all directions, differing based on the oil source: triglycerides in vegetable oils have shorter chains compared to those in animal fats or unsaturated oils.

The process begins with the reaction of triglycerides with an alcohol, typically methanol (CH_3OH), in the presence of a catalyst, such as sodium methoxide (CH_3ONa) or potassium hydroxide (KOH). This reaction breaks the carbon–carbon bonds responsible for the high viscosity of triglycerides, producing biodiesel and glycerol as a by-product.

The transesterification reaction (see Figure III.4.5) is exothermic, meaning it releases heat, which helps maintain the high operating temperature of around 100°C. Currently, the reaction takes place in an acidic environment, and an excess of methanol is required to ensure efficiency. While methanol is typically derived from fossil sources, it could potentially be produced from biomass, which would enhance the overall sustainability of biodiesel production.

Glycerol, a valuable by-product, finds extensive applications in cosmetic and pharmaceutical industries, contributing to the economic viability of the biodiesel production process.

After the chemical reaction, further operations are required because the product is not purely esters. The operational phases of biodiesel production are:

- **Reaction**: The oil is converted into esters, a process that can be carried out in batch (discontinuous) or continuous modes, depending on the production scale and facility design.

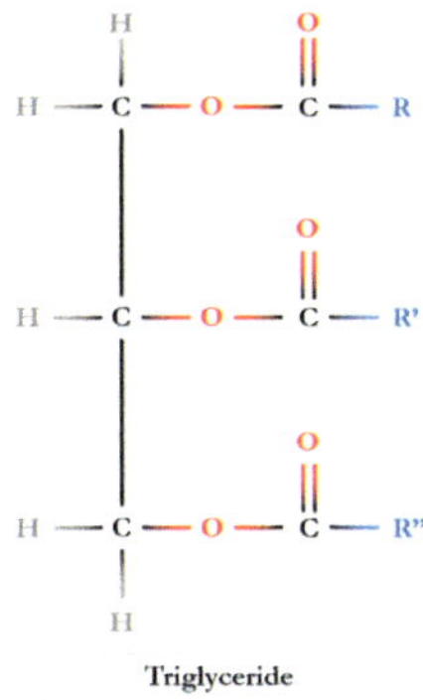

FIGURE III.4.4 Structure of a triglyceride.

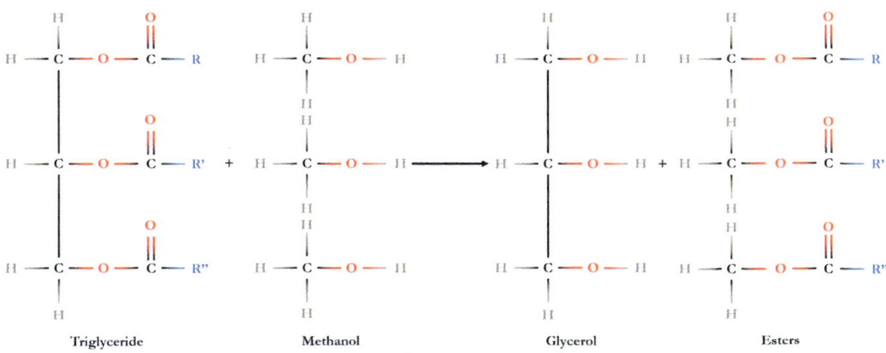

FIGURE III.4.5 Transesterification reaction.

- **Methanol Recovery through Distillation**: Excess methanol is recovered using vertical evaporators and condensers. Since methanol is highly volatile, it can easily be separated from other products through distillation, exploiting differences in boiling points.
- **Glycerine and soap separation** (the main causes of engine damage). In addition to glycerine, there are other reaction residues, commonly referred to as soaps (very dense and viscous). These must be removed to ensure that the biodiesel is engine compatible.
  - Decantation: Soaps and glycerine, being more viscous and denser, are separated by gravity, leaving the biodiesel below. However, this technique can leave traces of soap in the final product.
  - Centrifugal separation with water: Using rotation, the more viscous compounds settle on the walls of the separator and are collected. This technique is more efficient than decantation.
  - Separation with resins: Specialized resins are used to remove any residual soap and impurities, further improving the quality of the biodiesel.

III.4.4.4 Hydrotreated Vegetable Oil (HVO)

HVO is produced through the catalytic reaction of vegetable oils or animal fats with hydrogen in a process known as hydrogenation.

Hydrogenation (Figure III.4.6) leads to the breakdown of fats and oils, removing all oxygen atoms and unsaturated bonds. This process converts fatty acids into long-chain paraffins, while the glycerol component is transformed into propane gas. Oxygen binds with hydrogen to form water molecules.

HVO is an alternative fuel derived from various vegetable oils and animal fats containing triglycerides and fatty acids. It is also known by the acronym HEFA (hydrotreated esters and fatty acids). This fuel is often called renewable diesel or green diesel. It is generated through the hydrogenation and hydrocracking of paraffinic hydrocarbon forms of lipids, where the raw material consists of vegetable oil and animal fat.

Rapeseed is believed to produce the highest yield among all raw materials. Although HVO is obtained from raw materials similar to those used for biodiesel, it is produced via a refining process (unlike biodiesel, obtained through transesterification). The general reaction for HVO production is as follows:

$$\text{Tryglyceride} + H_2 \rightarrow \text{HVO} + \text{SAF} + \text{BioLPG}$$

The refining process is preferred because it helps improve the oxidative stability of HVO. This results in better resistance to bacterial growth compared to biodiesel, making it a superior solution.

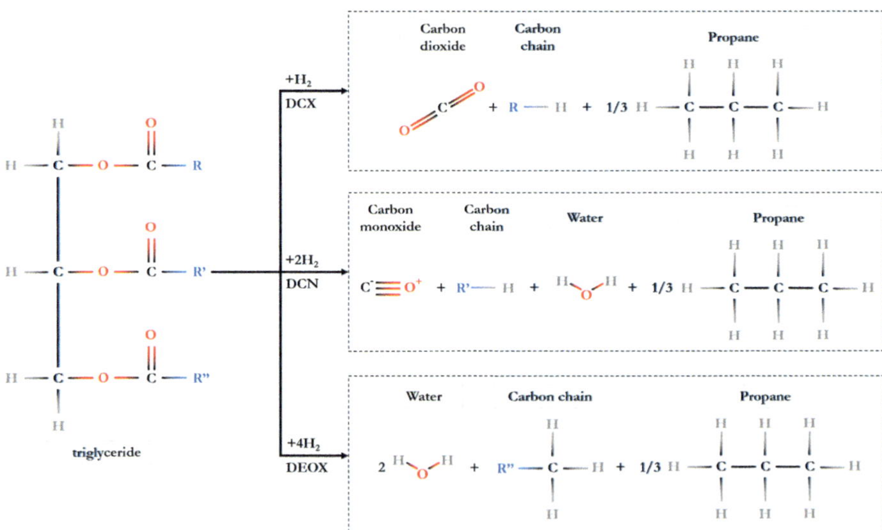

FIGURE III.4.6 Hydrogenation reaction.

The chemical characteristics of HVO are comparable to those of fossil diesel; they differ only in terms of density and energy content, both of which are lower. It has a high cetane number and is free of sulphur, oxygen, and aromatic hydrocarbons.

To date, it is the second-largest renewable diesel alternative and is blended with combinations of fossil diesel for use in the transportation sector.

Emerging processes, such as hydrogen production from microalgae (third-generation biofuels), are excluded from this analysis. These emerging technologies remain at a low level of technological maturity and have only been demonstrated in pilot plants or laboratory settings.

III.4.5 EXPERIMENTAL CHARACTERIZATION OF FUEL INJECTION

When introducing an alternative fuel in an internal combustion engine (ICE), understanding the dynamics of mixture formation is essential for optimizing combustion processes and reducing emissions. The characterization of a fuel jet, or spray, within an ICE is a fundamental step in achieving complete and efficient combustion.

Explaining all the physical processes involved is beyond the scope of this section. However, since the authors have conducted research in this specific field over the past decades, we briefly introduce two examples of experimental techniques used to characterize fuel jets for both liquid and gaseous fuels. Therefore, this short section only serves as a focused case study, offering readers insight into the complexity of the phenomena involved in ICE optimization.

Characterizing a fuel jet, in fact, permits obtaining information such as the morphology of the spray, the penetration of the jet, the size of the fuel drops, the speed of the individual particles, etc.

These quantities must be adequately studied in order to:

- Ensure the correct air/fuel mixing for subsequent combustion process.
- For liquid fuels, avoid the formation of liquid deposits on the piston head (piston-wetting) and on the cylinder walls (wall-wetting) which, not burning, are expelled as they are (carbon residues) or partially burned (presence of unburned materials in the exhaust) or can constitute hot-spots generating detonation phenomena.
- Ensure an adequate flow rate of injected fuel for each operating regime, thus allowing for different injection strategies. Optical investigation techniques allow the measurement of the characteristic parameters of a spray in a non-intrusive manner, i.e. without influencing the investigated process, and the analysis of transient phenomena with high spatial and temporal resolution. As real case studies, here the results of some experimental investigations performed by the authors for both high-pressure liquid and gaseous fuel jets are presented.

Both experimental activities were carried out by the authors at the Spray Laboratory of the Institute of Science and Technology for Energy and Sustainable Mobility (STEMS) of the CNR in Naples.

III.4.5.1 Liquid Fuel Jets – Flash Boiling Phenomena

When a liquid spray is injected into the combustion chamber of an engine, the fuel core must quickly break up into small droplets to accelerate the evaporation process and ensure proper mixing with the surrounding air.

The following experimental activity concerns the characterization of a high-pressure gasoline injector. The evolution of the spray was visualized inside a constant volume combustion chamber (CVCC), by means of optical techniques. Different experimental conditions were investigated, varying the injection pressure and the temperature and pressure in the CVCC, including flash-boiling conditions.

Isooctane was introduced inside the chamber by an eight-hole injector, and the jet spreading was observed by the means of the schlieren and Mie scattering techniques.

Indeed, the use of the schlieren technique alone does not allow discerning the vapour phase and the liquid phase, as it displays the density gradients, independently of the liquid phase. On the contrary, by introducing a high-intensity light source at the ends of the lateral optical window, it is possible to exploit the elastic diffusion of liquid drops to acquire, along the same measurement line, images of the liquid phase only.

The visualization of the liquid and vapour phases is, therefore, possible thanks to the combined use of the Mie scattering and schlieren techniques, with images acquired in an alternative but almost simultaneous manner. In this way, the liquid phase is extracted from the acquisitions with the Mie scattering technique while the global structure of the spray, including both the liquid and vapour phases, is visualized using the schlieren technique.

The optical diagnostic and acquisition system is shown in Figure III.4.7.

Once acquired, the Mie scattering images are superimposed on the corresponding schlieren ones to isolate the vapour phase of the spray, by means of a specific algorithm for processing the acquired images.

Finally, the acquired images were used to measure both the gaseous and liquid phases' penetration length of the jet, defined as the maximum axial penetration of any plume in the direction of the injector axis, measured from the tip of the injector itself. An example of the experimental results at different injection conditions is reported in Figure III.4.8.

III.4.5.2 Gaseous Fuel Jets – Mach Disk Formation

The experimental studies on gaseous fuels are mainly focused on direct injection techniques. In this case, the challenge lies mainly in the low energy density of natural gas, which entails, as stated earlier, the need for large injector nozzle outlet cross-sections and a slower penetration rate. These factors can be mitigated by maintaining high jet momentum through appropriately elevated injection pressures.

The following experimental activity concerns the characterization of a spray injected by means of a prototype injector for applications in heavy-duty direct injection engines of natural gas, under different operating conditions.

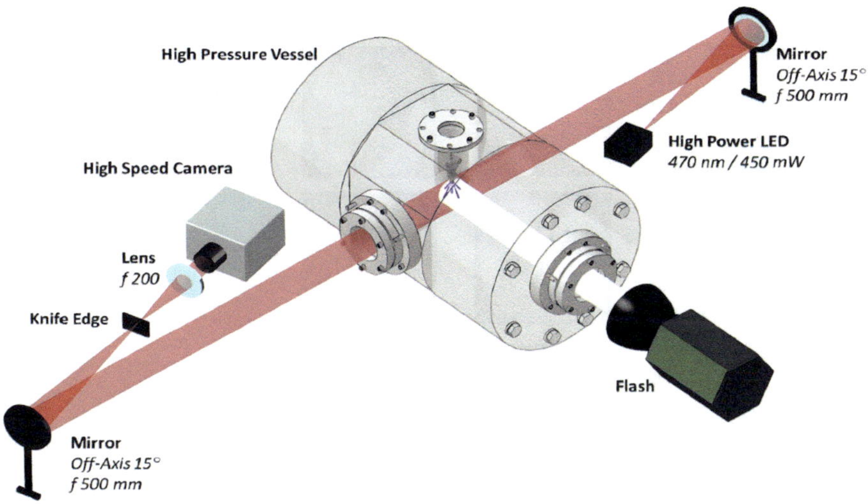

FIGURE III.4.7 Optical setup of the schlieren and Mie scattering techniques in the z-type configuration.

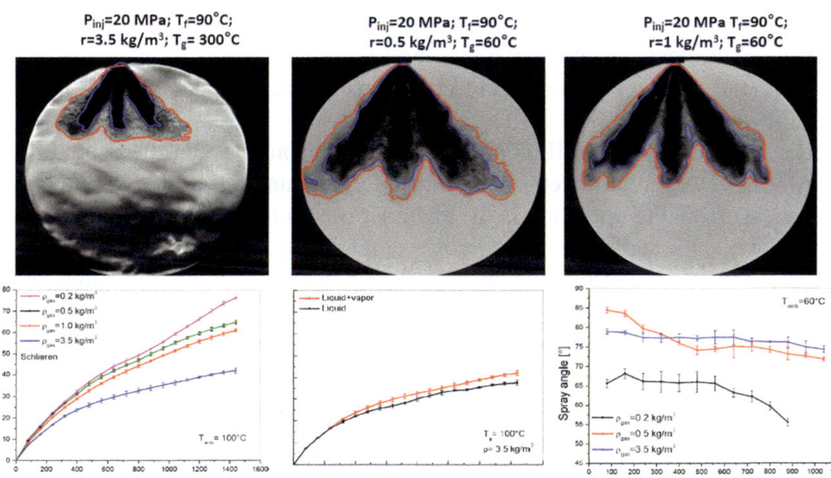

FIGURE III.4.8 Example of experimental measurements. The blue lines represent the liquid phase. The red lines represent the gaseous phase.

The evolution of the spray inside a CVCC was visualized by the sole optical schlieren technique in the "z-type" configuration, recording the images of the fuel evolution using a high-speed CMOS camera. Then, the macroscopic quantities of the jet were measured. The schlieren technique is, in fact, sensitive to the density gradient

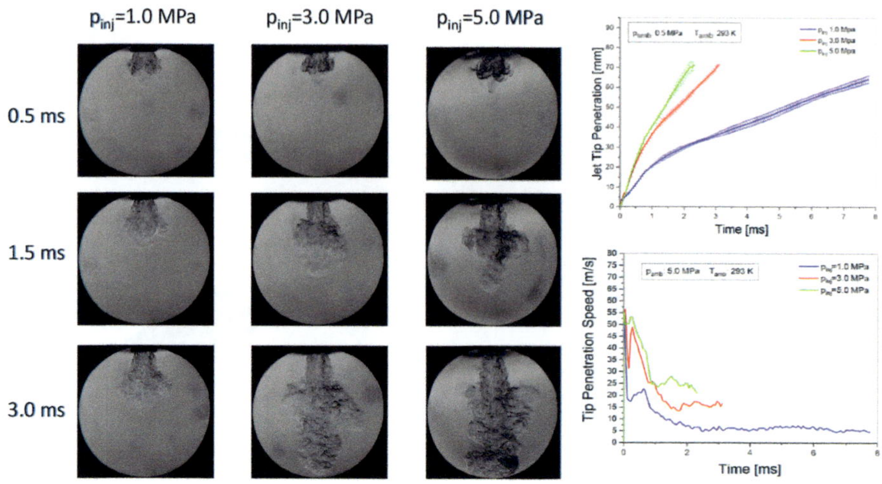

FIGURE III.4.9 Evolution of the gas spray morphology at different injection pressures and spray axial penetration trends.

along the optical path which, in turn, determines variations in the refractive index, making it particularly suitable for viewing transparent media such as gaseous fuels.

The experimental campaign was conducted to evaluate the effects of injection pressure and chamber backpressure on the evolution, morphology, and penetration of the spray.

The left part of Figure III.4.9 shows a comparison between the evolution of the spray at three different injection pressures (1.0, 3.0, and 5.0 MPa), maintaining constant pressure and temperature in the chamber (respectively p_{amb}=0.5 MPa and T_{amb}= 293 K), while the right part of the figure compares the values of the axial penetration of the jet as a function of time, averaged over five consecutive injections, for each test point.

In summary, alternative fuels for ICEs, eventually obtained by bio-sources and/or synthesized using electricity from renewable sources, can substantially increase ICE sustainability, in terms of both their toxic pollutant and GHG emissions. However, infrastructure development and cost reductions still pose significant challenges for their widespread adoption.

III.5 Vehicle Electrification

The concept of vehicle electrification often generates confusion in the literature due to differing definitions.

In this text, we define "vehicle electrification" as the process of powering the vehicle and all its energy requirements (both for traction and all the vehicle's auxiliaries) through electricity.

Therefore, vehicle electrification degree is defined as the fraction of whole vehicle energy requirements (both for traction and all the vehicle's auxiliaries), which is fed electrically. This definition applies regardless of whether the electricity is supplied externally or generated on-board via electric generators.

Then, increasing vehicle electrification means replacing any vehicle component that operates on a conventional energy source with another operating on electricity. Full vehicle electrification obviously also requires the complete electrification of its propulsion system, replacing ICEs with electric drives.

By this definition, conventional vehicles, whose propulsion is given through ICEs, have an electrification degree which is not 0%, because they commonly use a series of electrical components to fulfil a part of their energy requirements (e.g. ICE starter, lighting, electronic controls, …).

It is noteworthy to consider that the electrification degree of conventional vehicles rapidly grew in recent decades, since many components are now newly electrified in some vehicles (e.g. water and lubricant pumps, overcharging compressor, A/C compressor, …).

All the electric energy to feed these components is produced on board by an electric alternator and stored in a temporary BESS. Electric energy and power requirements of conventional vehicles depend on the vehicle's weight and ICE installed power, and also on its typical use (light-duty, heavy-duty, off-road vehicles) but are commonly in the range 1–5 kW peak power and 50–2000 Wh energy.

These specifications are normally fulfilled by lead-acid BESSs having an open circuit standard voltage of 12 or 24 V, and a Coulombic capacity ranging from 5 to 200 Ah. The vehicle electrification degree can reach 10% or more in some particular cases also for conventional vehicles, fully fed by chemical carriers (commonly gasoline or diesel) and driven by ICEs.

DOI: 10.1201/9781003631125-27

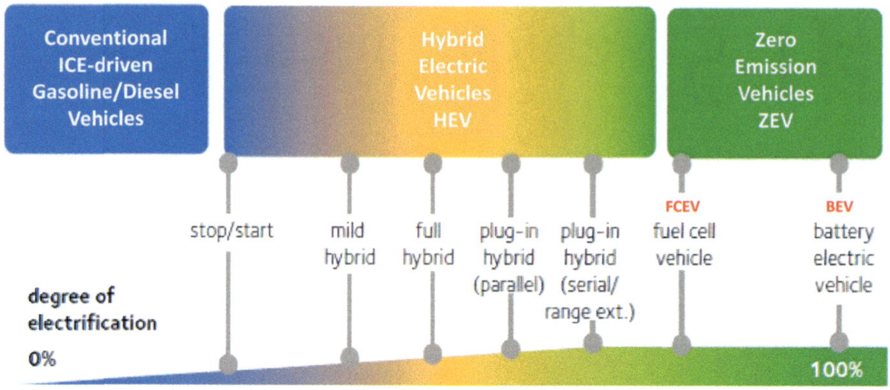

FIGURE III.5.1 Schematic representation of vehicle electrification degree.

Moreover, some fully electrified vehicles (EVs, having a 100% electrification degree) may not be connected at all to the electric grid, since they produce on-board all the electricity required by all vehicle's requirements through electric generation devices.

This is the case of both FCEVs (fuel-cell electric vehicles) producing electricity through chemically fed electrochemical devices (discussed in Chapter III.6), and of the so-called series hybrid ICE/electric vehicles (discussed in Chapter III.8 together with all other ICE/Electric HEVs – hybrid EVs).

On the other hand, some grid-connected vehicles may not have a 100% electrification degree, this being the case of so-called plug-in hybrid ICE/electric vehicles (also discussed in Chapter III.8).

Vehicles only fed by electricity, which is the only energy carrier stored on-board in a BESS, are normally referred to as BEVs (battery electric vehicles, discussed in Chapter III.7). They have a 100% electrification degree, but are not the only fully electrified vehicles.

Some authors, instead, prefer defining "vehicle electrification" only referring to the vehicle's traction requirements, and not to all energy requirements (both for traction and all the vehicle's auxiliaries). By this definition, all conventional vehicles have a 0% electrification degree.

All these considerations are depicted graphically in Figure III.5.1 schematical representation.

III.5.1 PASSENGER CAR MARKET EVOLUTION

Tables III.5.1 and III.5.2 present the results of a data elaboration performed by the authors on the EU car market in the light-duty passenger sector. This analysis is based on official data of the European Energy Agency.

The data show that in 2023 the vehicles sold in all 27 countries of the EU totalled 10.7 million, having an expected average CO_2 emission, measured on the WLTP cycle (as explained in Chapter III.2), of 106 g/km.

TABLE III.5.1
Market Analysis for Light-Duty Passenger Cars Sold in the EU in 2023: Mean Vehicular CO_2 Emissions, Mass, and Power Data for Different Countries

Data	Vehicles		CO_2		Mass		Power	
2023	**$*10^6$**	**%**	**g/km**	**Δ%**	**kg**	**Δ%**	**kW**	**Δ%**
Norway	0.1							
EU-27	**10.7**	**100%**	**106**	**0**	**1543**	**0%**	**115**	**0%**
Germany	2.7	26%	113	6%	1631	6%	134	17%
France	1.9	18%	97	–9%	1450	–6%	101	–12%
Italy	1.6	15%	120	13%	1407	–9%	93	–19%
Spain	1.0	9%	118	10%	1466	–5%	100	–13%
Belgium	0.5	4%	85	–20%	1651	7%	125	9%
Netherlands	0.4	3%	74	–30%	1587	3%	122	6%

TABLE III.5.2
Market Analysis for Light-Duty Passenger Cars Sold in the EU in 2023: Distribution of Vehicles in Categories for Different Countries

Data	ICEs	CI	SI	LPG-NG	®	BEV	PHEV
2023	%	%	%	%	%	%	%
Norway	**9%**	2%	7%	0%	**91%**	83%	8%
EU-27	**77%**	**56%**	**17%**	**3%**	**23%**	**16%**	**8%**
Germany	**75%**	52%	22%	0%	**25%**	19%	6%
France	**75%**	58%	14%	3%	**25%**	16%	9%
Italy	**91%**	61%	22%	8%	**9%**	4%	4%
Spain	**88%**	67%	18%	3%	**12%**	5%	6%
Belgium	**59%**	9%	50%	1%	**41%**	19%	21%
Netherlands	**56%**	0%	55%	1%	**44%**	31%	13%

The market penetration of electrified and more environmentally friendly vehicles (PHEVs and BEVs) already reached 23% (16% BEVs and 8% PHEVs). Amongst the four major EU markets (alone representing about 68% of EU-27), Italy is the less sustainable market, selling only 9% of "green" vehicles. Spain is a bit ahead (12%), while France and Germany are around 25%. Italy's performance in terms of average CO_2 emissions per car is correspondingly the worst among these four countries (120 g/km, 15% worse than the EU average), even though Italian cars are on average much smaller and less energy demanding (–9% average car weight with respect to EU-27, –14% with respect to Germany).

The best performing countries in the EU are Belgium and the Netherlands (respectively 41% and 44% of "green" vehicles sold) but together they only represent 7% of the EU-27 market.

Just outside the EU-27, Norway is a reference lighthouse country for vehicle electrification: in 2023, 83% of cars sold in Norway were BEVs, and only 9% were "conventional" ICE-based vehicles. As we have already said in Part I, however, it is noteworthy to consider that Norway paid most of the costs of this transition remaining outside the EU-27 not to follow its strict rules on sustainability issues and basing most of its economy on selling oil and methane to EU-27 countries. As already explained, this is only one of the enormous and evident contradictions to current global energy systems.

Let us now try to look at how the passenger car sector is evolving in terms of performances of the vehicle sold. As explained in Chapter III.2, car market evolution could lead to further contradictions if vehicles' electrification is not coordinated with the penetration of RESs in the electric sector. CO_2 emissions in Table III.5.1, in fact, are only TTW (those emitted from "Tank-to-Wheels"), and their reduction could become useless if vehicle weight correspondingly increases too much (mainly due to BESSs) leading to an intolerable increment of WTT ("Well-to-Tank").

So, the question is: how is the market evolving? Are we currently introducing into the market mainly small and lightweight EVs, mainly devoted to the urban context and applications (where the zero-local emissions is essential, and daily mileage is limited)?

To answer this question, in this paragraph we will report some results of our market analysis. These, unfortunately, show clearly that, up to now at least, the market is evolving towards the introduction of BEVs mainly on multi-purpose heavy-weight vehicles, not at all considering the issues we discussed in Chapter III.2. Most of the BEVs available on the market are definitely too heavy, too energy consuming, too

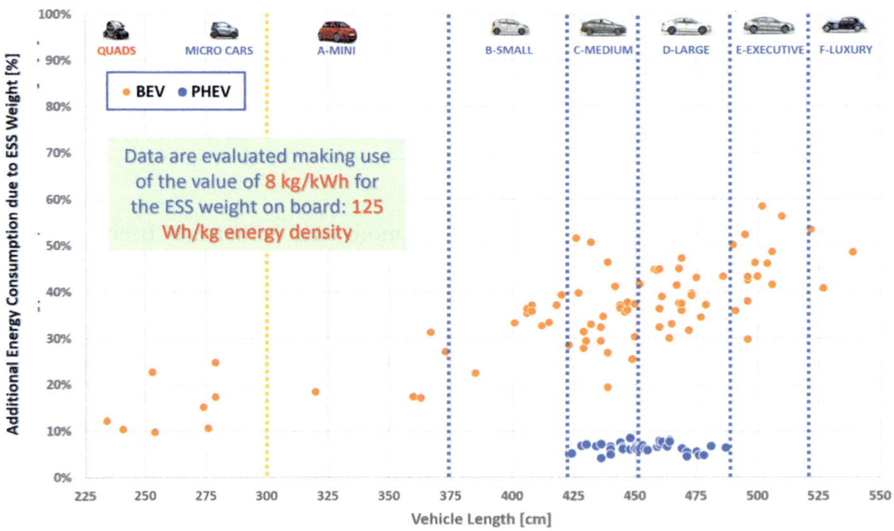

FIGURE III.5.2 Market analysis: Expected BEVs' and PHEVs' energy consumption percent penalty with respect to a conventional vehicle of the same size, as a function of vehicle length.

costly, and they grant a total mileage (500–600 km) which is completely useless for urban uses.

As already said, the authors realized a wide market analysis discussing these issues. Here, for brevity, only the cumulative results are reported in Figure III.5.2.

For each one of the BEVs (orange points) and PHEVs (blue points) currently (November 2023) available in the EU market, the figure reports the expected increment in energy consumption with respect to that of a conventional vehicle of the same size (same length).

The analysis clearly shows that BEVs' energy over-consumption ranges from 15% to about 60%, being higher than 30% for most of the vehicles on the market. This ratio significantly increases with vehicle size due to the additional mileage installed for the vehicles.

Installed vehicle mileage, in fact, is not at all based on transportation considerations. Bigger "executive" and "luxury" BEV cars are sold with more 600 km mileage, while small/medium-sized cars commonly have around 350 km mileage. As a result, small/medium-sized BEVs commonly weigh almost 2 tons, while bigger ones can weigh almost 3 tons.

From this point of view, a much better performance is obtained by PHEVs. Their additional weight and consequent energy over-consumption is much more limited with respect to BEVs and currently ranges between 4% and 8%. This is mainly due to the reduced mileage of these vehicles in pure electric mode (around 50–60 km) and to the correspondingly limited amount of BESS capacity installed on-board (commonly around 10 kWh).

III.6 Fuel Cell Electric Vehicles (FCEVs)

Throughout this text, we have cited fuel cells (FCs) as one of the most promising technologies for enhancing the sustainability of energy systems. But what is an FC and how does it work?

III.6.1 BATTERIES AND FUEL CELLS

A fuel cell is an electrochemical device that converts chemical energy from two reactive sources into electricity. Typically, these sources consist of a liquid or gaseous "fuel" undergoing oxidation at the anodic site and oxygen undergoing reduction at the cathodic site.

In a sense, FCs make a combustion reaction, which is a particularly endothermic redox reaction that occurs in two steps in two separate semi-reaction sites, which are connected through an ionic by-product which can be positively or negatively charged and must move from cathode to anode if negative, or backwards if positive to complete the redox reaction.

The movement of these ions happens through a suitable substrate (so-called electrolyte). This process itself "only" generates a charge separation and an electrostatic potential ΔV_{OC} (where *OC* stands for "open circuit", and is also usually referred as an electromotive force, *EMF*).

If anodic and cathodic sites are also connected by a conductive material outside of the device, the circuit is closed, so that the ions can return where they belonged and make the process potentially steady, generating a stable current on an external utilizer. From an energetic point of view, this is an energetic process converting at least part of the potential chemical energy of the fuel flow rate q (expressed, for unit mass, by its lower heating value, H_i) directly into electric energy, made available by the current (I) flowing under a voltage *DV*.

ΔV is $< \Delta V_{OC}$, because a part of the electrostatic potential is lost, when the circuit is closed, by the internal resistivity of the device (r), which should therefore be as small as possible, being ($\Delta V = \Delta V_{OC} - r\, I$).

FCs share several operational similarities with batteries. Therefore, many of the concepts we explain here will be used also in Chapter III.7, which is about battery electric vehicles (BEVs).

DOI: 10.1201/9781003631125-28

What really makes the difference between a battery and an FC is that the operation of an FC is based on continuous fuel supply, distinguishing it from an electrochemical battery. Therefore, while a battery can make available an energy quantity which strictly depends on the quantities of reactive material stored at its cathode and anode, an FC does not have an intrinsic capacity. The process can go on as long as the fuel and oxygen are fed to the cell, depending only on the capacity of the fuel tank (with oxygen normally being fed filtering external air based on its 21% volumetric oxygen content).

As a consequence, the EMF of a battery is an intrinsic characteristic of the device, while that of an FC depends on the flow rate of the fuel fed to the cell.

For both batteries and FCs, the nominal power output of the device is given by the kinetics of the semi-reaction sites, i.e., by the number of ions for unit time that can be created and moved through the electrolyte generating a stable process. This, in turn, obviously depends on the size of the reaction sites, and therefore of the overall devices, but for unit size and/or volume, basically depends on elements and the "chemistry" of the processes used to build the device. Some "chemistries", in fact, have an intrinsically higher power or energy density (for unit volume and/or mass) than others.

Both FCs and batteries can be invertible. In invertible electrochemical devices, the ions can flow "backwards" in the electrolyte, and the two semi-reactions can be inverted as well so that, basically speaking, the device transforms part of the flowing electric energy into chemical energy regenerating, respectively, oxygen at the cathode and new fuel at the anode. With his aim, ΔV should be $> \Delta V_{OC}$ because the device has a resistive behaviour expressed by its internal resistance in inverted operation (r_{inv}), which is normally higher than that in direct functioning, being ($\Delta V = \Delta V_{OC} + r_{inv} I$).

Reaction kinetics may also be very different in direct and inverse operation, so that inverse operation is much more critical than direct one, leading to much poorer performances.

From what we already said, it is clear that one of the main parameters influencing the efficiency of an electrochemical device is its operating point in terms of provided current: typically, the lower the current provided, the higher the efficiency (if very low currents are not provided).

Another important intrinsic difference between FCs and batteries is that FCs' performances are not influenced by the level of fuel at the tank, while batteries' behaviour is influenced by the remaining quantity of reactants, commonly referred to as their state of charge (SOC).

Batteries, in fact, have a limited number of reactions which can be realized, and therefore a limited number of charges (ions) that can flow through the electrolyte. Their limited capacity, in fact, is commonly expressed in coulombic terms (commonly in Ah, $1C$ being $1As$, therefore $1\ Ah = 3600C$).

When batteries or FCs are operated, they transform chemical energy into electricity (or vice versa, in their inverse operation). Their efficiency at a certain working point can be therefore defined as:

$$\eta = \frac{\Delta V \cdot I}{EMF \cdot I}$$

For batteries, charging and discharging efficiencies also can be defined, as well as their overall efficiency during a complete discharge–recharge process. In these phases they may lose energy (and efficiency) both for coulombic (some of the available charges do not take part in the reaction) and for energetic dissipations (Joule effect due to internal resistance being the most important of the energetic losses). Therefore, for a battery, both a coulombic and an overall efficiency can be defined: the coulombic efficiency of a lead-acid battery is particularly low (85–90%), while that of Li-ion batteries is higher than 99%.

III.6.2 FUEL CELLS

Various types of FCs may be suitable for automotive applications, each with distinct characteristics and advantages (see Figure III.6.1):

- **Proton Exchange Membrane Fuel Cells (PEMFCs)**
  PEMFCs are the most commonly used fuel cells in FCEVs. They operate at relatively low temperatures (around 70–100°C) and are known for their quick startup times, compact size, and high power density. PEMFCs rely on a polymer electrolyte membrane to transport protons from the anode to the cathode while electrons flow through an external circuit to generate electricity. Hydrogen and oxygen react at the cathode, producing water as the only emission. These features make PEMFCs ideal for vehicles, where fast response and compact design are crucial.

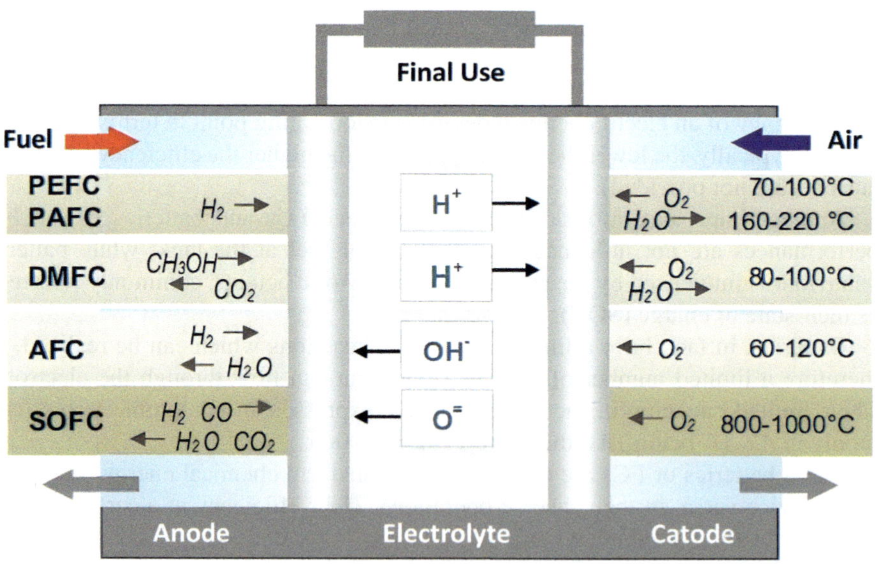

FIGURE III.6.1 Fuel cell types for automotive applications.

- **Phosphoric Acid Fuel Cells (PAFCs)**
  Operating at intermediate temperatures (around 160–220°C), PAFCs are robust and tolerant of impurities in hydrogen. However, their lower power density and bulkier design make them less attractive for automotive applications compared to PEMFCs.
- **Direct Methanol Fuel Cells (DMFCs)**
  Direct methanol fuel cells or DMFCs are a subcategory of proton-exchange fuel cells in which methanol (CH_3OH) is used as the fuel. Their main advantage is the ease of transport of methanol, an energy-dense yet reasonably stable liquid in all environmental conditions.

  This would mean storing methanol at the vehicle tank, overcoming one of the major challenges of H_2 vehicles, which is their extremely reduced mileage with respect to gasoline ICE-based vehicles.

  Unfortunately, efficiency is quite low for these cells, so they are targeted especially to portable applications, where energy and power density are more important than efficiency.
- **Alkaline Fuel Cells (AFCs)**
  AFCs use an alkaline electrolyte (such as potassium hydroxide) and are known for their high efficiency in using hydrogen and oxygen. They have been historically used in space applications, such as the Apollo missions. However, their sensitivity to carbon dioxide contamination still limits their practicality for widespread automotive use, as it requires extremely pure hydrogen and oxygen.
- **Solid Oxide Fuel Cells (SOFCs)**
  SOFCs operate at much higher temperatures (800–1000°C) and use a solid ceramic electrolyte to conduct oxygen ions. While their high operating temperature allows for greater fuel flexibility (e.g., using natural gas or biogas instead of pure hydrogen), it also leads to slower startup times and challenges with thermal management. SOFCs are generally more suited for stationary applications or range extenders in vehicles rather than primary propulsion systems.

III.6.3 FUEL CELL (HYBRID) ELECTRIC VEHICLES (FC(H)EVS)

Despite the significant advantages of fuel cell technology, FCEVs are often designed as hybrid systems, combining fuel cells with battery packs or other energy storage systems. These vehicles should be therefore more correctly referred to as fuel cell hybrid electric vehicles (FCHEVs). This hybridization addresses several challenges and enhances vehicle performance:

- **Energy Efficiency**
  Fuel cells operate most efficiently at steady loads. However, driving conditions require frequent acceleration, deceleration, and idling, which can cause inefficiencies in fuel cell operation. By integrating a battery or supercapacitor,

hybrid FCEVs can store excess energy during low-demand periods and use it during high-demand moments, stabilizing its operational point.

They also can much more easily stabilize and control their operational temperature, also avoiding overheating (which can damage components).

Water management, also, becomes a less restrictive issue. FCs, in fact, require precise control of water levels. Insufficient water can dry out membranes, while excess water can flood the system and block the reaction sites. Balancing water production and removal is particularly challenging in PEMFCs.

- **Regenerative Braking**
  Hybrid systems allow FCEVs to capture and store energy generated during braking. This energy is stored in the battery and can be used to supplement the fuel cell during acceleration or climbing, reducing hydrogen consumption.
- **Improved Performance**
  Batteries in hybrid FCEVs provide additional power for rapid acceleration or uphill driving, where the fuel cell alone might struggle to meet the demand. This ensures smoother and more responsive driving performance.
- **Cold Start Capability**
  Fuel cells, particularly PEMFCs, can take time to reach optimal operating temperatures in cold weather. Batteries in a hybrid system can provide the necessary power during startup, allowing the vehicle to function seamlessly until the fuel cell warms up.
- **System Longevity**
  Hybridization reduces the load on the fuel cell during peak power demands, leading to less wear and tear over time. This can extend the operational life of the fuel cell and lower maintenance costs.
- **Cost Reduction**
  Fuel cells remain an expensive technology due to the use of precious metals like platinum as catalysts. By hybridizing, manufacturers can downsize the fuel cell stack, relying on the battery for supplemental power. This reduces overall costs without compromising performance.

Fuel-cell electric vehicles offer a compelling solution for sustainable transportation, combining zero emissions with the convenience of faster refuelling and longer driving ranges as compared with BEVs.

The choice of fuel cell type significantly influences the vehicle's efficiency, performance, and practicality. PEMFCs have emerged as the leading technology for automotive use, thanks to their compact design and high power density.

However, hybridization with batteries or other energy storage systems is essential to address the inherent limitations of fuel cells, such as inefficiencies during variable loads and cold starts. By leveraging the strengths of both technologies, hybrid FCEVs represent a practical and efficient path forward in the transition to clean energy transportation.

Their operation, however, is fraught with challenges that span technical, economic, and infrastructural domains. Overcoming these obstacles requires advances in catalyst development, improvements in hydrogen production and storage technologies, and significant investments in infrastructure. Addressing these issues is essential to ensure that fuel cells can play a central role in the transition to a sustainable energy future.

III.7 Battery Electric Vehicles (BEVs)

III.7.1 ON ESS SPECIFICATIONS FOR HEVS AND BEVS

In Chapter III.2 we discussed vehicular energy and power requirements obtaining some information which is here recalled, for their implications for electrified vehicle design (both BEVs and HEVs).

When electricity is chosen as the energy vector to partially or fully feed mechanical energy to vehicle wheels, some specifications have to be satisfied both in terms of power and energy density of the on-board electricity generation system.

In both HEVs and BEVs, an ESS is chosen with electric energy as its direct output. In the case of BEVs, the ESS is an electrochemical battery which may be directly recharged, connecting it to the electric systems through a charging device; in HEVs, instead, different storage technologies can be used if they satisfy the required specifications.

So let us now recall briefly the main specifications we obtained in Chapter III.2. The discussions in Chapter III.2 and their main conclusions are summarized in Figure III.7.1.

In particular, the upper part of the figure recalls the main results about the trends of mechanical energetic at vehicle wheels:

- **Energy requirements:** About 100 Wh/km/ton (0.1 Wh/km/kg).
  - 15 kWh/100 km for a medium-sized car
  - 150 kWh/100 km for a medium-sized bus
- **Power requirements**
  - About 40 kW/ton for a medium-sized car (40 W/kg)
  - About 10 kW/ton for an urban bus
  - About 20 kW/ton for an extra-urban bus
- **Recoverable energy:** Up to 25–30%
  - Higher in the urban context
  - Higher in the context of complex topography
  - Much lower for heavy-duty vehicles that are mostly used in highway contexts

DOI: 10.1201/9781003631125-29

FIGURE III.7.1 Results of the analysis of vehicle energy and power demands made in Chapter III.2.

- **Storage requirements for recovery purposes:**
    - Up to 40 kW/ton (>20) are needed to cut-off transient requests (mostly due to inertia). Same magnitude in decelerating transients
    - About 100 Wh/ton for inertial phenomena
    - About 2–3 kWh/ton to include geodetic transients

The lower part of the figure indicates the major implications of previously detailed results on ESS design both for HEVs and BEVs, as discussed below.

III.7.1.1 HEVs

For hybrid applications, expectations are to cut-off power transients which mostly influence ICE polluting and consumption performance.

- **Specific Power**
    - At least 20 kW/ton of vehicle are needed.

    - Not to affect vehicle weight more than an additional 10%, ESS should have a minimum performance of 200 W/kg.
- **Specific Energy**
    - About 1 Wh/kg for inertial phenomena.
    - About 20–30 Wh/kg to include geodetic energy.

III.7.1.2 BEVs

For BEV applications, expectations are to run about 100 km in pure electric mode, which would make them perfectly suitable for pure urban applications.

- **Specific Power**
    - At least 20 kW/ton of vehicle are needed (same as for HEVs).
    - Not to affect vehicle weight more than an additional 10%, ESS should have a minimum performance of 200 W/kg.
- **Specific Energy**
    - About 100 Wh/kg for a 100 km mileage for an urban car (1 ton weight).
    - Regenerative braking is only possible if the ESS has sufficient capacity and a high specific power in regeneration: at least full-hybrid capabilities are required.

All these specifications are in terms of minimal ESS energy and power densities for HEV and BEV applications. They can be graphically represented in a power density vs. energy density log–log Cartesian plane.

This approach is commonly referred to as a Ragone Plot, after Davide Ragone who first proposed it in 1968. In the Ragone Plot, areas may be identified which are coverable by various available technologies, to simply identify which of them is suitable for an application to be designed.

With all this considered, Figure III.7.2 reports the Ragone Plot for HEV and BEV designs, making it possible to identify available technological solutions for their on-board ESSs.

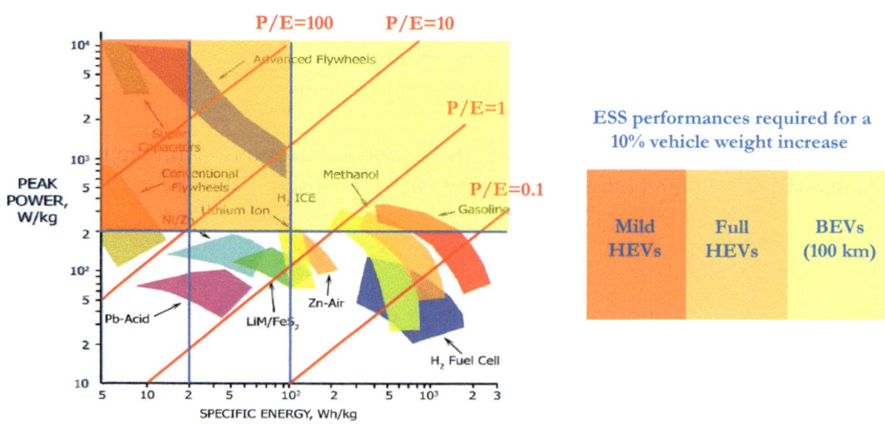

FIGURE III.7.2 Ragone Plot for HEV and BEV design.

All calculations are made with the hypothesis of an expected overall weight penalty below 10% for the vehicle due to its on-board ESS. For BEVs, the additional hypothesis is made to grant vehicles a mileage of 100 km on a WLTP driving cycle, which is enough for purely urban applications.

ESS technologies chosen for BEVs must be in the yellow-shaded zone in Figure III.7.2. The only ESS technologies which are touching that area and have an electric energy output are currently Li-ion and Zn-air batteries.

Among these, currently Li-ion batteries are the technology which has been proved to have higher technological, manufacturing, and commercial readiness levels (TRLs, MRLs, and CRLs). The next section will therefore be dedicated to these battery types.

ESS technologies chosen for HEVs should fall into the orange-shaded area (or in the red-shaded area for mild and minimal hybrid applications, which will be discussed in Chapter III.8). As shown, not only Li-ion batteries are available for these applications. Also advanced flywheels and super-capacitors have proven to be usable in some specific cases.

III.7.2 ON LI-ION BATTERIES AND THEIR MANAGEMENT

We already discussed some specific battery properties in Chapter III.6 explaining the major differences between batteries and fuel cells. Here, these concepts are recalled and specified for Li-ion electro-chemical devices.

III.7.2.1 Batteries' Typical Properties

The most important characteristics defining a battery are the following:

- ***C*** – This is battery capacity expressed in *Ah*, with 1 *Ah*=3600 Coulombs. *C* expresses the total number of unit charges available in the cell up to its complete discharge (from 100% to 0% state of charge – *SOC*). Battery capacity is typically measured under standardized conditions, differently defined depending on the battery type. Standards for Li-ion batteries are defined by the IEC 61960 regulation. The battery has to be discharged at a constant current. Tests must be performed at 25°C, with *C* being influenced by room temperature (as well as by cell operational temperature). The battery, in fact, tends to heat during its use due to the endothermic reactions occurring at its poles. The capacity-measuring test has to be stopped when the cell voltage reaches a minimal value (which is dependent on chemistry type). Standards like IEC 61960 for lithium-ion batteries define procedures for measuring capacity. These include discharge rates, cutoff voltages, and environmental conditions.
- $\boldsymbol{\Delta V_{OC}}$ – This is the battery nominal voltage at 100% SOC and open circuit. It is also referred to as the battery electromotive force (EMF).
- ***E*** – Measured in Wh, is the maximum theoretical energy available in a battery and is normally calculated as the product C*DV_{OC}.
- $\boldsymbol{r_{dis}}$ and $\boldsymbol{r_{ch}}$ – Measured in *W*, these are, respectively, the internal resistances of the battery during its discharge and charge.

- $I_{dis,max}$ – Measured in *A*, this is the maximum continuous discharging current deliverable at the battery poles.
- $I_{ch,max}$ – Measured in *A*, this is the maximum continuous charging current circulating inside the battery.
- $I_{dis,peak}$ – Measured in *A*, this is the maximum peak discharging current deliverable at the battery poles (commonly referred to as peak duration in seconds, usually 1 s, not more than 10 s).
- ***Cycle Life*** – This is the number of charge–discharge cycles a battery can undergo before its capacity drops below 80% of its original capacity.
- ***Self-Discharge Rate*** – This is the rate at which a battery loses its charge when not in use. Lithium-ion batteries typically have very low self-discharge, while nickel-based batteries, and lead-acid batteries have higher rates.
- ***Operating Temperature Range*** – This defines the temperature limits within which the battery can function optimally.

$I_{dis,max}$ and $I_{ch,max}$ are proportional to their capacitance *C*. It is therefore usual to define a non-dimensional power performance as a ratio of maximum currents (A) and capacity (in Ah).

This ratio, usually referred to as the C-ratio, also estimates the ratio between the total energy available in the cell and its maximum discharging and charging power (and therefore may be used to evaluate minimal durations of charging and discharging phases, which are therefore normally independent of cell size, and on the utilization and charging technologies (as long as their maximum power is sufficiently high).

III.7.2.2 Most Common Li-Ion Batteries

Lithium-ion batteries come with various chemistries, each tailored for specific applications based on energy density, safety, lifespan, and cost. Below is an overview of the most widely used types, along with their nominal voltage and typical C-rates (charge and discharge rates):

- **LFP**: Lithium Iron Phosphate ($LiFePO_4$)
  - Excellent thermal and chemical stability.
  - Long cycle life and improved safety.
  - Moderate energy density compared to other lithium-ion types.
  - ΔV_{OC} 3.2–3.3V.
  - Typical C-rates: Charge 1C–2C; Discharge 5C–25C (depending on design).
  - Widely used in electric vehicles (EVs), and renewable energy systems.
  - Enhanced safety and durability.
- **NMC:** Lithium Nickel Manganese Cobalt Oxide ($LiNiMnCoO_2$)
  - High energy density, decent power density; good cycle life.
  - Versatility due to customizable compositions (e.g., varying Ni, Mn, and Co).
  - ΔV_{OC} 3.6–3.7V.
  - Typical C-rates: Charge 0.5C–1C; Discharge 1C–3C.
  - Commonly used in BEVs, e-bikes, and grid energy storage systems.

- **NCA:** Lithium Nickel Cobalt Aluminium Oxide ($LiNiCoAlO_2$)
  - Exceptional energy density and power output.
  - Longer lifespan than many other lithium-ion chemistries.
  - ΔV_{OC} 3.6–3.7V.
  - Typical C-rates: Charge 0.5C–1C; Discharge: 1C–5C.
  - Primarily used in high-performance BEVs.
  - Higher cost and reduced thermal stability compared to $LiFePO_4$ or NMC.
- **LTO:** Lithium Titanate ($Li_4Ti_5O_{12}$)
  - Outstanding safety and fast charging capabilities.
  - Very long cycle life (up to 10,000 cycles or more).
  - Lower energy density compared to other types.
  - ΔV_{OC} 2.3–2.4V.
  - Typical C-rates: Charge 2C–10C; Discharge 10C–30C.
  - Suitable for industrial applications, electric buses, and military or aerospace uses where longevity and reliability are critical.
  - Can operate under extreme temperatures and demanding conditions.
- **LCO:** Lithium Cobalt Oxide ($LiCoO_2$)
  - High energy density.
  - Lightweight and compact design.
  - Limited thermal stability and shorter lifespan compared to other types.
  - ΔV_{OC} 3.6–3.7V.
  - Typical C-rates: Charge 0.5C–1C; Discharge 0.5C–2C.
  - Ideal for portable electronics, such as smartphones, laptops, and tablets.
  - Relatively low cycle life and safety risks under extreme conditions, such as overheating.

Summarizing the actual situation, we could say that each lithium-ion battery type serves a unique purpose, balancing trade-offs between energy density, safety, lifespan, cost, and performance.

For consumer electronics, LCO remains popular due to its compact size and high energy density. In contrast, LFP, NMC, and NCA dominate the electric vehicle and energy-storage markets, offering safety, durability, and scalability.

LTO, with its unmatched charging speed and cycle life, is ideal for industrial and specialized applications.

III.7.2.3 Single Cells and Battery Packs

Single Li-ion batteries (commonly also referred to as "cells") have a nominal capacity depending on their assembly and size. Four typical assemblies are possible, used referred to as "cylindrical", "prismatic", "pouch", and "button". Typical values for the capacity of single cells in the four assemblies are the following:

- **Cylindrical Cells**
  Their most common sizes are: diameter 18, 21, 26 mm; length 65 or 70 mm. Their capacity ranges from 1.5 to 6.5 Ah.

- **Prismatic Cells**
They commonly have a flat, rectangular design and a capacity range 20–200 Ah for larger cells used in electric vehicles and energy-storage systems. Smaller prismatic cells, used in consumer electronics, typically have capacities of 2–10 Ah.
- **Pouch Cells**
These have a flexible design, and commonly capacity is in the range 2–200 Ah, depending on size. They are often used in smartphones, tablets, drones, and EVs.
- **Button Cells**
These are currently only used in small portable electronic devices (watches, hearing aids, and other small devices) and have a very limited single cell capacity range: 10–300 mAh.

Summarizing, single Li-ion cells used for BEV applications have a capacity range of 2–200 Ah, and an EMF of around 4 V. Therefore, maximum energy stored in single cells may range from 8 to 800 Wh. All BEV applications, however, need at least an 8–10 kWh BESS (but easily overcome 100 kWh in vehicles nowadays) so that several cells have to be electrically connected somehow to give the energy needed by the vehicles.

Cell connections can be in parallel and/or in series. Single cells are normally only connected in series, increasing voltage, if the required energy is obtainable under an acceptably low nominal voltage which has to be set both according to safety regulations (which normally require maximum on-board voltage to be below 650–700 V) and further to design considerations for an efficient coupling with the electric drive.

If the single cell capacity is too small, however, the number of cells may increase up to thousands, making it impossible to connect all of them in series. In these cases, the cells are normally firstly connected in parallel in doublets, triplets, quartets, …, and then in series to reach the desired energy and nominal voltage output.

Moreover, consider also that Li-ion battery cell manufacturers no longer only sell single cells, but also pre-assembled groups of cells called modules.

Therefore, normally, car manufacturers buy and assemble these modules in greater elements to constitute the required BESS (also commonly referred to as a "battery pack") for their needs.

In Figure III.7.3 are three examples of BESSs for vehicles. They belong, respectively, to BMW, Tesla, and Hyundai, respectively equipping BMW i3, Tesla models S and X, and Hyundai models 5 and 6.

As is clearly shown, the design solutions are very different from one another, demonstrating that car manufacturers are still searching for the best available technology. They, in fact, are using different cell assemblies (prismatic, cylindrical, pouch), chemistries (NMC, NCA), single cell capacity (from 3 to 120 Ah cells), numbers (from 96 to 8256 single cells), and electric connections (only in series, parallelized in groups of 86 each, parallelized in doublets).

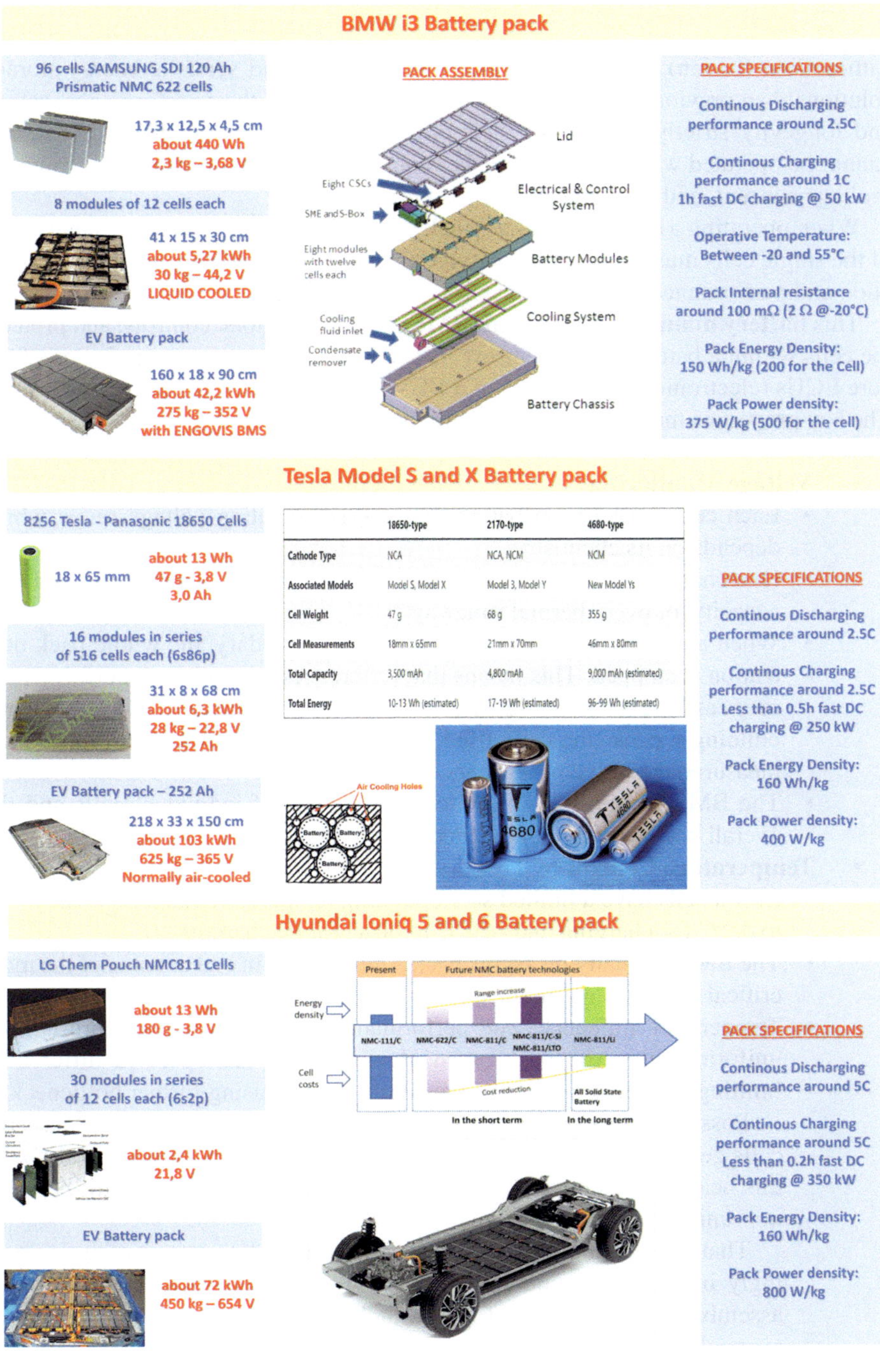

FIGURE III.7.3 Vehicle battery packs by BMW (upper); Tesla (middle), and Hyundai (bottom).

III.7.2.4 Battery Management System (BMS)

Lithium-ion (Li-ion) batteries are much more efficient and versatile energy-storage solutions than previous-generation lead-acid batteries, but their performance, safety, and longevity strictly depend on their precise management, so that a Li-ion cell cannot be operated without continuously controlling its main operational parameters (voltage, current, and temperature, if possible).

When operating a complete battery back made by several single cells, supervision of the single cells must be constantly assured, together with management of the operation and performances of the whole pack.

This **battery management system (BMS)**, which monitors, controls, and protects the cells within a battery pack to ensure its safe and efficient operation, is one of the core ECUs (electronic control units) of a BEV.

The key properties managed by a BMS are the following:

- **Voltage Monitoring**
  - Each cell in an Li-ion battery has a safe operating voltage range which depends on its chemistry.
  - Overcharging or deep discharging a cell can lead to degradation, reduced capacity, or even thermal runaway.
  - When a single cell reaches its operational boundary, the whole pack utilization is stopped. This means that differences in single cell performances may easily result in bottlenecks in the pack performances. Therefore, when building a pack, the cells used should be as uniform as possible in their measured operative performances.
  - The BMS continuously monitors the voltage of individual cells and the overall pack to maintain balance.
- **Temperature Control**
  - Li-ion cells have a limited safe operating temperature range, typically 0°C to 45°C for charging and –20°C to 60°C for discharging.
  - The BMS normally measures pack temperatures in one or some of the most critical points using thermistors or temperature sensors.
  - Temperature influences cell performances, so that if temperature is non-uniform within the pack, even identical cells are operated differently, limiting the performances of the whole pack, causing further bottlenecks.

    Possible temperature non-uniformities are easily created because the cells generate heat during their use, so that if the dispersion capabilities of this heat towards the environment are differently enabled for all the cells, non-uniformities arise.

    That is the main reason why battery pack cooling is becoming increasingly important. Liquid-cooled actively controlled packs are now being assembled, while most of the actually used packs are still passively air-cooled.
- **State of Charge (SOC)**
  - SOC represents the remaining charge in the battery as a percentage of its total capacity. The BMS calculates SOC using methods like voltage

measurements, coulomb counting (tracking and integrating current in/out), and advanced algorithms.
  - Accurate SOC estimation is crucial for optimizing performance, predicting range in electric vehicles, and controlling and optimizing BESS recharge.
- **State of Health (SOH)**
  - SOH indicates the overall condition of the battery compared to its original capacity and performance. The BMS tracks capacity fades, internal resistance changes, and cycle life to estimate SOH. This helps in determining maintenance needs and predicting end-of-life.
- **Current Regulation**
  - The BMS enforces current limits during charge and discharge to protect cells from excessive currents that can cause overheating or damage.
  - It monitors and controls the current to avoid conditions like overcurrent or short circuits.

As already said, due to manufacturing tolerances, individual cells may have slightly different capacities or internal resistances, leading to voltage imbalances during charging and discharging. Most BMSs therefore try to equalize these differences both to ensure all cells operate within their optimal range, and to reduce pack bottlenecks.

Cell balancing is one of the most critical functions of a BMS, as imbalances in voltage or capacity can reduce the overall performance and lifespan of the battery pack. The two primary equalization techniques are:

- **Passive Balancing**
  - Excess energy from higher-voltage cells is dissipated as heat through resistors.
  - This process equalizes the voltage across all cells by bringing higher-voltage cells down to the level of the lowest-voltage cell.
  - This is simple, cost-effective, and easy to implement. However, it is inefficient as energy is wasted as heat, and it requires heat management, especially in large packs.
  - It is still the most commonly used in small to medium-sized battery packs where cost and simplicity are prioritized.
- **Active Balancing**
  - Energy from higher-voltage cells is redistributed to lower-voltage cells using inductors, capacitors, or transformers. This ensures all cells are balanced without wasting energy.
  - This technique is more efficient than passive balancing. It extends the battery pack lifespan and enhances performance.
  - It is more complex and expensive to implement, because it requires additional components, increasing system size and weight.
  - It is now used in high-capacity battery packs for BEVs, renewable energy systems, and aerospace applications.

The BMS also should be able to actively cooperate on-board with all other ECUs, especially those controlling the vehicle propulsion system, vehicle cockpit (interacting with its user), and BESS recharging infrastructure.

Most BMSs communicate with external devices (e.g., EV controllers or grid inverters), with the CAN protocol being the most used standard.

During battery pack recharging, BMS is also responsible for charge management, eventually communicating with charging infrastructure and with the local electric grid, ensuring safe and efficient charging by controlling the charging current and voltage. BMS should implement charging algorithms, such as constant current/constant voltage (CC/CV), to optimize the charging speed and reduce degradation.

Summarizing, BMSs are indispensable for ensuring the safety, efficiency, and longevity of lithium-ion batteries. By managing key properties like voltage, temperature, SOC, SOH, and current, the BMS protects the cells from damage while optimizing performance.

Cell balancing techniques, whether passive or active, play a vital role in maintaining uniform cell behaviour, ensuring the entire pack operates at peak efficiency. As Li-ion batteries become more prevalent in applications ranging from consumer electronics to electric vehicles and renewable energy, the sophistication and capabilities of BMS technology will continue to evolve, driving advancements in energy-storage solutions.

III.7.3 ON ELECTRIC DRIVES AND THEIR PERFORMANCE

Several electric drive technologies are commonly used in BEVs, each with its own advantages and applications. The most widely used electric drive technologies are presented below.

III.7.3.1 Permanent Magnet Synchronous Motors (PMSMs)

Permanent magnet synchronous motors are among the most widely used motor types in BEVs due to their high efficiency, power density, and reliability. They are AC motors utilizing permanent magnets embedded in the rotor to create a magnetic field, which interacts with the stator's rotating field to produce torque.

They have high efficiency across a wide range of speeds and loads, making them suitable for both urban driving and highway conditions, not needing any gearbox to operate the vehicle.

They have high power density (output kW per unit weight), allowing for smaller and lighter motors. Their torque and speed control are fast, easy, and accurate. PMSMs are used by most car manufacturers including Tesla, Nissan, and Hyundai.

Their most critical issue is the dependence on rare earth materials (e.g., neodymium, dysprosium) for magnets, which can increase costs and raise supply chain concerns.

III.7.3.2 Induction Motors (IMs)

AC induction motors, also known as asynchronous motors, operate without permanent magnets. Instead, they rely on electromagnetic induction to create a rotating magnetic field in the rotor, which generates torque.

IMs are robust and durable and resistant to very high temperatures, where PMSMs could have some issues due to their potential demagnetization. Lower dependence on rare earth materials reduces material costs. They deliver high torque and power, especially at higher speeds.

For their robustness, they are commonly used in heavy-duty BEVs, such as buses and trucks. They have slightly lower efficiency compared to PMSMs, particularly at lower speeds, and they are slightly less compact than PMSMs.

III.7.3.3 Synchronous Reluctance Motors (SynRMs)

Synchronous reluctance motors are AC motors generating torque through the reluctance of the magnetic path in the rotor. They do not require permanent magnets, making them a cost-effective and sustainable option. SynRM fully eliminates dependence on rare earth magnets, reducing environmental and cost concerns. They have high efficiency, comparable to PMSMs in certain applications, especially with advanced control algorithms. Their simple rotor design improves thermal stability and reliability.

SynRMs are an emerging technology in BEVs; increasingly considered for cost-sensitive or sustainability-focused vehicles. However, they are relatively complex to control in order to optimize performance and have a lower power density than PMSMs.

III.7.3.4 Switched Reluctance Motors (SRMs)

Switched reluctance motors are AC motors relying on the principle of magnetic reluctance, where the rotor aligns with the magnetic field generated in the stator. The motor has a simple and robust design with no permanent magnets or windings on the rotor. They are extremely durable due to the lack of both magnets and rotor windings.

Simplified construction reduces manufacturing costs. They can achieve high torque and efficiency with advanced control systems. SRMs are ideal for industrial and commercial BEVs, such as forklifts, buses, and heavy machinery, but are nowadays less common in passenger BEVs, mainly due to noise and vibration challenges.

III.7.3.5 Brushless DC Motors (BLDCs)

Brushless DC motors are electronically commutated motors that offer high efficiency and reliability. They are similar to PMSMs but are optimized for simpler control and lower cost in certain applications. Their efficiency is comparable to PMSMs in specific operating conditions. They have no brushes or mechanical commutators, reducing wear and maintenance. BLDCs have high power density, suitable for smaller BEVs and space-constrained applications. Their utilization is, in fact, more common in smaller BEVs like electric scooters, motorcycles, and compact cars. They are also used in auxiliary systems such as pumps and fans in BEVs.

BLDCs have a normally lower torque performance compared to PMSMs and are reliant on rare earth materials for magnets.

III.7.3.6 Axial Flux Motors

Axial flux motors are a newer technology characterized by their "pancake-like" design, where the magnetic flux flows parallel to the AC motor's axis. This contrasts with the radial flux design of traditional motors. The focus is on power density increase to obtain an extremely compact and lightweight design. They have high efficiency due to reduced losses in the rotor and stator and are more easily integrated than PMSMs into compact vehicles.

They are still early-stage technology, and are expensive to manufacture due to precision requirements. Their utilization is growing in high-performance BEVs and sports cars due to their power density and lightweight characteristics, and they are also suitable for space-constrained applications like motorcycles and drones.

III.7.3.7 Hub Motors

Hub motors integrate the AC electric motor directly into the wheel hub, eliminating the need for a central drive system. They aim at saving space, also by eliminating the drivetrain, and at reducing mechanical losses for improved efficiency.

They are more frequently used in very small BEVs, electric bicycles, and concept vehicles, and they are gaining attention for urban mobility solutions and autonomous vehicles. Their most relevant challenge is the consequently increased unsprung weight in the wheels, which can affect vehicle suspension design and vehicle handling performances. There can also be heat dissipation and reliability issues due to the confined space in the hub.

Summarizing, the choice of electric drive technology for BEVs depends on the application, cost, and performance requirements. Permanent magnet synchronous motors (PMSMs) dominate the passenger BEV market due to their high efficiency and power density. The main characteristics of the currently most diffused alternatives are reported in Table III.7.1.

TABLE III.7.1
Most Used Electric Drives for BEVs and Their Main Characteristics

Technology	Efficiency	Power	Cost	Reliability	Applications
PMSM	High	High	Moderate/High	High	High-performance BEVs
IM	Moderate/High	Moderate	Moderate	High	Heavy-duty BEVs
SynRM	Moderate/High	Moderate	Low/Moderate	High	Cost-effective BEVs
SRM	Moderate	Moderate	Low	Very High	Off-road BEVs
BLDC	High	Moderate/High	Moderate	High	Small BEVs, e-scooters
Axial Flux	High	Very High	High	Moderate	High-performance BEVs
Hub	Moderate	Moderate	Moderate/High	Moderate	e-Bikes, urban BEVs

Induction motors (IMs) and switched reluctance motors (SRMs) excel in heavy-duty and industrial applications, while emerging technologies like axial flux motors and synchronous reluctance motors offer exciting possibilities for future BEV designs. As electric vehicle technology advances, continued innovation in motor design and control systems will drive improvements in efficiency, sustainability, and performance.

III.7.4 ON CHARGING DEVICES AND V2X TECHNOLOGIES

Battery electric vehicles (BEVs) rely on charging devices. Therefore, the efficiency, speed, and convenience of these devices play a critical role in the adoption and usability of BEVs. Charging devices may vary widely in terms of power delivery, design, and application, ranging from basic home chargers to advanced ultra-fast public charging stations.

They are designed to deliver direct current (DC) or alternating current (AC) electricity to the vehicle's battery. Depending on the charging device output, the conversion between AC (from the grid) and DC (used by the battery) can occur either inside the vehicle (via an onboard AC/DC charging converter) or within the charging station.

Chargers are mainly differentiated depending on their maximum power output (which determines charging speed and ranges) and their connector type (ensuring compatibility between the charger and the vehicle). Depending on these, current regulations distinguish three types (also referred to as "charging levels" of charging devices (graphically depicted in the upper part of Figure III.7.4):

- **Level 1 Chargers (AC Charging – Basic)**
  - Level 1 chargers use standard household outlets (120V in US; 230V in EU).
  - They typically deliver up to 2–3 kW of power, making them only ideal for overnight charging at home or in locations with long parking durations.
  - Charging Speed: Maximum 10–20 km of range per hour, for a 1-ton vehicle.
  - They are cost-effective and require no special installation, and fully accessible using standard power outlets.
  - Their very slow charging speed makes them unpractical for larger batteries or frequent BEV use.
- **Level 2 Chargers (AC Charging – Faster)**
  - Level 2 chargers operate on a 240V supply.
  - Typically deliver 2.5–20 kW of power, depending on the system and infrastructure.
  - They are commonly installed for home, workplace, and public charging.
  - Charging Speed: Adds approximately 10–100 km of range per hour, for a 1-ton vehicle.
  - Faster than Level 1 chargers and suitable for most daily charging needs.
  - Available with smart features like energy monitoring, scheduling, and load balancing. They require professional installation and may normally need an upgraded electrical panel if installed at home.

FIGURE III.7.4 Possible BEV charging levels (upper part) and charging connectors (lower part).

- **Level 3 Chargers (DC Fast Charging)**
  - Level 3 chargers deliver DC directly to the battery, bypassing the vehicle's onboard charger, being an AC/DC converter off-board, built-in in the charging device.
  - Power outputs range from 50 kW to 350 kW, though some chargers have a power output greater than 150 kW and are still not widely available.
  - Used in public charging stations for rapid energy replenishment, especially along highways and in urban centres.
  - Charging Speed: 200–1200 km of range per hour (15–20 km/min), for a 1-ton vehicle.
  - Ideal for long-distance travel and reducing charging downtime.
  - Suitable for high-capacity battery packs, like those equipping urban buses.

  - They are very expensive infrastructures, and also recharging costs for unit energy may be substantially higher (even 3–4 times higher than those for Level 1 and 2 domestic charging operations). They require advanced cooling systems to manage heat generated during high-power charging.

A further charging level is starting to be diffused:

- **Ultra-Fast and High-Power Charging (HPC)**
  - HPC is the latest generation of chargers delivering more than 350 kW.
  - They support advanced BEVs with high-voltage architecture (e.g., 800V systems like those in the Porsche Taycan or Hyundai Ioniq 5), or e-buses.
  - Positioned along major highways and in high-traffic areas to enable ultra-fast charging.
  - Charging speeds: a 1.5-MW charger may add up to 300–400 km of range in 10 minutes to a 2-ton vehicle.
  - Revolutionizes charging convenience, rivalling refuelling times for conventional vehicles, but substantially lowers charging efficiency, and largely increases recharging costs. Charging efficiency is also substantially reduced, increasing transportation energy consumption.
  - They require very robust electrical infrastructure and significant grid upgrades. They have a very limited compatibility with BEVs which are not specifically designed for high-power charging.

Chargers also require different connection cables, depending on location (North America, EU, Japan, EU, others) and on the power output level and type (Levels 1 and 2 in AC, Level 3 in DC). The most common standard connection plugs are depicted in the lower part of Figure III.7.4:

- **Type 1 (SAE J1772)**
  Widely used in North America and Japan for AC Level 1 and Level 2 charging. It may serve one-phase AC output with a maximum power delivery of 7.4 kW.
- **Type 2 (Mennekes)**
  This is the approved standard in Europe for AC Level 2 charging. It supports one-phase and three-phase power, with up to 22 kW for AC charging.
- **Combined Charging Systems Types 1 and 2 (CCS1 and CCS2)**
  Combine AC (Type 1 or 2) and DC fast charging capabilities in a single connector. They support up to 350 kW DC charging, making it a universal choice for many BEVs.
- **CHAdeMO**
  A DC fast charging standard developed in Japan, supporting up to 400 kW in newer versions. Gradually being phased out in favour of CCS1 and 2 in many markets.
- **Tesla Supercharger Connector**
  Proprietary connectors used by Tesla vehicles, though Tesla is increasingly adopting CCS connectors in certain regions. It is not supported in the EU. It delivers up to 250 kW.

FIGURE III.7.5 Schematic depicting wireless-inductive charging infrastructures.

It is clear that the installation of high-power charging infrastructures must be assisted by a correctly designed evolution of the electric grid, while low-power devices, especially Level 1, can certainly diffuse almost freely in homes and should be the preferred way of charging EVs in the near future.

In fact, developing and maintaining widespread charging infrastructure, particularly fast chargers, is expensive. High-power chargers place significant strain on electrical grids, requiring upgrades and smart grid technologies. Compatibility issues are also still critical, and these issues should all be solved in the near future to ensure charger interoperability across various vehicle brands, connector types, and charging networks, at least for those installed in public places.

Wireless Inductive Charging is also an option which is being investigated. Transferred power is still low (some kW at most) and electromagnetic compatibility issues are only solved through a very precise coupling (relative positioning) between the vehicle and the charger.

This concept also has the potential of being installed linearly in so-called "charging lanes", according to the schematic depicted in Figure III.7.5.

A further concept which is of great interest, especially for urban e-buses is robotized conductive charging. In these systems, the physical connection between the vehicle and the charger is robotized. It may also be assisted by pantograph-type connections, like those used in railway applications.

Charging phases robotization becomes of particular interest if charging phases can be very frequent and fast (e.g. at all bus stops for urban services, normally only 300–400 m from one to the next).

A 10-ton bus, in fact, only consumes around 1 kWh between two stops, so if at the stops it may rapidly recharge that energy, the bus could run its whole daily mission with a very limited BESS installed on-board, saving energy and bus CAPEX costs.

This concept (graphically depicted in Figure III.7.6) is usually referred to as opportunity charging, or "biberonage", a French word significantly referring to a baby who eats very frequently and very little, and has a very small stomach to store the energy they need.

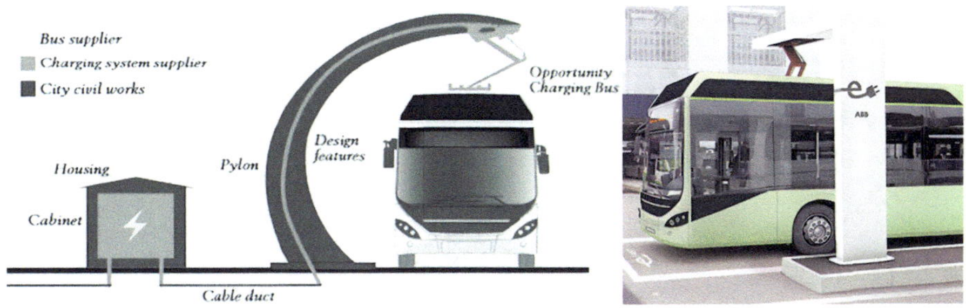

FIGURE III.7.6 Opportunity charging "biberonage" infrastructure.

III.7.4.1 V2X Technologies: Transforming Electric Mobility and Energy Systems

Vehicle-to-Everything (V2X) technologies represent a groundbreaking development in electric mobility and energy integration. These systems allow BEVs to communicate and exchange energy or information with various external entities, including homes (V2H), buildings (V2B), and the grid (V2G).

By leveraging the large energy storage capacities of BEV batteries, V2X technologies enhance energy management, grid stability, and sustainability, while offering new functionalities for vehicle owners and energy providers.

This extended discussion delves into the key V2X applications – Vehicle-to-Home (V2H), Vehicle-to-Building (V2B), and Vehicle-to-Grid (V2G) – outlining their functionalities, benefits, and challenges.

Vehicle-to-Everything (V2X) technologies enable two-way energy and information flow between BEVs and external systems. While traditional BEV charging systems are unidirectional, V2X systems are bidirectional, allowing BEVs to not only draw energy from the grid but also return it when needed. This requires specialized hardware, such as bidirectional chargers, and advanced communication protocols to manage energy exchange.

Algorithms have to be designed and installed in proper energy management systems (EMSs) to control the systems to optimize energy flows, and smart meters should measure and track energy usage and contributions to the grid. These activities are more easily applicable in home applications, by so-called Vehicle-to-Home (V2H) devices.

- **Vehicle-to-Home (V2H)**

  V2H allows a BEV to supply energy directly to a residential home (see Figure III.7.7). This system turns the vehicle into a mobile energy storage unit, capable of powering household appliances, lighting, and other systems.

  V2H is now starting to be diffused. It can:

  - Act as an emergency power source during outages, ensuring uninterrupted operation of essential systems like refrigerators, heating, or medical equipment (considering that a typical 60 kWh BEV battery can power an average household for 2–3 days).

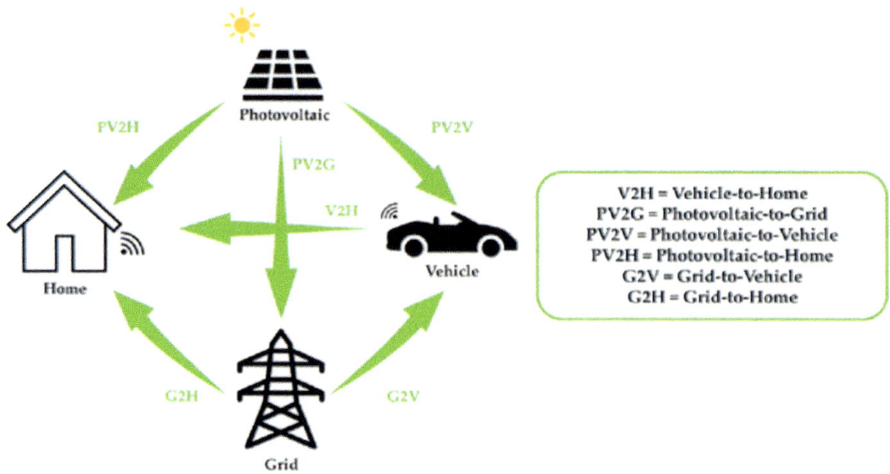

FIGURE III.7.7 f Sketch depicting the V2H technology concept.

- Charge BEVs only during off-peak hours when electricity is cheaper and use the stored energy during peak times to reduce utility bills.
- Store excess energy generated by rooftop solar panels during the day and supply it back to the home in the evening, reducing reliance on the grid.

- **Vehicle-to-Building (V2B)**

  V2B devices will be the next application in the V2X concept. They will make it possible to use vehicles' BESSs to supply energy to larger buildings, such as office complexes, factories, or schools. This application scales up the V2H concept to cater to the higher energy demands of commercial and industrial spaces.

  Buildings equipped with V2B systems can participate in demand response programs, where energy is drawn from BEVs during times of high grid demand in exchange for financial incentives. By integrating BEVs with renewable energy sources, V2B systems will also help businesses reduce their carbon footprint and achieve sustainability targets. Corporate fleets and employee vehicles can therefore act as distributed energy resources, contributing to building energy management while remaining available for transportation needs.

- **Vehicle-to-Grid (V2G)**

  V2G technology will further extend these concepts extending bi-directional chargers to public contexts. This application positions BEVs as a critical component of modern, flexible **smart grids**. BEVs could fully provide the ancillary services that the grid will increasingly require as much as non-programmable RESs will penetrate the electricity sector (such as frequency regulation and voltage support, to maintain grid stability during fluctuations in demand and supply).

  Therefore, BEV owners could earn money by participating in V2G programs, receiving compensation for the energy they supply to the grid.

The main challenges still to be solved for the wide diffusion of V2G technologies are:

- **Battery Degradation:** Frequent cycling of the battery may accelerate wear and reduce lifespan (advancements in battery chemistry are mitigating this issue).
- **Regulatory and Market Barriers:** Many regions lack regulatory frameworks and market mechanisms to support widespread V2G adoption.
- **Infrastructure Requirements:** High costs associated with installing bidirectional chargers and upgrading grid infrastructure.
- **User Participation:** BEV owners may initially hesitate to participate in V2G programs due to concerns about vehicle availability and battery health. Summarizing, the future of V2X technologies is certainly promising, with advancements in hardware, software, and policy paving the way for widespread adoption: Unified standards for communication protocols and connectors will simplify integration and enhance interoperability. Improved battery chemistry and management systems will reduce concerns about degradation from V2X applications.

The integration of V2X technologies will be certainly easier, especially in an urban context, if the MaaS Paradigm is implemented, as explained in Chapter III.1.

V2X technologies, encompassing V2H, V2B, and V2G applications, have the potential to revolutionize energy systems and mobility. By turning BEVs into mobile energy hubs, these systems enable more sustainable, efficient, and resilient energy management. While challenges remain in terms of costs, infrastructure, and regulation, the rapid pace of innovation and growing focus on renewable energy integration ensure that V2X will play a central role in the future of transportation and energy systems.

III.8 Hybrid Electric Vehicles (HEVs)

In Chapter III.2 we discussed vehicular energy and Power requirements. We saw, in particular, that the power request at the wheel is very rapidly varying.

In the following section, we show the effects of these fast variations on conventional vehicles' design and performance, which are far from being optimal. Therefore, the objective and role of propulsion system hybridization will become clear.

Then, the most common HEVs will be categorized and briefly discussed. Lastly, the main optimization techniques and technologies for HEVs will be explained.

III.8.1 VEHICLE POWER REQUIREMENTS AND ICE OPERATIONAL LIMITS

As explained in Chapter III.3, addressing both ICEs' thermodynamics and thermochemistry to explain their pollutant formation and abatement dynamics is definitely far beyond the objectives of this text and should be treated in separate academic courses. Here, instead, we will only briefly recall how the expected performance of ICEs can be described by the characteristic maps.

In conventional vehicles, the internal combustion engine (ICE) is mechanically connected to the wheels, so its rotational speed cannot be freely chosen. In fact, it has to be proportional to the vehicle's speed, according to the gear ratio selected in the transmission.

Unfortunately, ICE rotational speed can only vary in a limited range, with it not being possible to operate an ICE below 800–1000 rpm, and above 2500–3000 rpm or 3500–4000 rpm, respectively, for commonly used CI and SI ICEs.

As a consequence, the ratio between minimum and maximum engine speeds is in the range 3–4, while a vehicle should run at very low speeds 4–5 km/h as well as at 150 km/h, having an operative ratio range of 30–40 between minimum and maximum wheel rotation speeds.

This is the main reason why a variable power transmission device should be inserted in between the ICE and the wheels. This transmission system usually operates by discretely varying transmission ratios and is normally referred to as the gearbox.

Moreover, at a certain vehicle speed, a vehicle mission could require more or less power depending on the driver's actions on the pedals, requiring the vehicle to accelerate or decelerate.

DOI: 10.1201/9781003631125-30

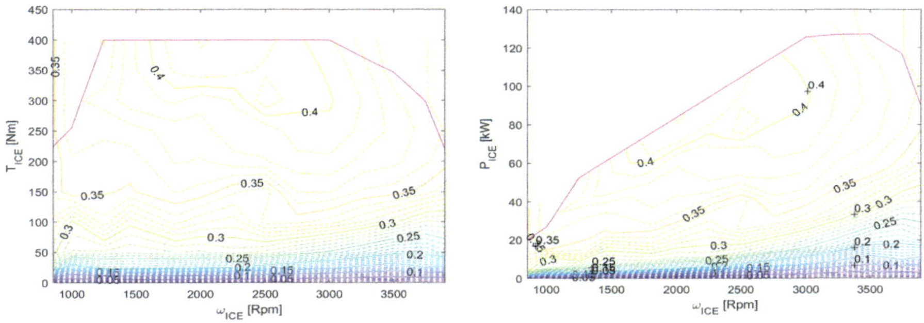

FIGURE III.8.1 Typical ICE torque and power operational maps.

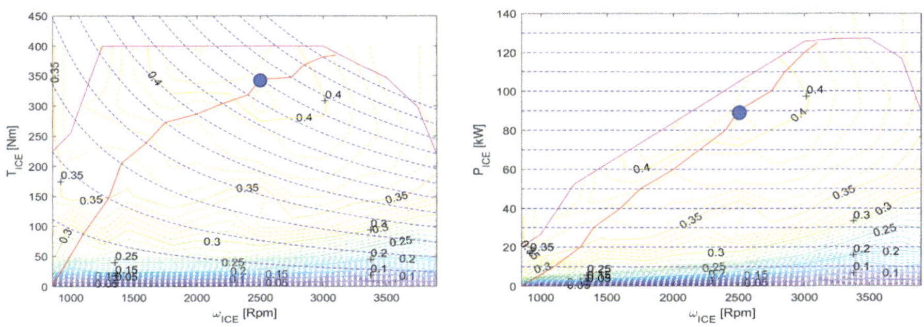

FIGURE III.8.2 Optimal ICE working points on ICE torque and power operational maps.

An additional parameter is therefore introduced in ICE operation which is normally referred to as engine load, this being the actual % power output of the ICE with respect to its maximum possible power output at the same engine speed.

All this considered, ICE performances can be fully described in a Cartesian plane having ICE speed (ω_{ICE}) in the x-axis and engine power (P_{ICE}), or engine torque (T_{ICE}), on the y-axis, with P_{ICE}, T_{ICE}, and ω_{ICE} being interdependent by $P_{ICE} = T_{ICE}\,\omega_{ICE}$.

Depending on the power transmission devices between the ICE and the wheels, it is only possible to vary the transmission ratio between the engine and wheel.

Figure III.8.1 illustrates an example of an ICE characteristic curve on these two planes introduced.

The magenta curves depict ICE maximum performance curves, i.e. the maximum power and torque values obtainable by the ICE for each given value of its rotation speed.

All the points below these magenta curves are "reachable" by the ICE, diminishing its load %. The iso-level curve depicts the efficiency values of the ICE in each of its possible operating points.

In Figure III.8.2 the same performance maps are reported, but some other points and curves are drawn within the ICE maps.

Blue curves are ICE iso-power output lines. These are obviously horizontal in the P_{ICE} vs. w_{ICE} plane, while they are equilateral hyperboles in the P_{ICE} vs. T_{ICE} plane, as $P_{ICE} = T_{ICE}\ \omega_{ICE}$.

Red curves are drawn connecting all the points having the maximum obtainable efficiency by the ICE at a given power output. On red curves, the blue points identify those ICE working points having a maximum ICE efficiency (amongst all the possible ICE working points delivering the same power output).

Now, the question to be addressed is how efficient the ICE is during its normal operation, and how far it operates (e.g. in a typical WLTP cycle) from its optimal working point.

In Figure III.8.3 trends are derived by simulating the ICE behaviour of a mid-sized vehicle on a WLTP cycle, equipped by the described engine.

The vehicle is equipped with a five-gear gearbox which is operated optimally choosing the best gear to be engaged for each operative condition (as shown by the trends in the upper-right plot).

The lower-right plot depicts the operational points where the ICE works during the WLTP cycle (each blue point is drawn representing mean engine speed and torque values in a second of operation).

As shown, operating points are mostly in the lower-left corner of the maps, showing that the ICE meanly works at a very low speed and load working conditions, with the resulting efficiency being very far from its optimum efficiency value.

At least a couple of questions arise:

- We described only how ICE energy performances vary depending on its operational point. But how does its environmental performance (toxic pollutant emissions) vary within the map?
  Unfortunately, the answer to this question is that they vary much more than efficiency, largely depending both on engine speed and torque. Moreover, toxic pollutant emissions are greatly influenced by the rapidity of the engine transients, i.e. the speed with which the operation points move on the maps.

  Therefore, an ICE whose engine conditions are varied smoothly can in fact emit less than a tenth the emission of a normally operated ICE on a very transient cycle like WLTP.
- Is it possible somehow to increase the mean ICE load and to make its rotational speed much more independent from vehicle speed, gaining somehow a degree of freedom to optimize its operative conditions, raising its efficiency, taking it much closer to its optimal value?
  Is it possible to smooth ICE transients making it pollute much less?

  The answer to these questions is YES. This is possible if the ICE is not the only on-board installed motor, usable to fulfil instantaneous power requirements at the wheels.

  With this aim, hybrid electric propulsion systems were designed and have started to be diffused in the last decade or more.

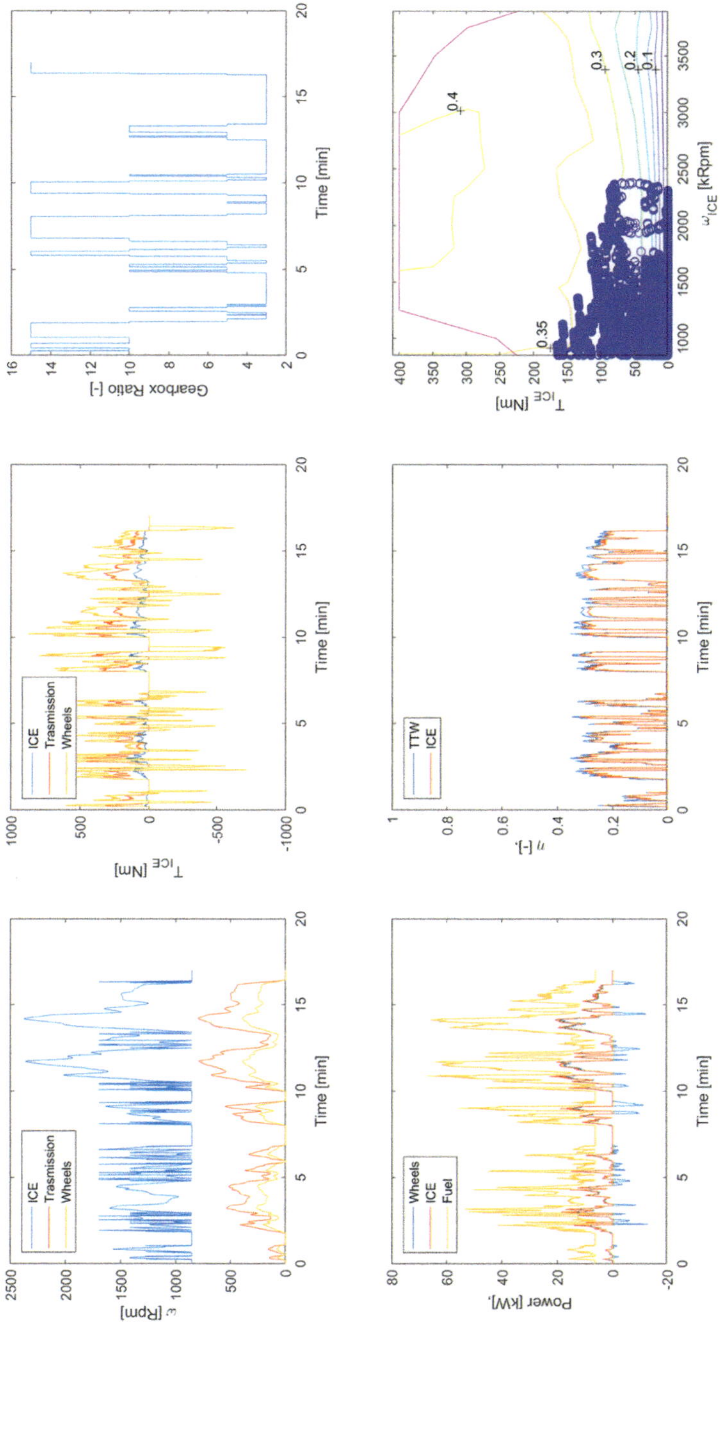

FIGURE III.8.3 Typical operation of a conventional vehicle on the WLTP cycle.

III.8.2 HEVS DESCRIPTION AND CATEGORIZATION

As already explained, HEVs represent a crucial transitional technology in the transportation sector to optimize ICE behaviour, making chemical carrier-based vehicles less energy consuming and also less toxic pollutant emitting (i.e. more compatible with air quality specifications, at least in extra-urban environments).

By combining an internal combustion engine with an electric motor and battery, HEVs may offer improved fuel efficiency, reduced emissions, and a smoother driving experience compared to traditional ICE vehicles.

The acronym HEV is usually referred to a vehicle whose traction may be powered by two distinct on-board motors:

- An internal combustion engine (ICE, typically powered by gasoline or diesel).
- An electric motor (EM, using electricity stored on-board in an ESS, and/or produced on-board instantaneously).

As detailed in the next section, all HEVs introduce a degree of freedom in the propulsion system which has to be optimized to increase both ICE and EM performances, relying on an advanced energy management system (EMS), and also depending on the driving conditions.

The interconnection between the various devices present on-board may define a series of different HEV topologies. Most of these can implement most of the key features which are expected by an HEV:

- **Regenerative Braking**
  HEVs use regenerative braking to convert kinetic energy into electrical energy, which is stored in the battery. This improves energy efficiency and reduces wear on traditional braking systems.
- **Start–Stop Technology**
  The ICE shuts off automatically when the vehicle is idling (e.g., at traffic lights), reducing fuel consumption and emissions.
- **Electric Assistance**
  During acceleration or climbing, the electric motor provides additional torque, reducing the load on the ICE and improving overall performance.
- **Vehicle Energetic and Environmental Optimization**
  An intelligent system that optimizes the interaction between the ICE, electric motor, and battery to ensure maximum efficiency and minimal emissions.

Typical HEV categorizations are based both on their degree of hybridization and on their drivetrain architecture.

Depending on the degree of vehicle hybridization, four categories of HEVs are identified:

- **Micro-Hybrid Vehicles**
  These vehicles do not use EM for propulsion but feature only start–stop technology and very limited regenerative braking.

The ICE powers the vehicle entirely, while the regenerative braking system recharges the battery, which powers some auxiliary systems like air conditioning or lighting.

They are obtainable simply by modifying a conventional vehicle and substituting starter and alternator electric machines with a single bidirectional EM of about 1–3 kW power output. This EM is commonly referred to also as an ISA (integrated starter alternator).

- **Mild Hybrid Vehicles**
  Mild HEVs feature a small electric motor that assists the ICE but cannot independently propel the vehicle. The electric motor provides torque during acceleration, reduces load during cruising, and assists in regenerative braking to recharge the battery. Typical sizes of EM for these applications are around 10 kW.
- **Full Hybrid Vehicles**
  Full HEVs feature a larger electric motor and battery, commonly allowing them to operate in three modes: electric-only, ICE-only, and combined. At least a 30–40 kW EM is installed on a mid-sized vehicle.

  Full HEVs commonly can start and drive at low speeds using electric power only. At higher speeds, when additional power is needed, or when BESS SOC is too low, the ICE takes over or works in tandem with the electric motor. Regenerative braking is also commonly used to recharge the battery.

  Significant fuel efficiency improvements may be obtained with respect to conventional vehicles (commonly around 20 on WLTP cycles).

  These vehicles can operate in pure-electric zero-emission mode for short distances (2–3 km).
- **Plug-In Hybrid Electric Vehicles (PHEVs)**
  Plug-in HEVs are full HEVs equipped with larger batteries that can be recharged on-board or by plugging into an external power source. They combine most of the benefits of HEVs and BEVs.

  PHEVs can operate purely on electric power for longer distances (typically around 50–60 km) before switching to hybrid mode, where the ICE and electric motor work together.

  In short-commute applications PHEVs mainly (or even only) operate in pure-electric ZEV mode, without consuming any fossil fuel on-board (zero TTW CO_2 emissions).

  PHEVs are more expensive than other HEVs due to their larger batteries (about €1000). A low-power Level 1 charging infrastructure is required for maximum benefit.

  PHEVs also extend vehicle on-board regenerative braking capabilities, for the increased specification of on-board BESS (both in power and energy terms).

Depending on the vehicle drivetrain architecture, HEVs, at least four HEV categories are commonly identified (see Figure III.8.4):

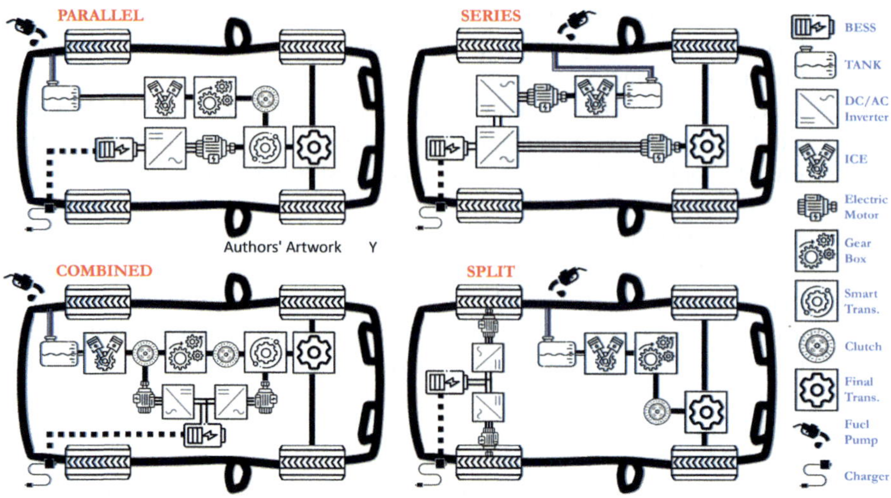

FIGURE III.8.4 Most common HEV topologies.

- **Series HEVs**
  In a series hybrid, the ICE is not connected to the wheels. It acts as a generator to charge the battery or power the electric motor, which drives the wheels.

  Series HEVs are highly efficient in stop-and-go urban driving, and have a much simpler mechanical design.

  Basically, these vehicles are derived from BEVs, adding an on-board electric generator to restore BESS SOC.

  Most HEVs for buses, last-mile trucks, and other heavy-duty urban applications are series HEVs.
- **Parallel HEVs**
  In a parallel hybrid, both the ICE and the electric motor are connected to the wheels and can drive the vehicle independently or together.

  They normally have higher efficiency in highway driving, where ICE directly gives mechanical energy to the wheels. They also commonly have improved power delivery and acceleration.

  Most passenger car HEVs nowadays are full parallel HEVs, some of them being PHEVs.

  Figure III.8.5 depicts all the possible operative modes of a parallel HEV. The way an HEV operates is defined by its EMS, depending on user request (who may specifically ask, for example, for a pure electric driving), engine speed and load (pedal positions), and BESS state of charge.

  This information is gained by the EMS interacting with the other vehicle ECUs through a standardized protocol (usually CAN), as explained already in Chapter III.7 about BEVs.
- **Combined HEVs**
  Combined HEVs pull together the features of both series and parallel hybrids, allowing the vehicle to switch between modes depending on the driving conditions.

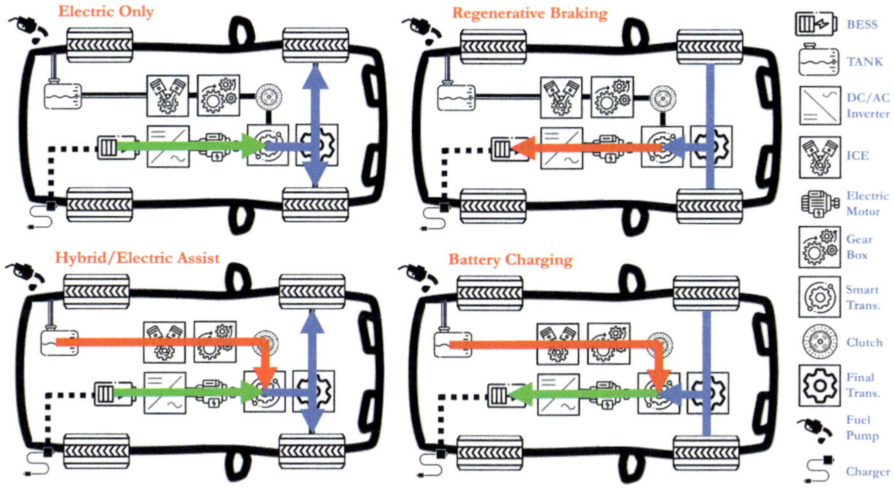

FIGURE III.8.5 Parallel HEV working modes.

They may optimize efficiency across a wide range of driving scenarios, but are more complex, doubling electric machines, and requiring a complex EMS.

- **Split HEVs**
  Split HEVs are used for four-wheel driving solutions using two separate propulsion systems, and also using hub motors for the electric propulsion systems. They combine the features of both series and parallel hybrids, allowing the vehicle to switch between modes depending on driving conditions.
  Their architecture is much simpler, but they normally implement reduced HEV features depending on the size of the various installed components. They fully rely on the efficiency and effectiveness of electronic controllers including for vehicle stability and traction/braking control.

Summarizing, most HEVs may grant substantial benefits with respect to conventional vehicles:

- **Fuel Efficiency**
  HEVs consume less fuel compared to ICE vehicles, resulting in lower running costs, which normally more than balance the increase in vehicle purchase costs.
- **Reduced Emissions**
  Hybrid operation reduces tailpipe emissions, contributing to cleaner air.
- **Smoother Driving Experience**
  The transition between an ICE and an electric motor is seamless, offering providing a quieter and smoother ride.

- **Energy Recovery**
  Regenerative braking recaptures energy that would otherwise be wasted, improving overall efficiency, more than balancing additional energy consumption due to incremental vehicle weight.
- **Lower Maintenance Costs**
  Electric motors and regenerative braking reduce wear and tear on traditional components.

HEVs are therefore a very versatile and efficient alternative to traditional ICE vehicles, offering significant benefits in terms of fuel savings and emissions reductions. With various types of HEVs catering to different needs, they are a key component in the transition to sustainable mobility.

Whether as a mild hybrid for incremental efficiency gains or a plug-in hybrid for extended electric driving, HEVs are shaping the future of transportation while complementing the broader adoption of fully electric vehicles.

III.8.3 HEVS CONTROL AND OPTIMIZATION

In HEVs, traction may be powered by two distinct on-board motors: an ICE (typically powered by gasoline or diesel) and an electric motor (EM, using electricity stored on-board in an ESS, and/or produced on-board instantaneously).

Therefore, each HEV should have a detailed supervising ECU (normally referred to as an energy management system, EMS) which has to control the vehicle deciding how to split actual power demand between the two (or more) available power sources, enabling the possible HEV modes.

With this aim, EMS has to dialogue with some of the vehicle's other ECUs through a standardized protocol (usually CAN).

In more detail, the EMS should:

- **Sniff data transmitted over the CAN network** are used to obtain the latest information on vehicle parameters necessary to determine the power split between the available motors. In particular, fundamental information includes:
  - Vehicle speed, and pedal positions are normally made available on CAN from the cockpit ECU.
  - BESS state of charge is normally made available on CAN by its BMS.
- **Elaborate these data implementing the designed control optimization algorithm**
  One of the more immediate ways to implement a control optimization algorithm on an EMS is using Matlab/Simulink®. This object-oriented programming code, in fact, permits to:
  - Read data from CAN about vehicle variables of interest.
  - Easily describe all possible dynamic algorithms governing the vehicle in all its possible modes, also describing the transitional rules between these modes (in a toolbox called Stateflow)

- Writing data on CAN about the power set-points of the two power sources. These set-points will be instantaneously read and implemented by the ECU governing the motors.
- Continuously read data from CAN to check if the set-points are correctly implemented, eventually implementing corrective actions.

Control-oriented algorithms can be synthesized and tested on a simple Windows Host-PC and then translated in C, and implemented and run on a target real-time OS-operated Target-PC running the supervising ECU of the EMS.

With this aim, two specific toolboxes of Matlab/Simulink have to be used, respectively named xPCTarget and Real-Time Workshop. Host and Target PCs can also remain in constant communication through fast and dedicated communication networks, including simply using Bluetooth or Wi-Fi.

In this way, virtual vehicle cockpits can be designed, which are very useful during the design phase of the EMS.

As an example, a sketch describing this whole design procedure implemented by the author on a prototypal vehicle by ENEA named Urb-E in 2010 is depicted in Figure III.8.6.

The procedure just described only explains how it is possible to implement a designed control optimization algorithm.

Let us now describe the most commonly used approaches to optimize HEV performances.

These strategies govern the operation of the ICE, electric motor, battery, and auxiliary systems, ensuring optimal energy utilization and seamless integration of the hybrid powertrain components.

III.8.3.1 Rule-Based Control Strategies

Rule-based control strategies rely on pre-defined logic and rules to determine the operation of the ICE and electric motor. These strategies are simple, computationally efficient, and widely used in commercial HEVs.

- **Thermostat Control**
  - Operates the ICE to maintain the battery's SOC within a specific range (SOC_{Min}–SOC_{Max}), similar to how a thermostat regulates temperature.
  - The ICE turns on when the SOC drops below a lower threshold (SOC_{Min}) and turns off when it exceeds an upper threshold SOC_{Max}.
  - This strategy is simple and easy to implement, especially for these vehicles having a sufficiently large BESS, because otherwise it can lead to frequent ICE start–stop cycles, reducing ICE efficiency and increasing wear.
- **Power Split Control (Load-Levelling)**
  - Distributes the load between the ICE and the electric motor based on the vehicle's power demand. Basically, a first-order filtering of estimated power requirements is implemented to lower the rapidity of ICE transients.

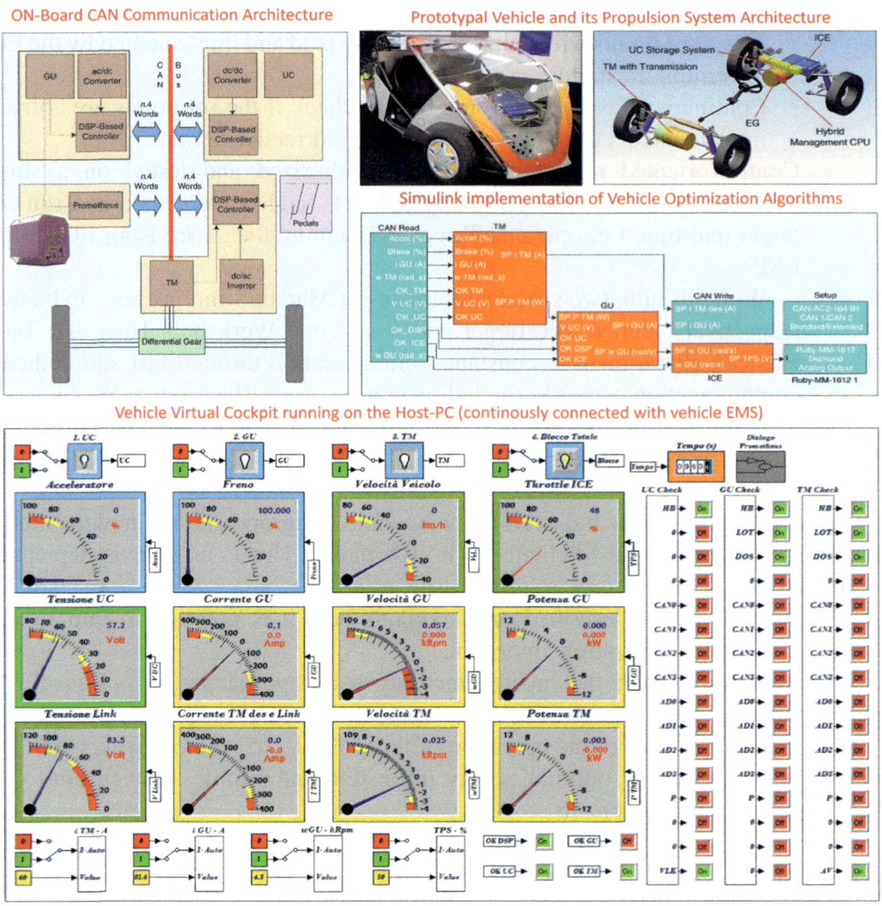

FIGURE III.8.6 EMS design process for the prototypal HEV "Urb-E" developed by the author for ENEA (Italian National Agency for New Technologies, Energy & Sustainable Development) in 2010.

Additionally, the electric motor is favoured if power demands are low and transient (e.g., city driving), while the ICE operates more during steady high-power demands (e.g., highway driving or acceleration).

- It improves fuel efficiency by operating the ICE within its optimal efficiency range, reducing emissions during low-power rapidly transient driving conditions. It requires careful tuning to balance power demand and energy efficiency.

- **Mode-Based Control**
  - Switches between different operating modes (e.g., electric-only, hybrid, or ICE-only) based on predefined rules. Modes are selected depending on factors like vehicle speed, battery SOC, and driver input.

    - Enables diverse operational scenarios, enhancing overall efficiency. This control approach is quite common in commercial HEVs.
    - Fixed-mode transition rules may not adapt well to dynamic driving conditions.

III.8.3.2 Optimization-Based Control Strategies

Optimization-based strategies aim to minimize fuel consumption, emissions, or operational costs by solving mathematical optimization problems. These strategies are more computationally intensive but usually provide superior performance compared to rule-based methods.

- **Equivalent Consumption Minimization Strategy (ECMS)**
    - Balances fuel consumption and electric energy usage by treating battery energy as an "equivalent" fuel. The control algorithm minimizes the total equivalent fuel consumption in real time.
    - It is highly adaptable to varying driving conditions, and provides near-optimal fuel efficiency. However, it depends on an accurate estimation of the equivalent fuel cost and is computationally demanding.
- **Model Predictive Control (MPC)**
    - It tries to predict future vehicle states (e.g., speed, power demand) based on a mathematical model and optimizes the control actions over a prediction horizon, updating decisions in real time as new data become available.
    - It can incorporate constraints (e.g., battery SOC limits, emissions), but has a very high computational complexity and strictly depends on the accuracy of predictive modelling.

III.8.3.3 Machine Learning-Based Control Strategies

With advancements in artificial intelligence, machine learning (ML) techniques are increasingly being applied to HEV control strategies. These strategies use data-driven models to optimize powertrain performance.

- **Reinforcement Learning (RL)**
    - RL algorithms learn optimal control strategies by interacting with the environment and receiving feedback in the form of rewards (e.g., fuel savings). The system continuously improves its performance over time.
    - This approach is highly adaptive to diverse driving conditions, because it can discover novel control strategies that are not explicitly programmed and are more suitable for the user driving a specific vehicle.
    - It has high computational demands, especially during training phases.
- **Neural Network-Based Control**
    - Uses neural networks to model complex relationships between inputs (e.g., vehicle speed, SOC) and outputs (e.g., power split decisions).

- Networks are typically trained both on historical data and on simulations. They can handle complex, nonlinear systems effectively. Their effectiveness improves over time with more data.
- This approach is computationally intensive.

Control strategies in HEVs are essential for optimizing performance, fuel efficiency, and emissions. Rule-based strategies offer simplicity and reliability, while optimization-based and machine learning methods enable higher adaptability and performance in dynamic conditions. Effective energy management, SOC balancing, and the integration of equalization techniques ensure that HEVs deliver their intended benefits over their lifecycle. As technology advances, more sophisticated control strategies will continue to enhance the efficiency and sustainability of hybrid vehicles.

Further Readings

1. Sustainable Mobility Issues

Banister, D., *Unsustainable Transport: City Transport in the New Century*. Routledge, 2018.

Geels, F. W., Regime Resistance against Low-Carbon Transitions: Introducing Politics and Power into the Multi-Level Perspective, Theory. *Culture & Society, 31*(5), 2020.

Gössling, S., & Choi, A. S., Transport Transitions in Copenhagen: Comparing the Cost of Cars and Bicycles. *Ecological Economics, 113*, 106–113, 2015.

Hickman, R., Givoni, M., & Cohen, T., *Handbook on Transport and Development*. Edward Elgar, 2019.

Tovey, M., Woodcook, A., Osmond, J., *Designing Mobility and Transport Services: Developing Traveller Experience Tools*. Routledge, 2017.

2. Vehicular Energy and Power Requirements

Bosch, R., *Automotive Handbook*. Bentley, 2020.

Crolla, D., *Automotive Engineering: Powertrain, Chassis System and Vehicle Body*. Elsevier, 2015.

Guzzella, L., & Sciarretta, A., *Vehicle Propulsion Systems: Introduction to Modeling and Optimization*. Springer, 2019.

Taylor, C. F., *The Internal-Combustion Engine in Theory and Practice*. MIT Press, 2018.

3. ICEs and Their Local and Global Impacts

Eastwood, P., *Particulate Emissions from Vehicles*. Wiley, 2016.

Heywood, J. B., *Internal Combustion Engine Fundamentals*. McGraw-Hill, 2018.

Khair, M. K., & Jääskeläinen, H., *Diesel Emissions and Their Control*. SAE International, 2016.

Reif, K., *Diesel Engine Management: Systems and Components*. Springer, 2019.

Stone, R., *Introduction to Internal Combustion Engines*. Palgrave Macmillan, 2017.

Turner, J. W. G., & Pearson, R. J., *Automotive Fuels Reference Book*. SAE International, 2020.

4. Alternative Fuels for ICEs

Dell'Aversano, S., & Villante, C., et al., *E-Fuels: A Comprehensive Review of the Most Promising Technological Alternatives towards an Energy Transition*, Energies, MDPI, 2024.

Demirbas, A., *Biofuels: Securing the Planet's Future Energy Needs*. Springer, 2017.

E-Fuels: A Technoeconomic Assessment of European Domestic Production and Imports Towards 2050. Concawe, 2024.

Ram, S. S., et al. *Biofuels: Production and Future Perspective*. Routledge, 2019.

Sharma, B. K., & Dunn, J. B., *Sustainable Biofuels: Opportunities and Challenges*. Elsevier, 2020.

Speight, J. G., *The Biofuels Handbook*. Royal Society of Chemistry, 2019.

Turner, J. W. G., & Pearson, R. J., *Automotive Fuels Reference Book*. SAE International, 2020.

5. Vehicle Electrification

Chan, C. C., & Chau, K. T., *Modern Electric Vehicle Technology*. Oxford University Press, 2017.
Ehsani, M., Gao, Y., & Emadi, A., *Modern Electric, Hybrid Electric, and Fuel Cell Vehicles*. CRC Press, 2018.
Husain, I., *Electric and Hybrid Vehicles: Design Fundamentals*. CRC Press, 2019.
Larminie, J., & Lowry, J., *Electric Vehicle Technology Explained*. Wiley, 2020.
Pistoia, G., *Lithium-Ion Batteries: Advances and Applications*. Elsevier, 2019.

6. Fuel Cell Electric Vehicles (FCEVs)

Barbir, F., *PEM Fuel Cells: Theory and Practice*. Academic Press, 2017.
Kordesch, K., & Simader, G., *Fuel Cells and Their Applications*. VCH, 2016.
Mench, M. M., *Fuel Cell Engines*. Wiley, 2018.
O'Hayre, R., *Fuel Cell Fundamentals*. Wiley, 2019.

7. Battery Electric Vehicles (BEVs)

Emadi, A., *Advanced Electric Drive Vehicles*, 2014.
Lu, L., Han, X., & Li, J., A Review on the Key Issues for Lithium-Ion Battery Management in Electric Vehicles. *Journal of Power Sources*, *226*, 272–288, 2013.
Pistoia, G., *Lithium-Ion Batteries: Advances and Applications*. Elsevier, 2019.
Spellman, F. R., *The Science of Electric Vehicles Concepts and Applications*. CRC Press, 2023.

8. Hybrid Electric Vehicles (HEVs)

Anderson, J., *Electric and Hybrid Cars: A History*. McFarland, 2017.
Chan, C. C., & Chau, K. T., *Modern Electric Vehicle Technology*. Oxford University Press, 2017.
Ehsani, M., Gao, Y., & Emadi, A., *Modern Electric, Hybrid Electric, and Fuel Cell Vehicles*. CRC Press, 2018.
Guzzella, L., & Sciarretta, A., *Vehicle Propulsion Systems: Introduction to Modeling and Optimization*. Springer, 2019.
Miller, J. M., *Propulsion Systems for Hybrid Vehicles*. IET, 2018.

Index

C

D

E

F

G

P

Q

R

S

V

W

Y

9781041050469